DISCARD
FCPL discards materials
that are outdated and in poor condition.
In order to make room for current,
in-demand materials underused materials
are offered for public sale

Raspberry Pi Electronics Projects for the Evil Genius™

Evil Genius Series

15 Dangerously Mad Projects for the Evil Genius

22 Radio and Receiver Projects for the Evil Genius

25 Home Automation Projects for the Evil Genius

30 Arduino Projects for the Evil Genius, Second Edition

30 BeagleBone Black Projects for the Evil Genius

50 Awesome Auto Projects for the Evil Genius

50 Model Rocket Projects for the Evil Genius

51 High-Tech Practical Jokes for the Evil Genius

101 Spy Gadgets for the Evil Genius

101 Outer Space Projects for the Evil Genius

123 PIC® Microcontroller Experiments for the Evil Genius

123 Robotics Experiments for the Evil Genius

Arduino + Android Projects for the Evil Genius

Bionics for the Evil Genius: 25 Build-It-Yourself Projects

Electronic Circuits for the Evil Genius, Second Edition: 64 Lessons with Projects

Electronic Gadgets for the Evil Genius: 28 Build-It-Yourself Projects

Electronic Games for the Evil Genius

Electronic Sensors for the Evil Genius: 54 Electrifying Projects

Fuel Cell Projects for the Evil Genius

Mechatronics for the Evil Genius: 25 Build-It-Yourself Projects

MORE Electronic Gadgets for the Evil Genius: 40 NEW Build-It-Yourself Projects

PC Mods for the Evil Genius: 25 Custom Builds to Turbocharge Your Computer

Programming Video Games for the Evil Genius

Raspberry Pi Projects for the Evil Genius

Solar Energy Projects for the Evil Genius

Raspberry Pi Electronics Projects for the Evil Genius™

Donald Norris

New York Chicago San Francisco Athens London Madrid
Mexico City Milan New Delhi Singapore Sydney Toronto

McGraw-Hill Education books are available at special quantity discounts to use as premiums and sales promotions, or for use in corporate training programs. To contact a representative, please visit the Contact Us pages at www.mhprofessional.com.

Raspberry Pi Electronics Projects for the Evil Genius™

Copyright ©2016 by McGraw-Hill Education. All rights reserved. Printed in the United States of America. Except as permitted under the Copyright Act of 1976, no part of this publication may be reproduced or distributed in any form or by any means, or stored in a database or retrieval system, without the prior written permission of publisher, with the exception that the program listings may be entered, stored, and executed in a computer system, but they may not be reproduced for publication.

McGraw-Hill Education, the McGraw-Hill Education Publishing logo, Evil Genius® and TAB™, and related trade dress are trademarks or registered trademarks of McGraw-Hill Education and/or its affiliates in the United States and other countries and may not be used without written permission. All other trademarks are the property of their respective owners. McGraw-Hill Education is not associated with any product or vendor mentioned in this book.

1 2 3 4 5 6 7 8 9 0 RHR 21 20 19 18 17 16

ISBN 978-1-25-964058-2
MHID 1-25-964058-2

Sponsoring Editor
Michael McCabe

Editorial Supervisor
Donna M. Martone

Project Manager
Patricia Wallenburg, TypeWriting

Acquisitions Coordinator
Amy Stonebraker

Copy Editor
James Madru

Proofreader
Claire Splan

Indexer
Claire Splan

Production Supervisor
Pamela A. Pelton

Composition
TypeWriting

Art Director, Cover
Jeff Weeks

Information has been obtained by McGraw-Hill Education from sources believed to be reliable. However, because of the possibility of human or mechanical error by our sources, McGraw-Hill Education, or others, McGraw-Hill Education does not guarantee the accuracy, adequacy, or completeness of any information and is not responsible for any errors or omissions or the results obtained from the use of such information.

To the highly trained staff at

The Works Family Health & Fitness Center, Somersworth, NH

and especially to Samantha

who provided me with professional emergency treatment.

About the Author

Donald Norris has a degree in electrical engineering and an MBA specializing in production management. He is currently teaching undergrad and grad courses in the IT subject area at Southern New Hampshire University. He has also created and taught several robotics courses there. He has over 35 years of teaching experience as an adjunct professor at a variety of colleges and universities. Mr. Norris retired from civilian government service with the U.S. Navy, where he specialized in acoustics related to nuclear submarines and associated advanced digital signal processing. Since then, he has spent more than 20 years as a professional software developer using C, C#, C++, Python, Node.js and Java, as well as 5 years as a certified IT security consultant. Mr. Norris started a consultancy, Norris Embedded Software Solutions (dab NESS LLC), that specializes in developing application solutions using microprocessors and microcontrollers. He likes to think of himself as a perpetual hobbyist and geek and is always trying out new approaches and out-of-the-box experiments. He is a licensed private pilot, photography buff, amateur radio operator, and avid runner.

Contents

Acknowledgments xiii

Foreword xv

Introduction xvii

1 Introduction 1
- Brief Raspberry Pi Background 1
- Chapter 1 Parts List 1
- Raspberry Pi GPIO 4
- Establishing a RasPi Development Station 7
- Setting Up the RasPi Software 11
- Linux Users, Privileges, and Permissions 21
- Summary 22

2 Touchscreens 23
- PiTFT Touchscreens 23
- Chapter 2 Parts List 24
- Touchscreen Background 24
- RasPi Touchscreen Installation 29
- Framebuffer 41
- Touchscreen Project 43
- Summary 53

3 Arduino Coprocessor 55
- What Is a Coprocessor? 55
- Chapter 3 Parts List 55
- Communication Implementation Techniques 55
- Arduino Board and Arduino IDE 65
- Lidar Demonstration Project 69
- Demonstration Project 82
- Summary 84

4 RGB LED Matrix Display 85
- 32 × 64 RGB LED Matrix 85
- Chapter 4 Parts List 86
- How the RGB LED Matrix Works 87
- RasPi Interface Board 89
- Software to Drive the RGB LED Matrix 92
- Using Python with the RGB LED Matrix Display 104
- Summary 106

5 Raspberry Pi Supercomputer Cluster 107
Brief Supercomputer Discussion and History 107
Chapter 5 Parts List 107
RasPi Cluster 110
RasPi Cluster Software 112
Software Setup 115
Pi Calculations 126
Unique Functions for Cluster Operations 130
Basic MPI Operations 132
Monitoring Cluster Network Traffic 135
Summary 136

6 RasPi-to-RasPi Communications Using MQTT 137
Paho and Eclipse.org 137
Chapter 6 Parts List 137
RasPi MQTT Publisher-Client System 139
RasPi Subscriber Client 152
MQTT Two-Phase Thermostat 155
Summary 162

7 Software-Defined Radio 163
Basic Radio Concepts 163
Chapter 7 Parts List 163
SDR Dongle 167
rtl-sdr and GNU Radio Software Installation 169
Receiving Aviation Data Signals 174
Spectrum Analyzer 183
Summary 188

8 BrickPi Python Robot 191
BrickPi 191
Chapter 8 Parts List 191
The CasterBot 195
Software Installation and Configuration 199
Summary 208

9 Python-Controlled Robotic Arm 211
Background for Robotic Arms 211
Chapter 9 Parts List 211
Degrees of Freedom 212
Robotic Arm Classifications 213
SainSmart Robotic Arms 215
Robotic Arm Software 221
Summary 235

10 Gigapixel Camera System **237**
Chapter 10 Parts List 237
Stack and Stitch 238
DSLR Camera 253
Summary 262

11 Nighttime Garden Monitor **263**
Pi Noir Camera 263
Chapter 11 Parts List 263
Installing the Camera 264
Installing and Configuring the Camera Driver Software 264
Physical Monitoring System 266
Mounting All the System Components 269
Setting Up the Trip Beam 270
Software Installation and Configuration 272
Sample from a Capture Video 274
Summary 274

Index 275

Acknowledgments

I THANK KAREN for putting up with all my experiments and enduring all the "discussions" about the book projects.

Thanks to Michael McCabe for his fine support as my editor and also to Amy Stonebraker for her support as editorial assistant.

Thanks also goes out to Patty Wallenburg for her fine work as the Project Manager.

Finally, I like to thank all the folks at the Raspberry Pi Foundation for creating the boards and getting them to the marketplace.

Foreword

This is the second book I have written that cover a series of interesting Raspberry Pi (RasPi) projects. The first book, *Raspberry Pi Projects for the Evil Genius* is more introductory in the sense that the projects are somewhat less complex as compared to this book's projects. This book's projects utilize the latest features of this single board computer first introduced by the Raspberry Pi Foundation in 2013. The creator of the board, Dr. Eben Upton, originally thought it might sell upwards of 10,000 units; however, there have been over 5 million RasPis in various versions manufactured to date.

I have tried to present a range of projects in this book, which should intersect with the interests and hobbies of my readers. There is one on photography, another on radio communication, two on robots and a variety on computer-related subjects, including one on how to build your own computer cluster for those who wish to take on an ambitious project.

Foreword

Introduction

This project book is about building a series of interesting projects and also about providing an education regarding the underlying project technologies. I am positive that my extensive experience as a college educator required me to ensure readers not only could build the projects but also understand why they function as designed.

Building a successful project is a reward unto itself, but additionally understanding why and how it functions is far more important. The reader should expect a great increase in experience and knowledge with the Raspberry Pi if a commitment is made to complete most of these projects. I personally always learn a great deal while designing and finishing them. Often, things work out just fine while at other times they are fraught with problems. However, that's what I consider the joy of experimenting. As the renowned Professor Einstein once stated, "Anyone who has never made a mistake has never tried anything new."

The joy of learning about building projects are the key concepts within this book. I designed and built all of the projects and along the way gained a lot of knowledge about Linux and how it really shines as an operating system for embedded development platforms.

The complexity of the book projects are relatively equal and you should read each one that interests you and decide if you want to tackle it. Be assured that I have built and tested every project and can unequivocally state that they will function as designed if you do not deviate from the chapter instructions. You should also gain experience and confidence in dealing with the Linux OS, Java, and Python languages as you proceed through these projects.

Experienced Linux developers should feel free to jump into any of the projects, however, there are useful hints and techniques sprinkled throughout the book that might be missed by taking too selective an approach to reading the book. I have also tried to point out constraints and limitations of the Raspberry Pi as I encountered them when designing and building the projects. Just keep in mind, a $35 computer simply cannot meet all expectations.

Most of the book projects may be expanded and modified as desired. I tried to point out areas where you can make changes to suit your preferences and to suit your own particular situation. I strongly recommend that readers try to experiment and modify as this only enhances the learning experience. It has been stated that "The ability to experiment has been described as one of the key attributes that modern employers are looking for in twenty-first century employees."

Software downloads available at:
www.mhprofessional.com/RaspPiEvilGenius

Raspberry Pi Electronics Projects for the Evil Genius™

CHAPTER 1

Introduction

This is a Raspberry Pi book in which I will provide 10 interesting projects. However, before I start discussing the projects, I think it is important to provide some background and configuration information so that you gain an understanding of options available in the extensive Raspberry board series and how they compare with each other.

Brief Raspberry Pi Background

The Raspberry Pi (RasPi) had been in existence for three years at the time this book was written. More than 5 million RasPis have been produced since it was introduced, which is not too shabby considering that the creator, Eben Upton, originally thought that about 10,000 boards would be sold. I will not go into extensive detail about the origins, history, and structure of the RasPi because I have already covered that subject in my original Evil Genius book, *Raspberry Pi Projects for the Evil Genius*. However, I will reiterate some key RasPi concepts that are critical to your success in building the RasPi projects in this book, and it is always convenient to have the data immediately available and in one place.

I also believe that it is important to have an understanding of all the different RasPi models and revisions that have been manufactured to date. Having this knowledge should enable you to select that best RasPi for a particular project. It is also important to realize that the core features of the RasPi have been relatively unchanged from the original models to the latest ones. The RasPi still remains a rather inexpensive single-board computer that is capable of running a full-featured Linux operating system (OS). Differences in models and revisions arise mainly with regard to supported peripherals, on-board memory, and central processing unit (CPU) clock speed.

Chapter 1 Parts List

The following is a composite list of parts used throughout all the following chapters in this book. Not all the listed parts are used in every chapter, but they are listed for your reference.

Item	Model	Quantity	Source
Light-emitting diodes (LEDs), various colors available	Commodity	Varies	mcmelectronics.com adafruit.com digikey.com mouser.com farnell.com
270-Ω, 0.25-W resistor	Commodity	Varies	mcmelectronics.com adafruit.com digikey.com mouser.com farnell.com
2N3904 transistor	Commodity	1	mcmelectronics.com adafruit.com digikey.com mouser.com farnell.com
Solderless breadboard, 830 tie points	Commodity	Varies	mcmelectronics.com adafruit.com digikey.com mouser.com farnell.com
5-V power supply, micro-USB connection	Commodity	1	mcmelectronics.com adafruit.com digikey.com mouser.com farnell.com
HDMI-to-HDMI cable, 1 m	Commodity	1	mcmelectronics.com adafruit.com digikey.com mouser.com farnell.com
Jumper-wire package	Commodity	1	amazon.com
Pi Cobbler, RasPi 2	2029	1	adafruit.com
USB keyboard	Commodity	1	amazon.com
USB mouse	Commodity	1	amazon.com
Pluggable USB hub	7 port	1	amazon.com
HDMI monitor	Commodity	1	amazon.com
USB WiFi adapter	Commodity	1	amazon.com

Table 1-1 lists the key features of all RasPi models and revisions arranged by date of introduction. The original model B Raspberry Pi is shown in Figure 1-1.

The latest and greatest Model B Raspberry Pi 2 is shown in Figure 1-2 for your comparison with the original Model B.

What is most remarkable about these two boards, which are separated by more than three years between their dates of introduction, is that the RasPi 2 Model B is six times faster than the original Model B, has twice the dynamic memory, and yet costs exactly the same as the original.

TABLE 1-1 Key Features for All RasPi Models and Revisions

Model	Revision	CPU Speed/ No. of Cores	Memory/ No. of GPIO Pins	Ethernet Jack	No. of USB Ports	SD Card Type	Remarks
B	1	700/1	512 MB/26	1	2	Std	Initial release, very early B's had 256 MB
A	1	700/1	256 MB/26	0	1	Std	Initial release
B	2	700/1	512 MB/26	1	2	Std	Poly fuses removed
A	2	700/1	256 MB/26	0	1	Std	Poly fuses removed
Computer module	1	700/1	512 MB/0	0	0	N/A	4-GB electronic multimedia card (eMMC)
B+	1	700/1	512 MB/40	1	4	Micro	Audio +
A+	1	700/1	256 MB/40	0	1	Micro	Audio +
B gen 2	1	900/4	1 GB/40	1	4	Micro	Audio +

Figure 1-1 Model B Raspberry Pi board.

Figure 1-2 Model B Raspberry Pi 2.

Cheaper models, such as A and A+, are available, but they do not have an onboard Ethernet port, and they have fewer USB ports and only half the memory of the corresponding RasPi 2 Model B or the original Model B. Interestingly, none of these two constraints would prevent you from using Model A or Models A+, but you would need to provide a wireless USB adapter for Internet connectivity, and the diminished memory certainly would slow down the RasPi applications while they were running. The Raspberry Pi Foundation's (RPF's) design intent for the A models was that they would be used in strictly embedded applications where user interaction is typically not required, such as monitoring and controlling a home automation application.

Most real-world projects involve using some type of digital input and/or output (I/O) to interface with sensors and actuators. These digital I/Os are generally referred to as *general-purpose input-output* (GPIO). The RasPi GPIOs have specific maximum voltage and current limits that most be followed to prevent damage to the board. Unfortunately, this damage is irreversible, rendering the board nonoperative or useless.

Raspberry Pi GPIO

The RasPi Models A+, B+, and B gen 2 use a 40-pin connector designated as J8 for its GPIO. This connector on the Model B gen 2 is shown in Figure 1-3 with the first two beginning and ending pin numbers annotated for orientation and reference.

The 40-pin connector is a change from the previous 26-pin connector used on Models A and B. The 26-pin connector from a Model B is marked as P1 and is shown in Figure 1-4.

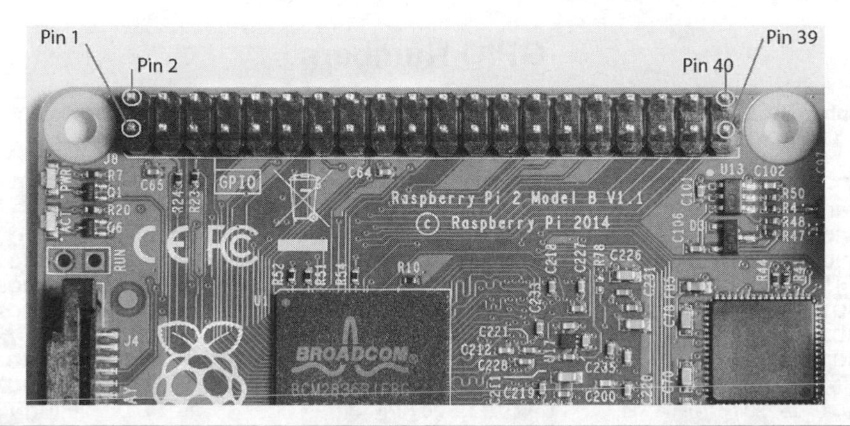

Figure 1-3 Model B gen 2 GPIO J8 connector (40 pins).

Figure 1-4 Model B GPIO P1 connector (26 pins).

The reason for the change was to make nine additional GPIO pins available for project use. These additional pins are clearly shown in Figure 1-5, which also contains the 26-pin configurations for both the RasPi Models A and B.

Many GPIO pins in the P1 and J8 connectors have multiple functions that extend beyond simple digital input and/or output, as shown in the Figure 1-6, which is for the B+, but the pin descriptions hold true for all the RasPi models except for the Compute Module.

You should also note that the first 26 pins on the 40 pin J8 connector are both physically and logically identical to those of the P1 26-pin connector. This means that GPIO connectors designed for earlier RasPis may be used with the 40-pin connector without any problems as long as the pin-out numbers are followed, that is, 1 to 1, 2 to 2, and so on.

GPIO Numbers

Raspberry Pi B Rev 1 P1 GPIO Header

	Pin No.		
3.3V	1	2	5V
GPIO0	3	4	5V
GPIO1	5	6	GND
GPIO4	7	8	GPIO14
GND	9	10	GPIO15
GPIO17	11	12	GPIO18
GPIO21	13	14	GND
GPIO22	15	16	GPIO23
3.3V	17	18	GPIO24
GPIO10	19	20	GND
GPIO9	21	22	GPIO25
GPIO11	23	24	GPIO8
GND	25	26	GPIO7

Raspberry Pi A/B Rev 2 P1 GPIO Header

	Pin No.		
3.3V	1	2	5V
GPIO2	3	4	5V
GPIO3	5	6	GND
GPIO4	7	8	GPIO14
GND	9	10	GPIO15
GPIO17	11	12	GPIO18
GPIO27	13	14	GND
GPIO22	15	16	GPIO23
3.3V	17	18	GPIO24
GPIO10	19	20	GND
GPIO9	21	22	GPIO25
GPIO11	23	24	GPIO8
GND	25	26	GPIO7

Raspberry Pi B+ B+ J8 GPIO Header

	Pin No.		
3.3V	1	2	5V
GPIO2	3	4	5V
GPIO3	5	6	GND
GPIO4	7	8	GPIO14
GND	9	10	GPIO15
GPIO17	11	12	GPIO18
GPIO27	13	14	GND
GPIO22	15	16	GPIO23
3.3V	17	18	GPIO24
GPIO10	19	20	GND
GPIO9	21	22	GPIO25
GPIO11	23	24	GPIO8
GND	25	26	GPIO7
DNC	27	28	DNC
GPIO5	29	30	GND
GPIO6	31	32	GPIO12
GPIO13	33	34	GND
GPIO19	35	36	GPIO16
GPIO26	37	38	GPIO20
GND	39	40	GPIO21

Key

Power +	UART
GND	SPI
I²C	GPIO

Figure 1-5 GPIO pin configurations for RasPi Models A, B, A+, B+ and B gen 2.

Raspberry Pi J8 Header (Model B+)

GPIO#	NAME			NAME	GPIO#
	3.3 VDC Power	1	2	5.0 VDC Power	
8	GPIO 8 SDA1 (I2C)	3	4	5.0 VDC Power	
9	GPIO 9 SCL1 (I2C)	5	6	Ground	
7	GPIO 7 GPCLK0	7	8	GPIO 15 TxD (RS232)	15
	Ground	9	10	GPIO 16 RxD (RS232)	16
0	GPIO 0	11	12	GPIO 1 PCM_CLK/PWM0	1
2	GPIO 2	13	14	Ground	
3	GPIO 3	15	16	GPIO 4	4
	3.3 VDC Power	17	18	GPIO 5	5
12	GPIO 12 MOSI (SPI)	19	20	Ground	
13	GPIO 13 MISO (SPI)	21	22	GPIO 6	6
14	GPIO 14 SCLK (SPI)	23	24	GPIO 10 CE0 (SPI)	10
	Ground	25	26	GPIO 11 CE1 (SPI)	11
	SDA0 (I2C ID EEPROM)	27	28	SCL0 (I2C ID EEPROM)	
21	GPIO 21 GPCLK1	29	30	Ground	
22	GPIO 22 GPCLK2	31	32	GPIO 26 PWM0	26
23	GPIO 23 PWM1	33	34	Ground	
24	GPIO 24 PCM_FS/PWM1	35	36	GPIO 27	27
25	GPIO 25	37	38	GPIO 28 PCM_DIN	28
	Ground	39	40	GPIO 29 PCM_DOUT	29

http://www.pi4j.com

Figure 1-6 Raspberry Pi J8 Header (Model B+).

These multipin connectors will be the gateway through which the RasPis will interface with real-world devices. As you are probably aware, software drivers must be loaded that provide the logical interface between the control program, operating system (OS), and GPIO pins. The particular type of driver depends primarily on the programming language used to develop the control program. I will be using the C, Python, and Java languages in this book to develop control programs, so a separate set of drivers will be loaded to accommodate each development environment.

I will not review all these pin functions at this time but will discuss them when they are encountered while building a project. Incidentally, none of the projects connect directly with the P1 pins but instead rely on the use of a Pi Cobbler that is plugged into a solderless breadboard. Figure 1-7 shows the Pi Cobbler adapter plugged into a solderless breadboard with the 26-conductor ribbon cable plugged into a Model B P1 connector.

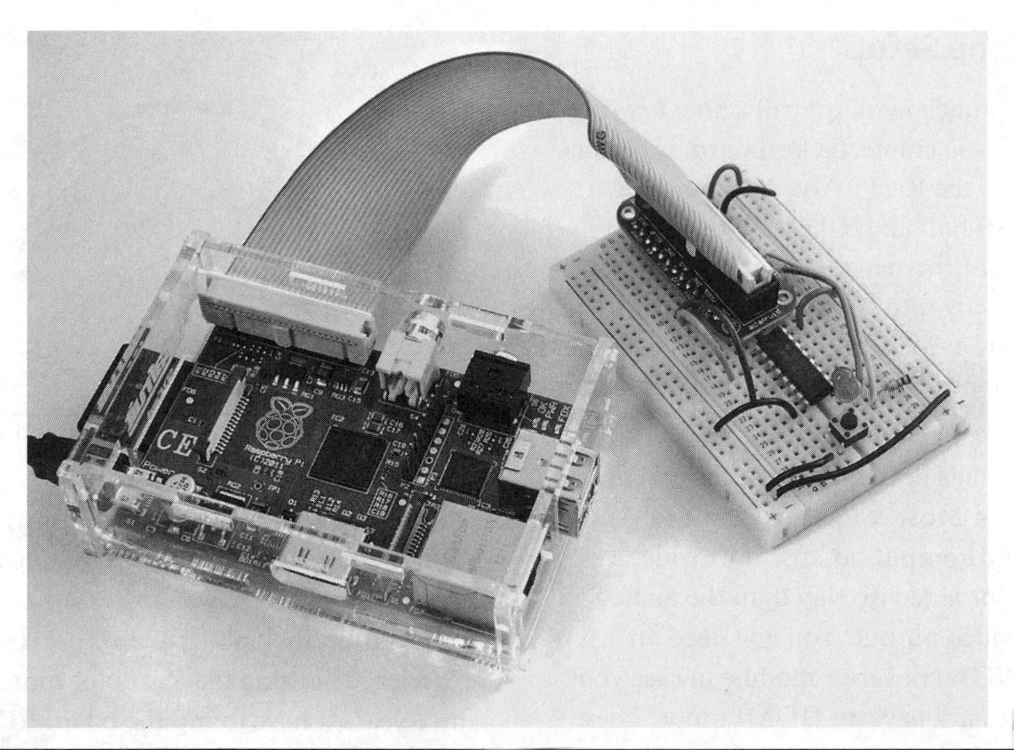

Figure 1-7 Pi Cobbler.

Both 26- and 40-pin Pi Cobblers are available from a variety of suppliers, such as Adafruit Industries and MCM Electronics. You can purchase one either fully assembled or as a kit, which you will have to assemble by soldering a connector to the printed circuit board (PCB). This task, which is not too difficult, allows you to practice your soldering skills. Just don't add too much solder to the connector pins because they are close together, and it is easy to form a solder bridge that might be disastrous to the RasPi when you connect the Pi Cobbler to it.

I prefer to use manufactured jumper wires, as shown in Figure 1-8, when connecting components on a solderless breadboard. These jumpers are very sturdy and can be easily inserted into the breadboard without the bending or crinkling that affects ordinary precut wires. Inexpensive jumper wire kits are also typically available from the same Pi Cobbler suppliers.

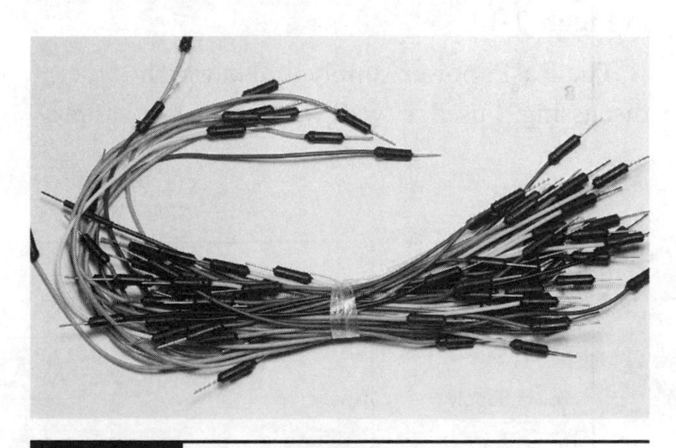

Figure 1-8 Manufactured jumper wires.

Establishing a RasPi Development Station

There are several ways to set up a RasPi development station, each with its own pros and cons. I will cover two approaches that will likely fulfill the needs of most users.

Stand-Alone Setup

The first approach is what I call a *stand-alone setup*, where you connect a keyboard, monitor, and mouse to the RasPi. You will also need a powered USB hub and either a wireless WiFi adapter or an Ethernet patch cable that you can plug directly into your router or network switch. Figure 1-9 is a block diagram showing all the components needed for a stand-alone workstation.

The RasPi has both composite and HDMI video outputs. Most readers will elect to use the HDMI output because it provides a much superior video display than the analog composite video output. You will need an HDMI-to-VGA converter module in case your monitor does not have an HDMI input. These converters are relatively inexpensive, with a typical module, available from Adafruit, shown in Figure 1-10.

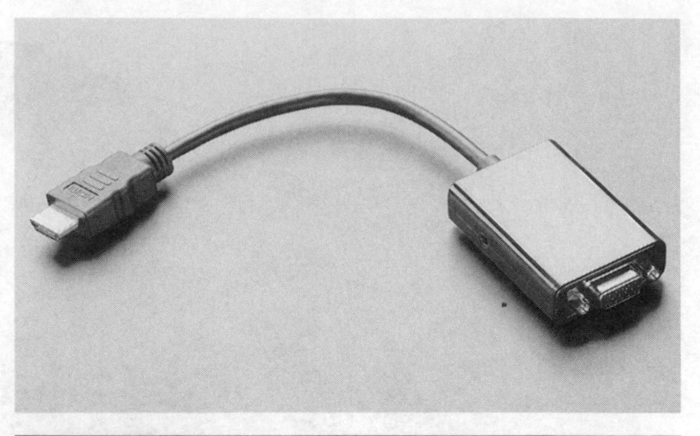

Figure 1-10 HDMI-to-VGA converter module.

The RasPi power supply is also worth discussing. I used a "wall wart" 5-V, 1-A supply, which is more than adequate for providing sufficient current to the RasPi as long as you do not attempt to power any external USB devices from the onboard USB ports. From my experience in using the RasPi for more than three years, I have found the board to be a bit "sensitive" to the quality and level of the 5-V supply. Strange and frustrating events happen if the power supply droops to 4.75 V or less, which is only a 5 percent drop. Often, simply swapping the power supply clears up mysterious and

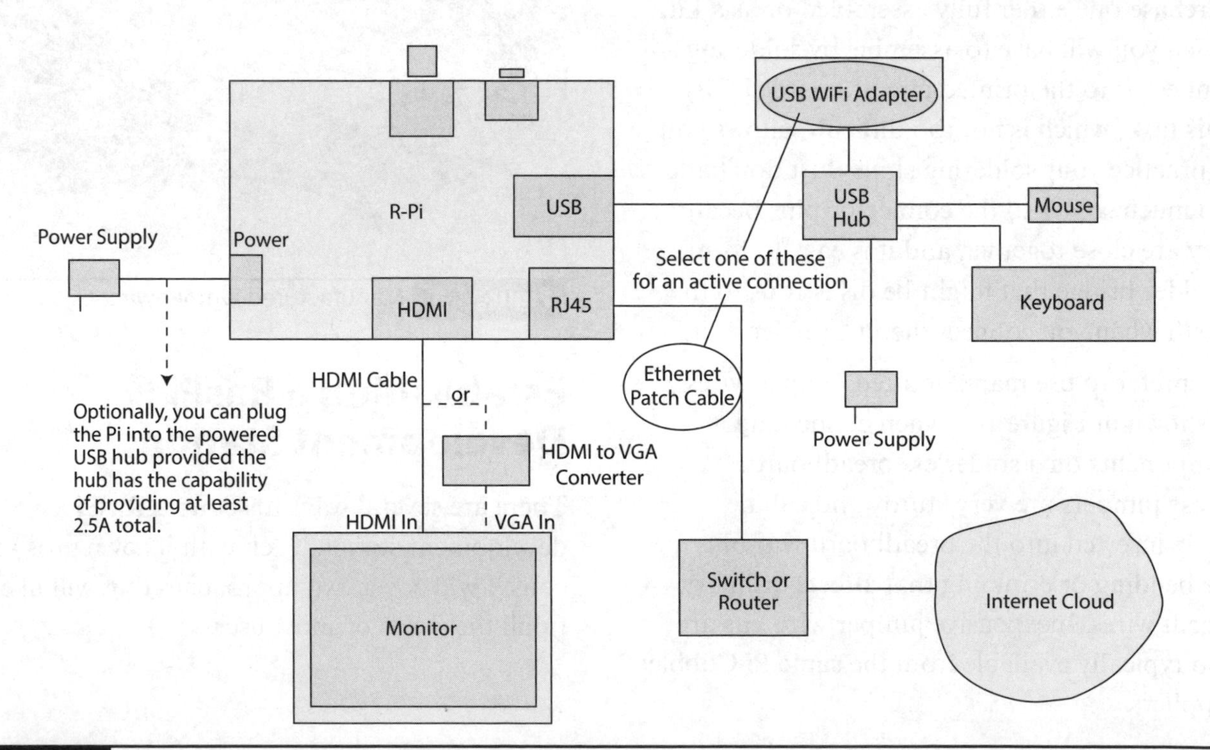

Figure 1-9 RasPi stand-alone workstation block diagram.

Figure 1-11 Pluggable-powered USB hub.

intermittent operational issues, which can lead to unproductive and hair-tearing development sessions. In Figure 1-9, I have included a note that mentions that you can also power the RasPi directly from the hub using a micro-USB/USB cable as long as the hub power supply is rated for a minimum of 2.5 A. I have used the Pluggable series of powered hubs to do this in the past, one of which is shown in Figure 1-11.

Any USB keyboard and mouse combination, whether wired or wireless, will suffice for user input. However, I did find the wireless Logitech K400 keyboard/mouse device to be a very handy and flexible combination. There were no issues with the RasPi detecting this device and installing the proper driver. The K400 is inexpensive and is shown in Figure 1-12. I highly recommend this keyboard/mouse unit.

I would like to mention the wireless WiFi adapter that I have used successfully for a number of projects. It is the EDIMAX EW-7811Un and is shown in Figure 1-13. It is very inexpensive and seems to perform quite well for the relatively low-bandwidth projects with which I have used it.

Figure 1-12 Logitech K400 wireless keyboard/mouse unit.

Figure 1-13 EDIMAX EW-7811Un USB WiFi adapter.

You should note that it is rated at a maximum of 150 Mps, which is somewhat lower than other, more expensive brands. However, none of the projects in this book requires very high bandwidth, so why spend the money for performance you will not require.

Headless Setup

The second approach is not a gruesome RasPi decapitation, as the name suggests, but rather a network-centric configuration to control a RasPi remotely. For this approach, you will only need a networked RasPi and another computer. It doesn't matter if the RasPi is connected wired or wirelessly to your network. All you really need is the Internet Protocol (IP) address that your router assigns to the RasPi when it discovers it on initial startup. Note that no keyboard, mouse, monitor, or powered hub is required for this setup, just a RasPi, a power supply, and either an Ethernet cable or a wireless WiFi adapter are needed. Figure 1-14 is a block diagram showing all the headless components and their interconnections.

The secret to the simplicity of the headless setup is the software running both on the RasPi and on the computer used to communicate with the RasPi. This software will be one of the items discussed in the following software section.

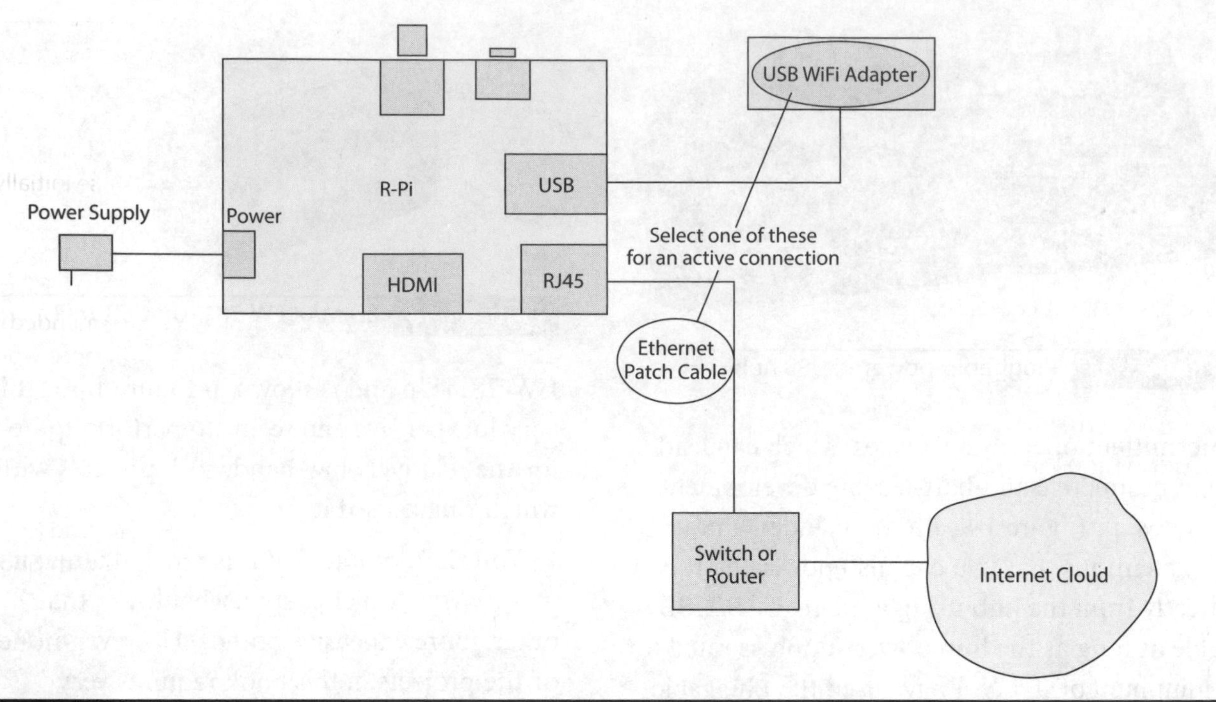

Figure 1-14 RasPi headless workstation block diagram.

The last hardware item that needs to be discussed is the SD card that stores the software the RasPi needs to function. A standard 4-GB SD card is the minimum required for RasPi operations, but I strongly suggest you use at least an 8- or 16-B card so that you have enough space for all this book's projects without having to delete any one of them. It is fairly easy to add software whose memory requirements can quickly add up to the point where RasPi operations could be adversely affected. However, don't be deterred if you purchased a RasPi starter kit that came with a prebuilt image 4-GB SD card. It will be sufficient for all this book's projects, but you might have to delete some early project files to ensure there is space for later projects.

SD cards are also rated for speed with a class number. Table 1-2 shows the various classes and the associated minimum data transfer speeds.

TABLE 1-2 SD Card Class Designations

Class	Minimum Performance
Class 2	2 MB/s
Class 4	4 MB/s
Class 6	6 MB/s
Class 10	10 MB/s

Using an SD card with a higher class number in the RasPi allows for much better performance. Just be mindful that SD cards with high class numbers are more expensive than ones with lower numbers, but the cost differential seems to be getting smaller as time progresses. I strongly suggest that you purchase at least a Class 4 or higher card; anything less and you will be disappointed in how slow your RasPi responds.

Finally, don't be worried about how to create an operational RasPi SD card. I will show you in the software section how to download and store the latest software image on a blank SD card. It really is quite easy, and you will feel like an expert after a few downloads and stores.

Setting Up the RasPi Software

I will begin this section by assuming that you are starting out using a stand-alone workstation with a blank SD card. Your first step is to set up the SD card with a suitable OS image from which to boot the RasPi. Go to the RPF's download website at http://www.raspberrypi.org/downloads and download the file named NOOBS_v1_5_0.zip, which was current at the time of this writing. I am sure that a later revision will be available when you visit the site, which is okay. The name NOOBS is short for "New Out Of Box Software" and is a recent revision to the way the RasPi images have been traditionally made available by the RPF. This is a compressed file that should be extracted directly to the SD card that must be inserted into the computer that holds the downloaded NOOBS file. You must ensure that the SD card is properly formatted before you extract or unzip the file. The easiest way to format the SD card is to use the SD Card Association's formatting tool, SDFormatterv4.zip, which also may be downloaded from the same RPF website mentioned earlier. Of course, the formatting tool also must be extracted before use.

The freshly formatted and NOOBS-loaded SD card has been designed to boot the RasPi into a clever menu that allows you to select one of four operating systems. To boot the RasPi, first ensure that the workstation is set up as shown in Figure 1-7 without the power supply attached to the RasPi board. It is okay to power up the USB hub, provided that the hub is not directly powering the RasPi.

Next, insert the NOOBS SD card into the RasPi, and then connect the power supply to the RasPi. If everything has been done properly, you will see the NOOBS menu selection displayed. The NOOBS revision menu selection has eight choices that are detailed in Table 1-3.

TABLE 1-3 Initial NOOBS Selection Menu

Menu Selection Name	Description
Raspian (*recommended*)	This is the recommended Linux distribution you should use initially. It is based on the Debian "Wheezy" distribution and has the LXDE desktop built in.
Arch	This distribution is recommended for experienced users or those who wish to learn Linux using a text-based interface. There is no built-in desktop.
OpenELEC	This is the Ubuntu 13.04 distribution customized for the RasPi. Users should try this one after using Raspian. It usually comes with Gnome desktop.
Pidora	This is a fully functional distribution based on Fedora 18 using the Xfce desktop. *NOTE:* Future open-source support for Pidora is "problematic," meaning that it will not likely be updated.
RISC OS	This is a British developed and maintained OS especially designed for ARM processors such as the one that runs the RasPi. It is compact (memory-wise) and very efficient. *NOTE:* It is *not* Windows or Linux but its own open-source OS.
RaspBMC	This is a distribution that is media-centric in that it blends Raspian with the XBOX media center. It is fairly mature now with many of the initial problems resolved. It turns the RasPi into a very nice media center that supports most media formats.
Raspian (Boot to Scratch)	This Raspian variant is set to boot directly into the Scratch application. It is useful for users who wish to focus on learning and using the Scratch app. Readers of this book will not find this OS useful.
Data Partition	In my humble opinion (IMHO), this is an odd selection to add to the menu. It sets up a 512-MB partition to the SD card. Using the gparted application is much more useful and general purpose. This selection is *not recommended*.

I strongly recommend you select the first menu item, which is to install the latest Raspian distribution. The top menu Install button will become active after you click on any selection. Simply click on Install to commence the installation process.

A dialog box asking you to confirm that the pending install will delete any existing data from the SD card will appear next. This is the last time you can avoid the serious mistake of overwriting an SD card that you didn't intend to use. Click on OK assuming that everything is proper and that you are indeed using the desired SD card.

Next, a series of screens will be displayed commencing with a Welcome message that also contains a progress bar indicating how much of the installation has been completed. The install will take a while depending on the size of the distribution and the data transfer speed of the SD card you are using. The initial installation portion will be complete when the Raspian banner is displayed. My only comment on this screen is that it contains, in part, this statement: "based on Linux and optimized for the Raspberry Pi." This is all true, but it really should have included an indication that it is based on the Debian Linux distribution because there are significant differences between Linux distributions, which can be seen from the descriptions in Table 1-3.

The next screen that appears in the installation sequence is very important. Figure 1-15 shows both the username and password that you will need when you attempt to run the Raspian OS on the RasPi. Every Raspian distribution that is downloaded from the RPF website has the same username and password. Obviously, this is not a very secure situation if you connect the RasPi to the Internet. However, never fear, I will show you later on how to change both the username and password to establish much better security for your RasPi installation.

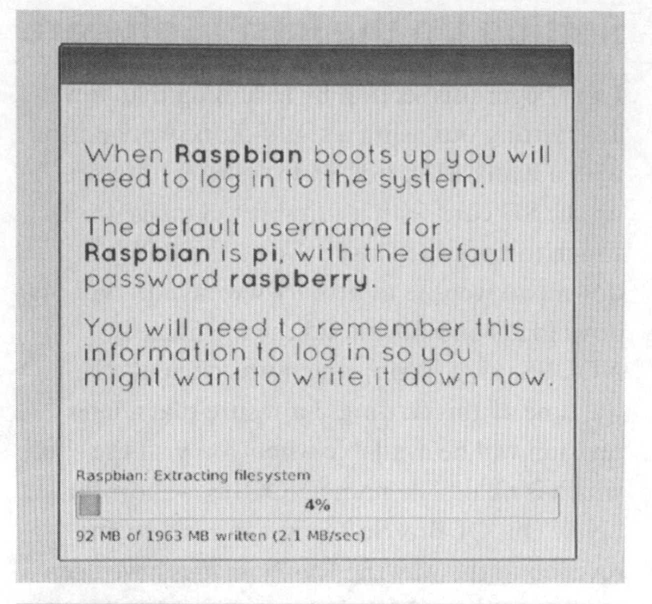

Figure 1-15 Default username and password screen.

You will next see the raspi-config introduction screen. The raspi-config main menu is automatically shown the first time you boot up the Raspian OS. Its purpose is to easily allow you to configure your OS to address your needs and requirements. I will shortly discuss the raspi-config application in detail.

The last display screen shown indicates that the Raspian OS has been successfully installed. Notice near the bottom center of the screen is the phrase, "For recovery mode, hold Shift." What this means is that you can get back to the NOOBS opening selection menu by holding down the keyboard's SHIFT key while powering up the RasPi. At this point, you can reinstall an old OS or select a new one. This is very useful if and when you corrupt your existing OS, which is likely to happen with all the experimenting we will be doing. Now, you must be very aware that any data files stored on the NOOBS SD card will be deleted when a reinstallation happens. This is why it is very important to copy and store any and all data files either to a network drive or to removable media such as a thumb drive. Neglecting to do constant backups will cause you distress when you realize that you have corrupted the OS and consequently lost all your data. Also realize that any applications that you might have loaded above and beyond the core Raspian installation will be lost. This is not a problem because you can reload and reconfigure using the same procedures you followed to install them initially. The data, however, are another story, and any data likely will remain gone unless you have done the backups as suggested.

Clicking the OK button on this last screen will reboot the RasPi and eventually bring you to a command-line prompt where you will enter the username (pi) and the password (raspberry). The raspi-config menu screen should now appear.

Table 1-4 shows all the raspi-menu selections arranged by menu number along with descriptions and my recommendation as to what you should do with a particular selection. I believe that you should initially follow my recommendation. You can always change at a later time.

There are also seven Advanced Options menu selections that I discuss in Table 1-5.

Click on the Finish button after you have entered all the raspi-config menu selections. You should be returned to the command line.

Next, enter the following command to check whether you have installed the desktop graphical user interface (GUI) successfully:

```
startx
```

You should see the Desktop screen appear, as shown in Figure 1-16, if the OS installed correctly.

TABLE 1-4 raspi-config Menu and Recommendations

raspi-config Menu Number	Description	Recommendation
1	Expand Filesystem	Do nothing. The SD card will already have been expanded because of the NOOBS installation.
2	Change User Password	Leave it as "raspberry" for now.
3	Enable Boot to Desktop/Scratch	Do nothing. You will need to do most of the configurations at the command line.
4	Internationalization Options	Personalize your language and set your time zone.
5	Enable Camera	Enable the camera option if you have one or are planning on buying the Pi camera.
6	Add to Rastrack	It's your choice if you wish to register your Pi at this private website. There is no advantage either way regarding this book's projects.
7	Overclock	Not recommended. Overclocking adds additional heat stress to the Pi CPU and is not needed for any of this book's projects.
8	Advanced Options	This is a series of additional choices that are discussed later.
9	About raspi-config	A simple credits screen.

TABLE 1-5 raspi-config Advanced Options Menu and Recommendations

raspi-config Advanced Options Menu Number	Description	Recommendation
A1	Overscan	Removes black bars from display. Use it if you need it.
A2	Hostname	Do nothing. It sets a network hostname for the Pi. I will show you how to do this from the command line.
A3	Memory Split	Do nothing. This changes the amount of memory allocated to the graphics processor unit (GPU). The default split is fine.
A4	SSH	Enable SSH. SSH is short for the "Secure Shell Protocol." You will need it for headless operation.
A5	SPI	Enable SPI. SPI is short for the "serial peripheral interface bus," which is used in one of this book's projects.
A6	Audio	Do nothing. There are three modes: 0: Auto (default) 1: Force audio to headphone jack 2: Force audio to HDMI connector
A7	Update	Do nothing. I will show how to both update and upgrade from the command line.

Figure 1-16 Desktop GUI.

This screen is the LXDE Desktop, which is the default Raspian OS GUI interface. For informational purposes only, LXDE is short for "Lightweight X11 Desktop Environment" and is built on the X-Window System. X-Windows has nothing to do with Microsoft Windows but instead is based on a windows framework created at Massachusetts Institute of Technology during the mid-1980s. X-Windows is independent of any particular OS, which means that developers must create appropriate interface software for it to function with a specific OS.

Clicking on the LXDE icon button located in the lower-left corner of the screen pops up a menu with four choices:

- Shutdown
- Reboot
- Logout
- Cancel

Shutdown turns off the RasPi (as the name implies). Reboot causes the RasPi to cycle through a complete restart and presents you with a command-line login prompt after it is done. Logout stops the GUI and brings you right back to a command-line prompt. No reboot or resetting is involved with this command. The Cancel command brings you back to the GUI screen.

You will now have a complete Raspian OS up and running if you have followed all the preceding steps successfully. Before proceeding to any more advanced instructions, I would like to show you how to set up the RasPi using a complete Raspian OS image that may be downloaded from the RPF's website.

How to Set Up the RaspPi OS Using an Image File

This section shows you how to set up a RasPi with a raw image file. This was the only way you could create an operating OS prior to introduction of the NOOBS software. It is important to understand this procedure because it allows you to load any OS image and not be limited to the ones contained in NOOBS.

The first step is to download the desired image file from the RPF's download website. This is the same site mentioned earlier where you downloaded the NOOBS software. The image software is located further down on the website listing from the NOOBS section. At the time of this writing, the current Raspian image was listed as 2015-02-16-wheezy-raspian.zip. It will need to be unzipped or extracted before being further processed.

You cannot simply unzip the file onto an SD card. It won't work because the image must be transferred in a very specific manner for it to boot and function properly as an OS. There is a free open-source program named Win32DiskImager that you would use on a Windows computer to transfer the unzipped image to a formatted SD card. This program is available from the sourceforge site at http://sourceforge.net/projects/win32diskimager/files/latest/download. The program download is in a zipped format that must be extracted to a convenient location prior to use. Figure 1-17 is a screen capture of the Win32 Disk Imager program in action downloading the latest Raspian image to a Class 10 SD card.

Notice the more than 17 MB/s transfer rate shown in the figure. You will quickly appreciate using high-speed SD cards because they allow read and write operations to occur an average of two to three times faster than the much more common Class 4 SD cards.

All you need to do next is put the newly imaged SD card into the unpowered RasPi and apply power to start the boot process. This is what I did, and I saw absolutely nothing on the monitor screen. This was certainly discouraging because I was sure that I had done everything as described in the RPF's instructions. It turns out that this raw Raspian image caused the RasPi to default to the analog video output instead of

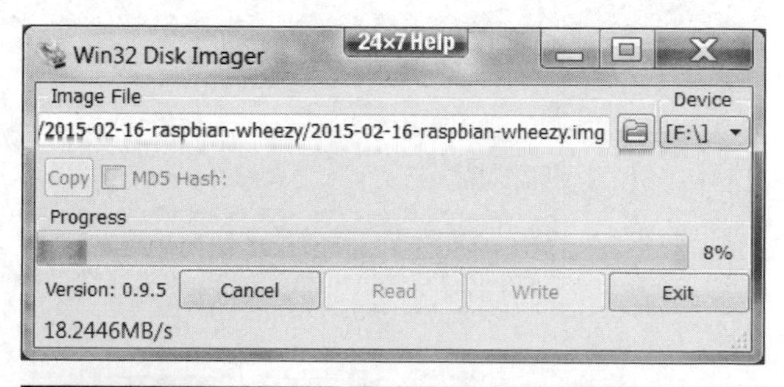

Figure 1-17 Win32 Disk Imager program executing.

using the HDMI output to which my monitor was attached. This was not the case with the NOOBS installation, which apparently defaults to the HDMI video output. In any case, it is fairly easy to remedy this situation. Figure 1-18 is a listing of all the files that are installed on the SD card after the Win32 Disk Imager finishes executing. Note that this screen shot is from the laptop that I used to create the SD card and not from the RasPi.

Shown in the list near the bottom is a file named config.txt, although the .txt extension is not shown in the file name list because of my Windows folder configuration. This file must be edited for the video display to appear on the HDMI video output. Figure 1-19 shows this file's content using the Notepad editor.

Name	Date modified	Type	Size
overlays	2/15/2015 1:07 PM	File folder	
bcm2708-rpi-b.dtb	2/15/2015 10:20 A...	DTB File	5 KB
bcm2708-rpi-b-plus.dtb	2/15/2015 10:20 A...	DTB File	5 KB
bcm2709-rpi-2-b.dtb	2/15/2015 10:20 A...	DTB File	6 KB
bootcode.bin	2/15/2015 10:20 A...	BIN File	18 KB
cmdline	2/15/2015 1:08 PM	Text Document	1 KB
config	2/15/2015 1:08 PM	Text Document	2 KB
COPYING.linux	2/15/2015 10:20 A...	LINUX File	19 KB
fixup.dat	2/15/2015 10:20 A...	DAT File	6 KB
fixup_cd.dat	2/15/2015 10:20 A...	DAT File	3 KB
fixup_x.dat	2/15/2015 10:20 A...	DAT File	9 KB
issue	2/16/2015 1:57 PM	Text Document	1 KB
kernel	2/15/2015 10:20 A...	Image Files	3,904 KB
kernel7	2/15/2015 10:20 A...	Image Files	3,877 KB
LICENCE.broadcom	2/15/2015 10:20 A...	BROADCOM File	2 KB
LICENSE.oracle	9/25/2013 9:57 PM	ORACLE File	19 KB
start.elf	2/15/2015 10:20 A...	ELF File	2,581 KB
start_cd.elf	2/15/2015 10:20 A...	ELF File	542 KB
start_x.elf	2/15/2015 10:20 A...	ELF File	3,516 KB

Figure 1-18 Raspian raw image file listing.

You will need to uncomment the line

```
#hdmi_force_hotplug=1
```

All you need to do to uncomment the line is to delete the # symbol from the line's beginning, then save the file, and exit Notepad. The SD card should now be all set to display the boot sequence from the HDMI port.

Booting a raw image will bring you to the raspi-config screen, as was the case for the NOOBS installation. All the recommendations made for that installation hold true for this one, with the addition of expanding the file system. The NOOBS installation does this automatically, but this is not the case for this more manual install.

Updating and Upgrading the Raspian Image

The NOOBS software and raw-image OS should be updated and upgraded to have the

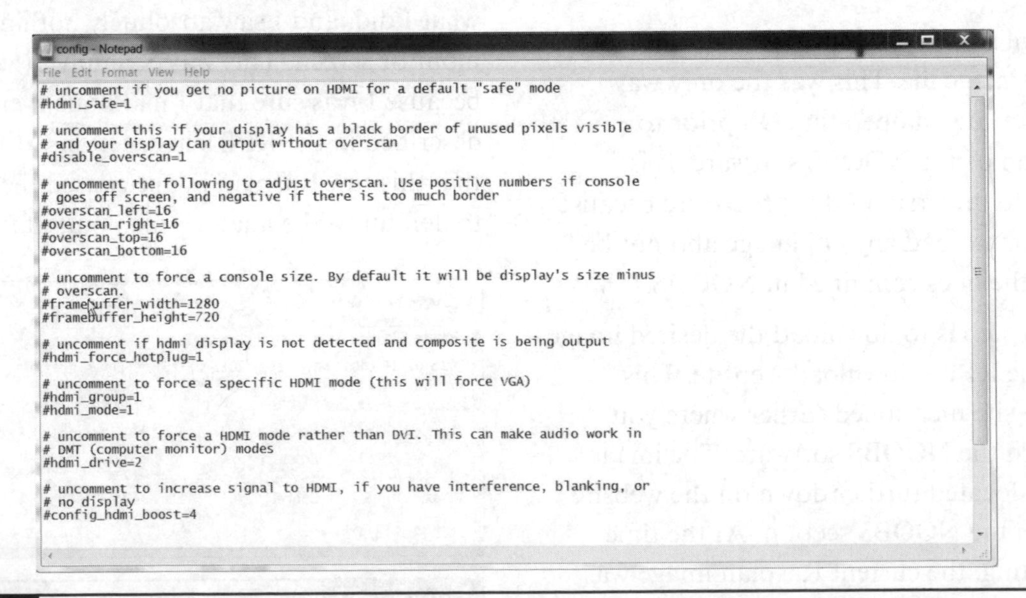

```
# uncomment if you get no picture on HDMI for a default "safe" mode
#hdmi_safe=1

# uncomment this if your display has a black border of unused pixels visible
# and your display can output without overscan
#disable_overscan=1

# uncomment the following to adjust overscan. Use positive numbers if console
# goes off screen, and negative if there is too much border
#overscan_left=16
#overscan_right=16
#overscan_top=16
#overscan_bottom=16

# uncomment to force a console size. By default it will be display's size minus
# overscan.
#framebuffer_width=1280
#framebuffer_height=720

# uncomment if hdmi display is not detected and composite is being output
#hdmi_force_hotplug=1

# uncomment to force a specific HDMI mode (this will force VGA)
#hdmi_group=1
#hdmi_mode=1

# uncomment to force a HDMI mode rather than DVI. This can make audio work in
# DMT (computer monitor) modes
#hdmi_drive=2

# uncomment to increase signal to HDMI, if you have interference, blanking, or
# no display
#config_hdmi_boost=4
```

Figure 1-19 Config.txt file contents.

latest software revisions and patches in place. The update should be done first by entering the following at the command-line prompt:

```
sudo apt-get update
```

This will normally take several minutes depending on how out of date the OS image was at install time versus the number of updates issued from the OS image publication date. Incidentally, I want to explain this command a bit further for those of you without much Linux command experience: `sudo` instructs the OS to execute the commands that follow as if an administrator issued them. Linux is constructed by privilege layers with the admin layer at the top or having the least number of restrictions. I discuss this in much further detail later in this chapter. `apt-get` update simply instructs the OS to get the update.

At this point, you should have a fully functional and updated RaspPi running the Raspian Linux distribution after completing either the NOOBS or raw-image installation. This must be in place before proceeding with any of the following RaspPi projects.

Headless Configuration

Unfortunately, it is one of those catch-22 situations (apologies to younger readers who don't know what this means—Google it!) where you need a fully configured SD card to run a headless configuration. But you can't configure it without a stand-alone workstation, as described earlier, or a preimaged SD card. My strong recommendation would be to but a preimaged SD card if you already know beforehand that you want to run headless.

A headless configuration was shown in Figure 1-13, which indicated that you only need to connect to a RasPi in a network using either an Ethernet cable or a wireless WiFi adapter. The network router will automatically provide an IP address to the RasPi using what is known as the Dynamic Host Configuration Protocol (DHCP), which is normally the default setup in most home or business wireless routers. What you need to do is attach another computer to your network that runs a program that can connect to the RasPi and remotely run it. For Windows computers, that program is called PuTTY and is freely available for download at http://www.chiark.greenend.org.uk/~sgtatham/putty/download.html.

PuTTY uses the SSH Protocol to communicate with the RasPi, and this protocol must be enabled on the RasPi to allow the communication link to function. You will need to determine the RasPi's IP address to establish this link. The following procedure is usually successful in determining the RasPi's IP address:

1. Open a browser session on the computer you wish to use to control the RasPi.
2. Go to the admin IP address for the router that is the DHCP server for your network. Often it is at 192.168.0.1.
3. Enter the username and password to get to the control webpage. These are normally shown in the instructions that came with the router, but they are also readily available by doing an Internet search for your specific router model.
4. Click on Attached Devices or some similar menu selection that displays the IP addresses of all devices attached to the network, whether through wires or wireless.
5. Look for the entry labeled Raspberry Pi. This is the IP address you will need for PuTTY.

It is a simple matter to connect to the RasPi through your computer once you have the RasPi's IP address. I will also show you another way to determine or confirm the RasPi's IP address a little later.

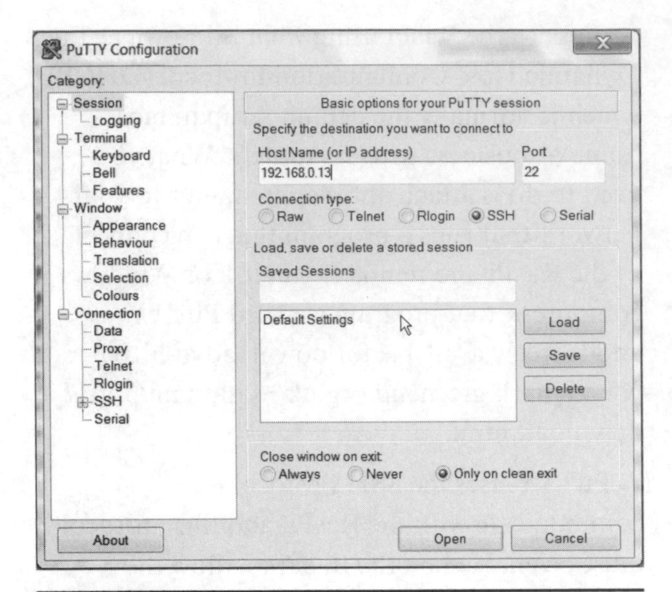

Figure 1-20 Initial PuTTY screen.

Start PuTTY, and enter the IP address in the Host Name (or IP Address) text block. Leave all the other selections and text blocks alone. Figure 1-20 shows the initial PuTTY screen with my RasPi's IP address entered into the Host Name text block.

Your RasPi's IP address likely will be different from the one I entered into the Host Name block. Also check that the port number is set at 22, which is the default for the SSH Protocol. You should see the RasPi's command-line opening screen after you click on the Open button located at the bottom of the PuTTY opening screen. Figure 1-21 shows the RasPi login screen being delivered to the remote computer from the RasPi via SSH.

Figure 1-21 Opening the RasPi login screen over the network.

Just enter the default username (pi) and the default password (raspberry), and you will see the normal command-line prompt appear, as shown in Figure 1-22.

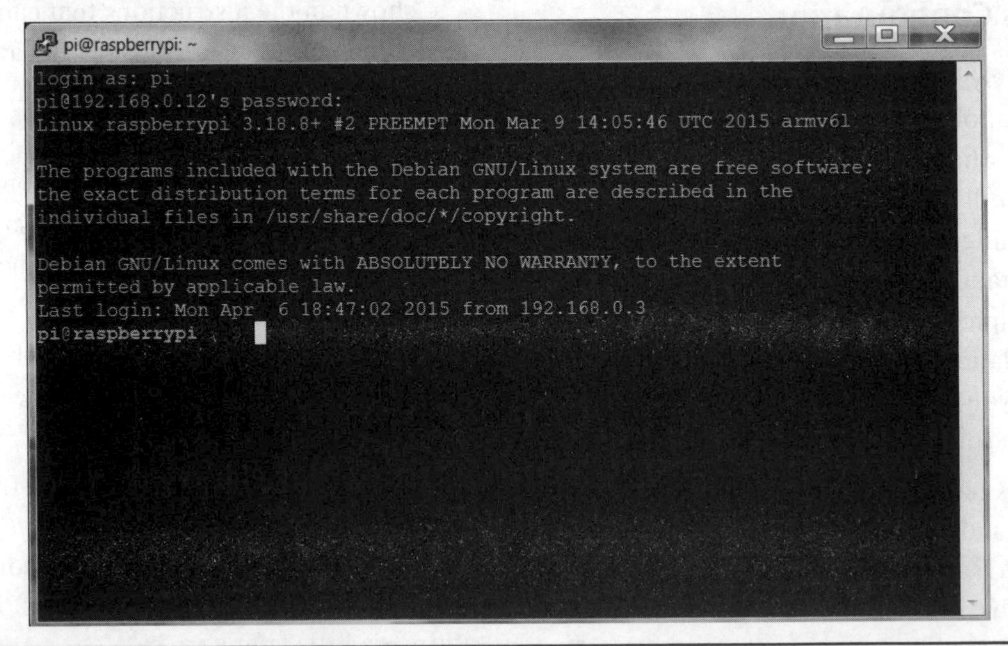

Figure 1-22 RasPi command-line prompt.

You will now be able to interact with the RasPi in exactly the same way as if you were sitting in front of a stand-alone workstation. The one major limitation with the SSH Protocol is that it is text only and you cannot open a GUI desktop. This would be fine for most operations, but it would prevent you from running any program with graphics, which, in my opinion, is a big constraint. However, there is a great solution to this situation that I will discuss in the next section.

Headless Operation with Graphics

Linux has a wonderful program suite named xrdp, which stands for "X11 Remote Desktop Protocol." I first mentioned the X11 server in the earlier discussion of the LDXE desktop GUI. This is the same server engine used in this software suite. xrdp also contains a virtual networking connection (VNC) server called tightvncserver that functions in a similar manner to SSH except that it handles both text and graphics. Type the following command to install xrdp on the RasPi:

```
sudo apt-get install xrdp
```

This program suite takes only a few minutes to install and takes up about 11 MB of file space. You start the VNC server by entering the following at the command line:

```
vncserver
```

Every time you start the VNC server, you will use the following line:

```
New 'X' desktop is raspberrypi:x
```

where the lowercase `x` represents a number. The first time you start, the number should be a `1`. You need to remember this number because it is an important parameter when you run the Windows client on the remote computer. Also, at the first startup, you will be prompted to enter a password that can be up to eight characters in length. You will need to input this password when you authenticate the remote computer with the RasPi. This is all that is required on the RasPi or server side; it is now time to focus on the Windows or client side.

You will need to download a free VNC suite from http://tightvnc.com/download.php. This download includes both server and client VNC packages, but only the client package is needed for this configuration. The website has two Windows installers (.msi files), one for 32-bit machines and another for 64-bit machines. Select the appropriate one for your computer, and install it.

Go into the Start menu, Program Files, and find the TightVNC folder. Click on it, and then double-click on the TightVNC Viewer menu item. You should see the screen shown in Figure 1-23.

Enter your RasPi's IP address in the Remote Host text box, as seen in the figure. Also append a colon to the number that you saw when you started the RasPi's VNC server. In this case, I added `:2`. Yours will be different. Then click the Connect button next to the text box. If everything goes smoothly, you should see the RasPi's VNC server authentication dialog box appear, which is shown in Figure 1-24.

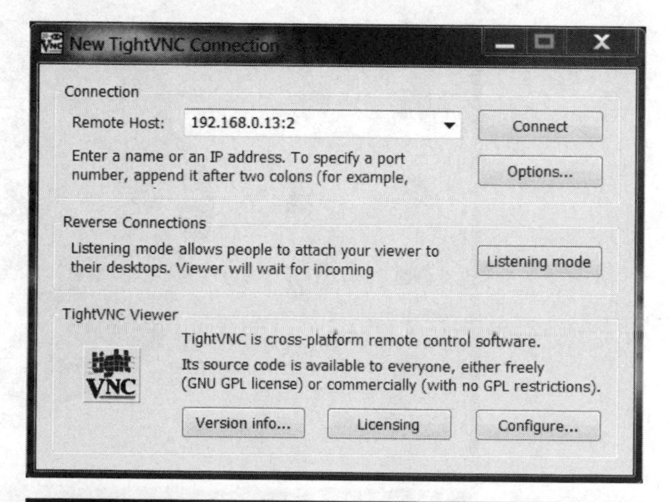

Figure 1-23 Opening screen for the TightVNC Viewer.

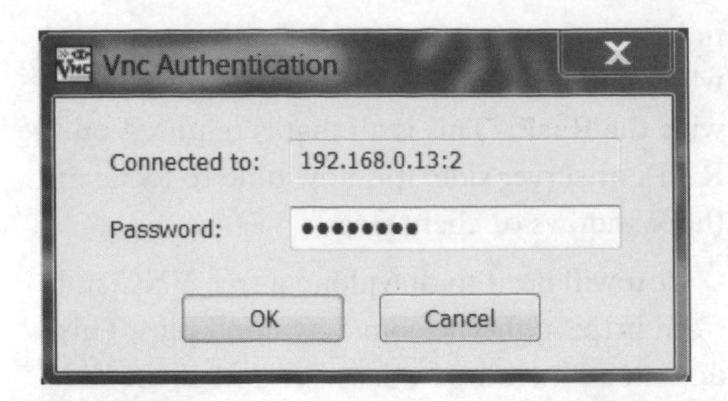

Figure 1-24 The Raspberry Pi VNC authentication dialog box.

Enter the password you created when you first configured the RasPi's VNC server. Click the OK button, and you should see the classic Raspian LXDE GUI desktop, as shown in Figure 1-25.

There is absolutely no difference in using this desktop GUI versus interacting with the stand-alone desktop GUI. I also launched a terminal window, shown in Figure 1-26, to demonstrate that everything responds as it should, even though it is a remote desktop connection.

This really is very cool technology when you consider what has taken place. I am remotely controlling the GUI desktop of an extremely inexpensive Linux computer using a completely separate Windows computer, all with free, open-source software. I guess that it's just the geek in me surfacing to appreciate this setup. I hope you also appreciate it.

I will close this chapter with a discussion of Linux users, privileges, and permissions because they will be important considerations on how all the software is run for this book's projects.

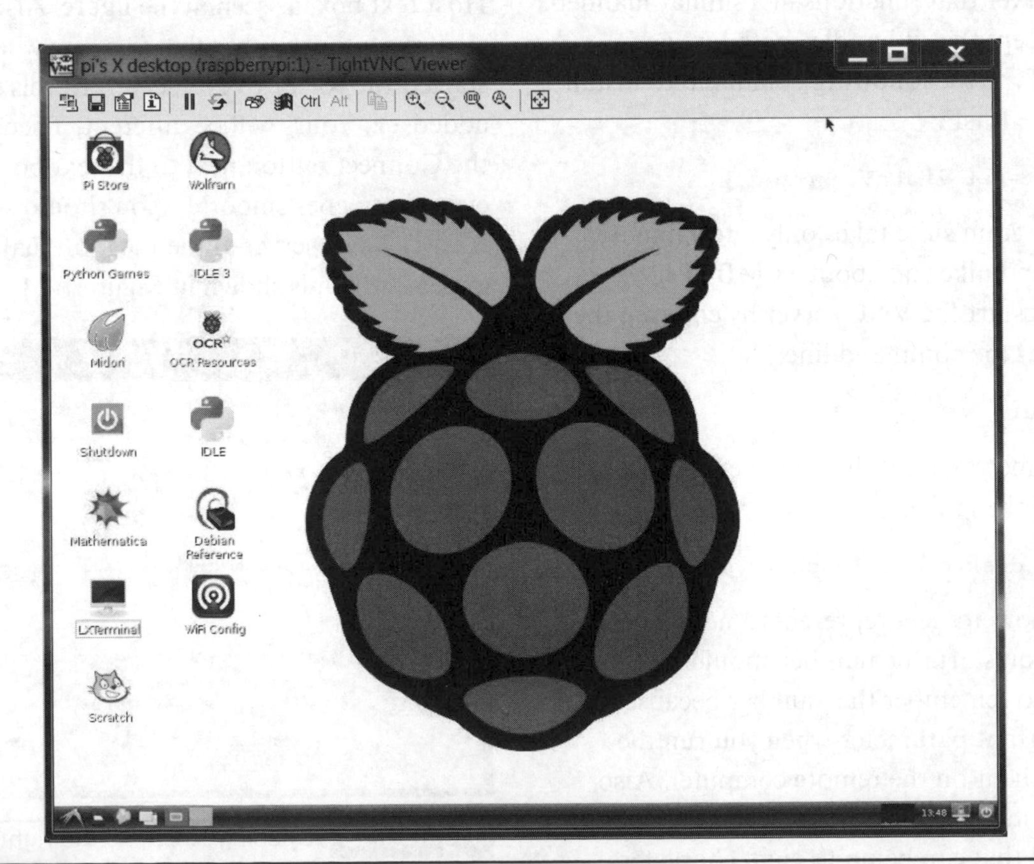

Figure 1-25 Raspian LXDE desktop served by the TightVNC connection.

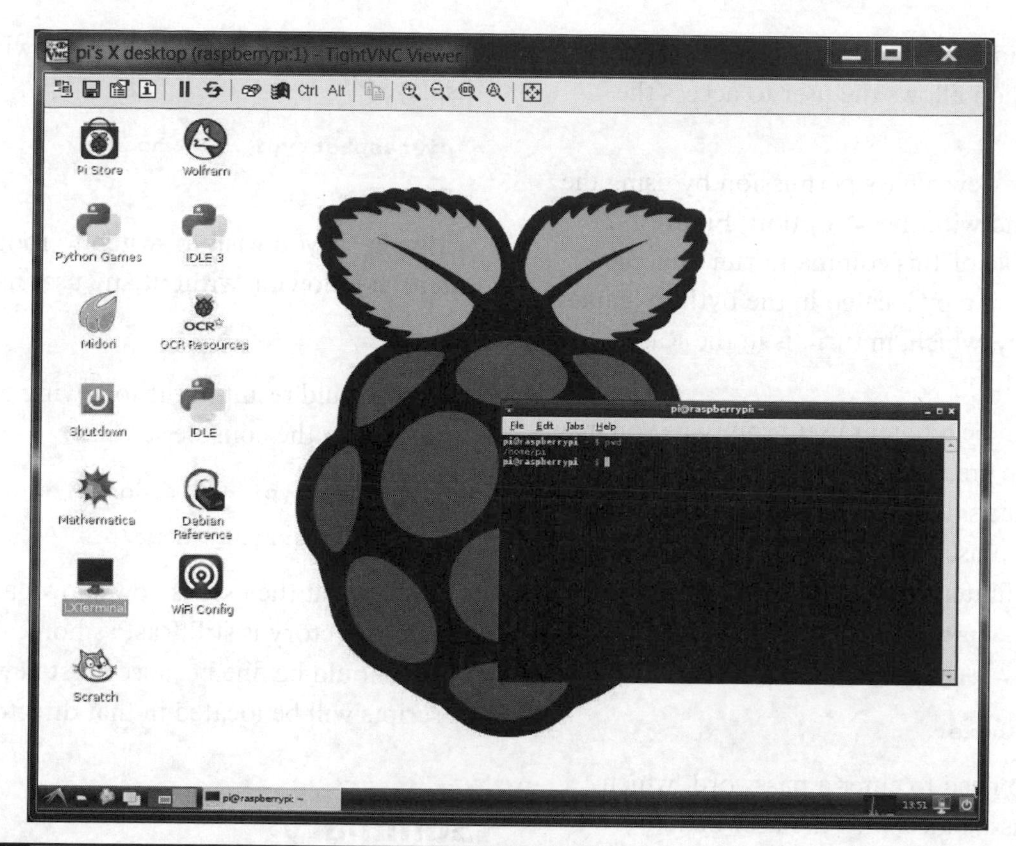

Figure 1-26 Launching a terminal window in the desktop.

Linux Users, Privileges, and Permissions

Not all Linux users are created equal; one named *root* has administrative powers that enable the root user to access all files and directories and to perform any possible operation allowed within the Linux OS. Another name for root is the *super user*, which pretty much describes the overarching powers of this user. You will often need to act as root to perform certain operations, such as opening kernel-level files and creating new files using a text editor. Fortunately, there is an easy way for a regular user named *pi* to act as root, and this is to invoke the `sudo` command. sudo is really an application that is already included in the Wheezy Debian distribution because it is used so often. For example, to create a new Python file named test.py using the nano editor, you would enter

```
sudo nano test.py
```

The Linux OS also has user privileges that date back to the original Unix OS design principles that the OS should simultaneously support multiple users and that these users would not interfere or somehow disrupt each other. This is where the concept of permissions came into being. A *permission* may be thought of as a right to interact with a file or directory, and remember that almost everything in Linux is considered to be a file from the OS perspective. There are three permission types assigned to every file:

- `read`: This allows a file's contents to be accessed but not altered.
- `write`: This allows a file's contents to be modified.
- `execute`: This permits a file to be run or executed. Of course, the file must be in

an executable format. For directories, this permission allows the user to access the directories.

You may view a file's permission by using the `ls` command with the `-l` option. Figure 1-27 is an example of this command run on a file named drawing.py located in the python_games subdirectory, which, in turn, is in the RasPi home directory.

There can be different user groups, as you may see from the preceding figure. For most of this book's projects, you will create and run project software as a user in a directory both named pi. You can definitely change this by adding another user with the `useradd` command. For instance, to create a new user named boxer, you would enter:

```
useradd boxer
```

You also need to enter a password, which would be associated with the user boxer:

```
passwd peachy123
```

To change to the new user, just enter the `su` command, which is short for "switch user":

```
sudo su boxer
```

You will then be prompted to enter the password you created for this new user. Enter the following to check for the current user name:

```
whoami
```

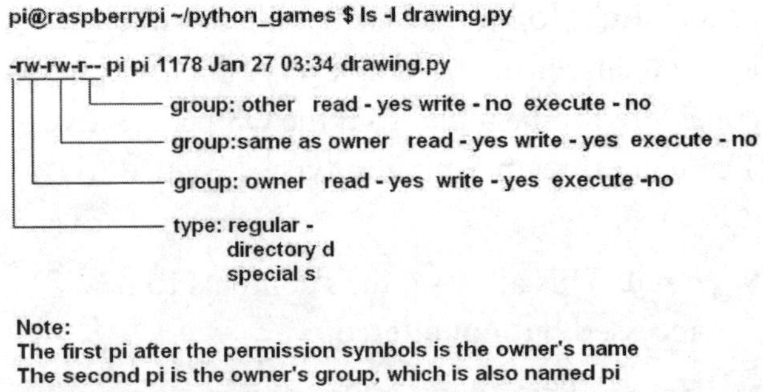

pi@raspberrypi ~/python_games $ ls -l drawing.py

-rw-rw-r-- pi pi 1178 Jan 27 03:34 drawing.py

group: other read - yes write - no execute - no

group:same as owner read - yes write - yes execute - no

group: owner read - yes write - yes execute -no

type: regular -
directory d
special s

Note:
The first pi after the permission symbols is the owner's name
The second pi is the owner's group, which is also named pi

Figure 1-27 Displaying file permissions.

This command results in the following being displayed:

```
pi@raspberrypi ~ $ whoami
pi
```

Finally, if you wish to switch to root, simply enter the following without any user name:

```
sudo su
```

This should result in the following being displayed on the console screen:

```
pi@raspberrypi ~ $ sudo su
root@raspberrypi:/home/pi#
```

Notice that the user is now shown as root, but the directory is still RasPi's home directory, which should be fine because most new programs or scripts will be located in that directory.

Summary

This was an introductory chapter in which I discussed all the available RaspPi models. This book's projects use RasPi Models B, B+, A+, and B gen 2. I showed you how to create bootable SD card images that you could use in both stand-alone and headless configurations. The GPIO digital pin connections also were discussed for all the RasPi models. The GPIO connector is the primary means used to interface a RasPi with sensors and actuators. The chapter concluded with a brief discussion of Linux users, privileges, and permissions.

CHAPTER 2

Touchscreens

In this chapter, I will show you how to install, connect, and program a mini-sized touchscreen display to a RasPi Model B. I will also explain the SPI bus and how you can use a Linux file to control a GPIO pin because both these functionalities are required for this project.

PiTFT Touchscreens

A touchscreen in is just an ordinary LCD device that has an additional human-interface device (HID) that allows users to touch the screen to further interact with the RasPi in accordance with an underlying control/display program. I will be using the 2.8-inch (diagonal) PiTFT, distributed by Adafruit Industries and shown in Figure 2-1.

This touchscreen is very compact and is specifically designed to be plugged into either a Model A or B RasPi. Incidentally, the figure shows the touchscreen mounted on a Model B RasPi, which happens to be running the classic Blender video *Big Buck Bunny*. I will

Figure 2-1 Adafruit PiTFT Model 1601.

Figure 2-2 PiTFT mounted on a Model B RasPi.

demonstrate how to display this video later on in this chapter.

Figure 2-2 shows a side view of the touchscreen mounted on a Model B RasPi, where it is plugged directly into P1, which is the 26-pin GPIO connector.

I will now provide you with a bit of background on the touchscreen technology now that I have shown you the actual touchscreen that will be used for this chapter's project.

Chapter 2 Parts List

Item	Model	Quantity	Source
PiTFT module	1601	1	Adafruit.com
RasPi Model B	998	1	Adafruit.com
RasPi PiBow case	1723	1	Adafruit.com
Tactile button	1489	Package	Adafruit.com

Touchscreen Background

I will begin this discussion by giving credit to Steve Kolokowsky, of the Cypress Semiconductor Corp., whose white-paper content I freely used in this chapter. Incidentally, *Planet Analog* published Steve's white paper entitled "Touchscreens 101" in 2009. Touchscreens may be classified as either resistive or capacitive. Their functions are identical, but how they achieve their functionality is completely different. I will begin by describing the resistive touchscreen because this is the type used in this project.

Resistive Touchscreens

Figure 2-3 shows a layer-view cutaway of a resistive touchscreen.

The touchscreen is made up of six layers, as you can see from the figure. The top layer is a protective flexible, clear-plastic layer, which is the one you press your fingertip or plastic stylus against. Immediately underneath the plastic overlay is the ITO *X*-layer. ITO is an acronym for "indium-tin-oxide," which is a mildly conductive metal alloy. The ITO layer is extremely thin and transparent. It can be thought of as a two-dimensional resistor that is used to detect the *X*-axis touchpoint coordinate. I will discuss how this is accomplished after I

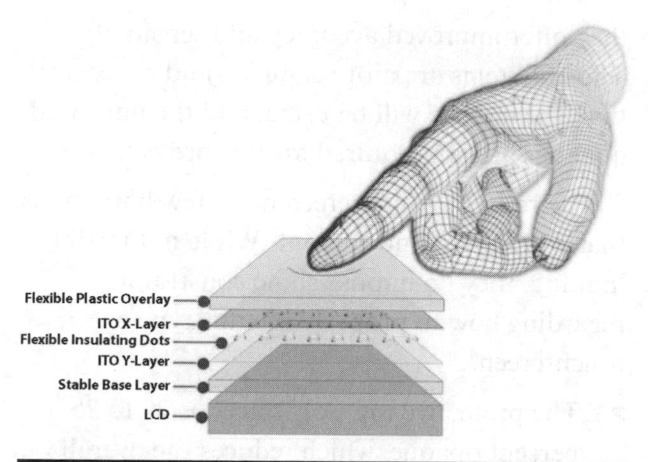

Figure 2-3 Cutaway view of a resistive touchscreen.

introduce the remaining layers. Next follows a layer of nonconductive transparent dots that provide an air gap between the ITO *X*-layer and the ITO *Y*-layer, which is located immediately under the dot layer. The ITO *Y*-layer is made up of the same material as the *X*-layer, and it is used to detect the *Y*-axis touchpoint coordinate. A thin glass layer lies directly under the ITO *Y*-layer, and it provides a stable platform for all the layers above it. Finally, a LCD display with a backlight is positioned under all the layers just discussed. It should be noted that the LCD used in touchscreen design should be very quiet from an electrical noise perspective so as to minimize potential interference with the touchsensor elements.

Figure 2-4 is a block diagram illustrating all the interconnected components necessary to detect the *X*-axis touch-coordinate point.

There are three flexible metalized traces printed on the *X*-axis ITO layer; two are vertical bus lines, and the third is a horizontal sense line. The left-hand bus line has +5 VDC applied to it from the touchscreen controller chip. The right-hand bus line is at ground or 0-VDC potential. These bus lines will cause a distributed current flow across the ITO layer because the ITO alloy acts more like a resistor than a perfect conductor.

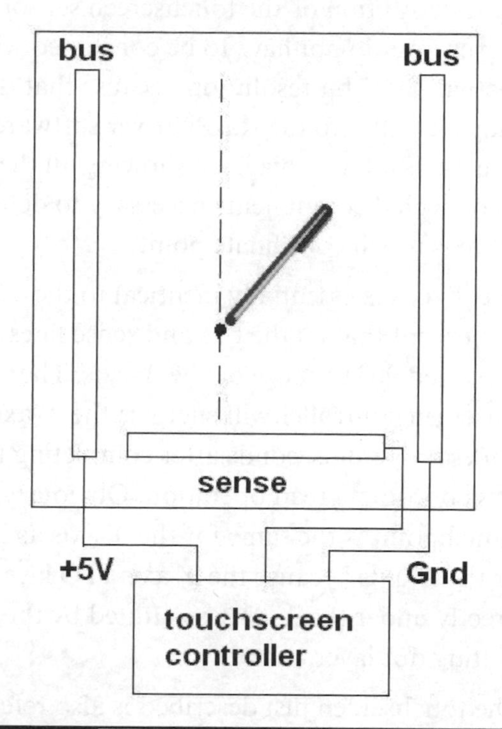

Figure 2-4 Block diagram of the *X*-axis touch-sensor layer.

The ITO layer will deform when a fingertip or a plastic stylus, as shown in the figure, is pressed onto the outermost poly layer. This deformation causes a slight disruption in current flow, which can be detected by the sense line as a voltage between 0 and 5 V. The deformed ITO layer acts as a virtual potentiometer in which the sense line is the virtual center tap, as shown in Figure 2-5.

The sense line is connected to an analog-to-digital converter (ADC) within the touchscreen controller, and the digital equivalent to the sensed voltage is then sent to the RasPi for further processing. The ADC typically has either

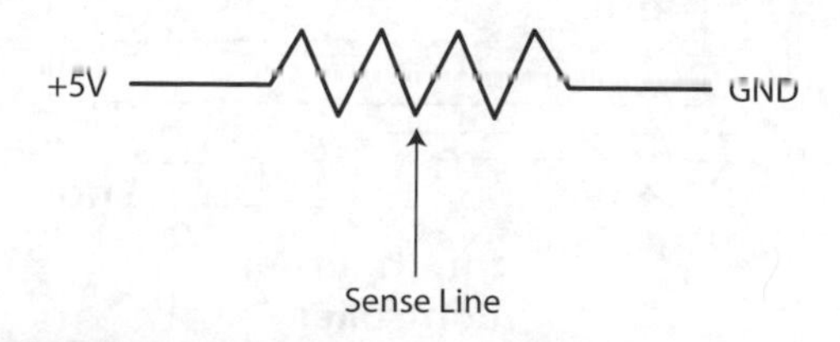

Figure 2-5 Deformed ITO layer as a virtual potentiometer.

10 or 12 bits of resolution depending on the size and resolution of the touchscreen sensor. However, you do not have to be concerned with the actual ADC bit resolution because that is already figured into the RasPi driver software. Figure 2-6 is a block diagram showing all the interconnected components necessary to detect the *Y*-axis touch coordinate point.

The *Y*-axis is essentially identical to the *X*-axis except that all the bus and sense lines are oriented 90° to those of the *X*-axis. The touchscreen controller will energize the *Y*-axis bus lines only milliseconds after completing the *X*-axis coordinate read operation. Obviously, the touchpoint is the same for the *Y*-axis as it is for the *X*-axis because the *Y*-axis ITO layer is directly under the *X*-axis, separated by the insulating dot layer.

The touchscreen just described is also referred to as a *four-wire touchscreen*, as the number refers to the interconnecting wires between the ITO layers and the controller. Five- and eight-wire resistive touchscreens are also available, as

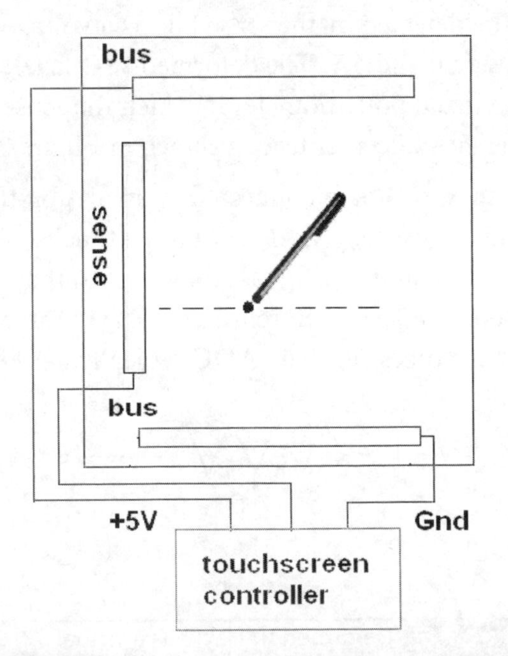

Figure 2-6 *Y*-axis touch sensor layer block diagram.

they offer improved accuracy and sensitivity. Those systems are more complex and expensive than the unit we will be using, and the improved qualities are not required for this project.

The resistive touchscreen has a few limitations that you should know about. While not terribly limiting, they do impose some constraints regarding how to properly use this type of touchscreen.

- The protective top poly screen is 20 to 25 percent opaque, which reduces the overall LCD luminance. This might cause problems when the touchscreen is used outdoors under a bright sun.
- The ITO layers can be a bit nonuniform, which leads to nonlinear resistance within the layer. This translates to inaccurate coordinate position sensing.
- Resistance touchscreens often require more recalibrations than capacitive touchscreens. This might be a nuisance if the touchscreen is remotely deployed and development tools not readily available to accomplish the recalibration.

I also should mention that both the *X* and *Y* digital coordinate values are sent to the RasPi using the SPI Protocol, which I will describe following discussion of the capacitive touchscreen. Although this project uses the resistive version, I believe that it is important to understand how both the resistive and capacitive touchscreens function and when to select one or the other for use in a specific project.

Capacitive Touchscreens

You are probably quite familiar with capacitive touchscreens because they are used universally in just about all modern smart phones and tablets. The capacitive touchscreen functions in a somewhat similar fashion to the

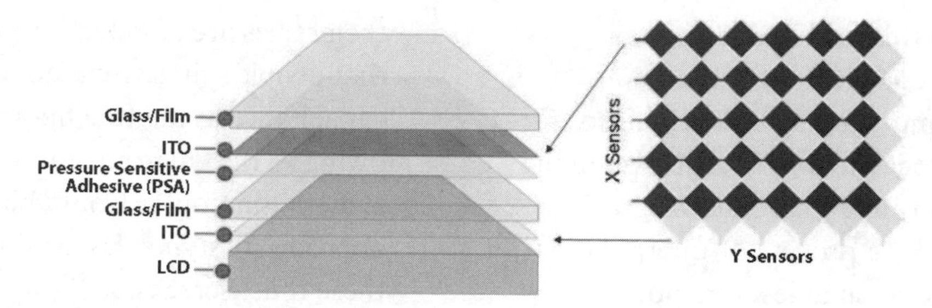

Figure 2-7 Cutaway view of a capacitive touchscreen.

resistive version, but it uses a somewhat different layer structure, as can be seen in Figure 2-7.

This touchscreen uses a glass top layer, which provides 100 percent clarity. The next layer is an ITO layer consisting of an array of closely spaced dots, which are one part of the capacitive sensing scheme. Following this layer is a pressure-sensitive adhesive (PSA) that closely couples the two ITO layers while providing an insulator between them. The next layer consists of more ITO dots, similar to the layer above. The two ITO layers make up a fine grid of microcapacitors, with the PSA layer acting as a dielectric. The display LCD is placed beneath the second ITO layer, completing the touchscreen assembly.

The essence of touch sensing is that when a fingertip or appropriate stylus touches the top glass cover, it will slightly affect the capacitance of one or more of the microcapacitors beneath the touchpoint. The touch effect is not much, generally about a 4 to 5 percent change, but the touchscreen controller can detect this small change in capacitance. The touchscreen controller also uses individual wires to connect to each row and column of the microcapacitor array, and it will poll or sense these wires to detect the capacitors that have changed value as a result of the touch.

As you can imagine, this sensing scheme is quite stable and requires little to no recalibration once it is set up. There is, however, one issue regarding coordinate resolution, which you can see from Figure 2-8. The problem is that placing a fingertip on the top glass screen will affect more than one or two microcapacitors, thus hindering accurate sensing of the desired touchpoint.

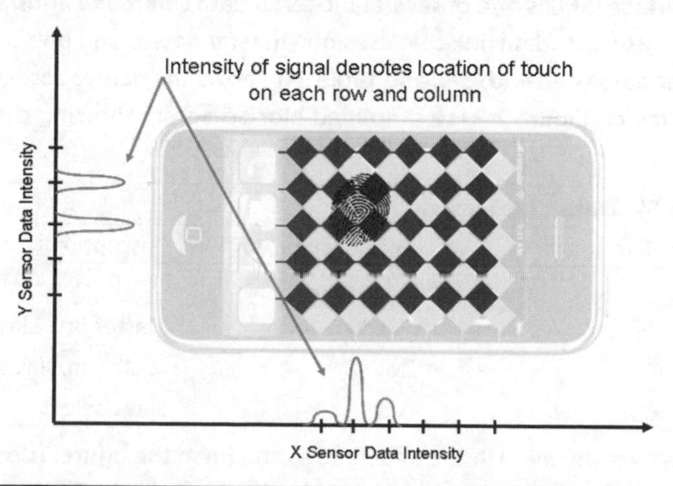

Figure 2-8 Touching a capacitive screen.

The solution to this problem is to employ a technique known as *projected capacitive sensing*. This technique uses all the changed values of the microcapacitors in and around the touchpoint to determine the center of gravity of the touchpoint by a process of interpolation. Note that no physical changes were done to the ITO dots to implement this technique. It's all accomplished by software residing in the touchscreen controller.

NOTE As of February 2015, Adafruit Industries offered a capacitive touchscreen for sale. It is Model 1983 and is a little more expensive than the resistive model I am using for this project. All the software that I describe in this chapter can be used on either type of touchscreen without any modifications as to the type.

There are a few limitations or constraints that you should know before using a capacitive touchscreen in a project. For example:

- Wearing a glove likely will not result in enough of a capacitance change to be detected. You probably already know this if you tried to use a gloved hand with your smart phone.
- Extreme weather also can affect these touchscreens. It is not that the screens themselves are affected; rather, it's just that people's fingers cannot change the capacitance at the touchpoint sufficiently owing to the environmental conditions. In such a situation, it would be wise to use an appropriate stylus designed to work with these touchscreens.
- Some people have a condition, colloquially known as "zombie fingers," and cannot reliably operate a capacitive touchscreen. Their only alternative is to use a stylus.

I will next discuss how a resistive touchscreen interfaces with a RasPi using the SPI Protocol. This interface is typical, but it is not mandatory for connecting touchscreens to RasPi-type controllers.

PiTFT Touchscreen Interface

The touchscreen controller sends and receives data from a RasPi using the SPI Protocol, which the RasPi supports in the Wheezy Linux distribution. SPI is a bit-serial protocol, which I explain in the following sidebar if you wish to understand it further. However, you can skip reading the sidebar without any loss of continuity regarding this project.

The serial peripheral interface (SPI) is one of several bit-serial data communication channels that the RasPi supports. It is a synchronous serial data link that uses one master device and one or more slave devices. A minimum of four data lines are used with SPI, and Table 2-1 shows the names associated with the master (RasPi) and slave (1601) device. Figure 2-9 is a simplified block diagram showing the principal components used in a SPI data link.

TABLE 2-1 SPI Data-Line Descriptions

Master device: RasPi	Slave device: 1601	Description
SCLK	CLK	Clock
MOSI	MISO	Master out, slave in
MISO	MOSI	Master in, slave out
CS/SHDN	CS	Slave select

Usually, two shift registers are involved in the data link, as shown in the figure. These registers may be hardware or software depending on the devices involved. The RasPi and PiTFT both implement their shift

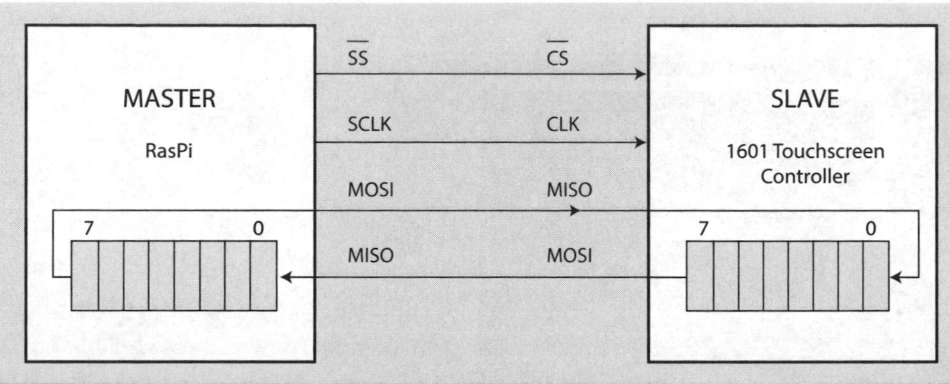

Figure 2-9 Simplified block diagram of an SPI data link.

registers in software. The two shift registers form what is known as an *interchip circular buffer arrangement* that is the heart of the SPI.

Data communication is initiated by the master by first selecting the required slave. The RasPi selects the 1601 by bringing the SS line to a low state or 0 VDC. During each clock cycle, the master sends a bit to the slave, which reads it from the MOSI line. Concurrently, the slave sends a bit to the master, which reads it from the MISO line. This operation is known as *full-duplex communication*, that is, simultaneous reading and writing between master and slave.

The clock frequency used depends primarily on the slave's response speed. The touchscreen controller in the PiTFT can run at 32 MHz, which easily should be compatible with the RasPi's computational rate. The PiTFT clock speed may be adjusted if the display becomes erratic or sputters. I show how this is done in the installation instructions that follow.

The preceding sidebar completes the background on how touchscreens function. It is now time to discuss how to install an Adafruit resistive touchscreen with a Model B RasPi.

RasPi Touchscreen Installation

I have separated the installation process into four sections as follows:

1. Hardware installation
2. Software installation
3. Configuration
4. Calibration

Hardware Installation

The hardware installation is quite simple: simply plug the matching 26-connector into the Model B RasPi. This is all that is needed if want the touchscreen installation to be barebones as far as the physical setup. However, I wanted it to be a bit fancier, so I elected to install the RasPi and PiTFT in a custom case. The case I used was the PiBow case with a PiTFT extension available from Adafruit Industries as Model 2779. Figure 2-10 shows the final installation, and I believe that it really looks like a finished product instead of an experimental project.

One modification to the PiBow case is necessary if you wish to use the Pi Cobbler with the RasPi/PiTFT combination. I have made this modification, as you can clearly see in the figure. Figure 2-11 is a close-up of the modification showing where the Pi Cobbler ribbon cable exits the PiBow case.

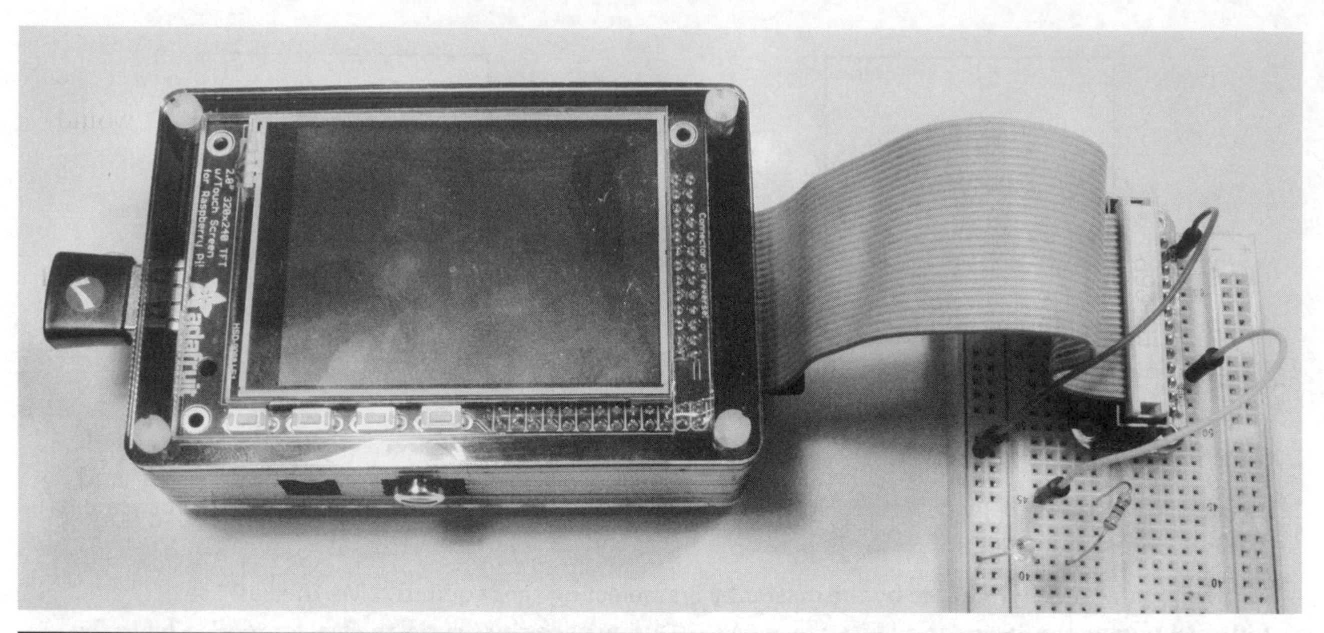

Figure 2-10 PiTFT with RasPi installed in a PiBow case.

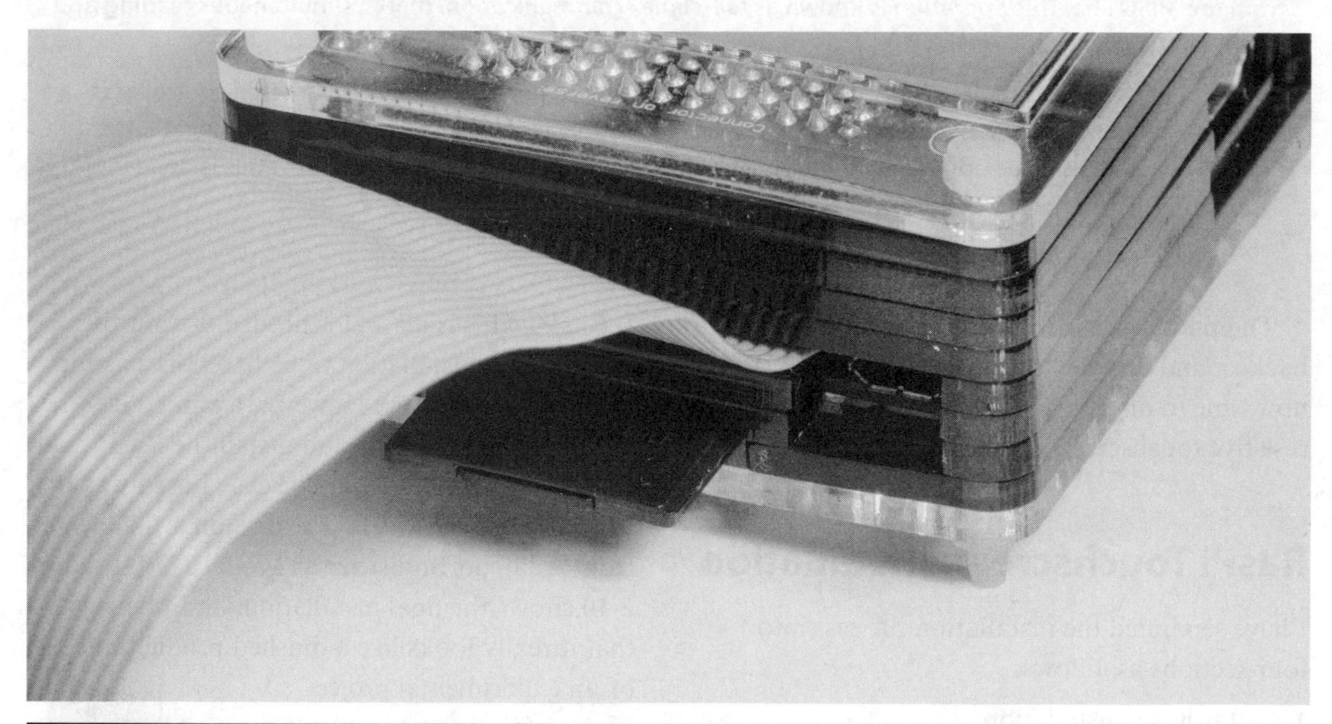

Figure 2-11 Pi Cobbler ribbon cable pass-through in the PiBow case.

The modification involves cutting off a piece of plastic from stack layer four, as shown in Figure 2-12, to allow room for the Pi Cobbler ribbon cable to pass through the PiBow case.

The Pi Cobbler ribbon cable then can be plugged into the 26-pin extension socket as shown in Figure 2-13. The ribbon also must be carefully folded back to allow it to pass over the socket and out of the PiBow case.

You also should note how the flexible ribbon cable from the PiTFT display is inserted into the touchscreen controller socket. Be very careful to

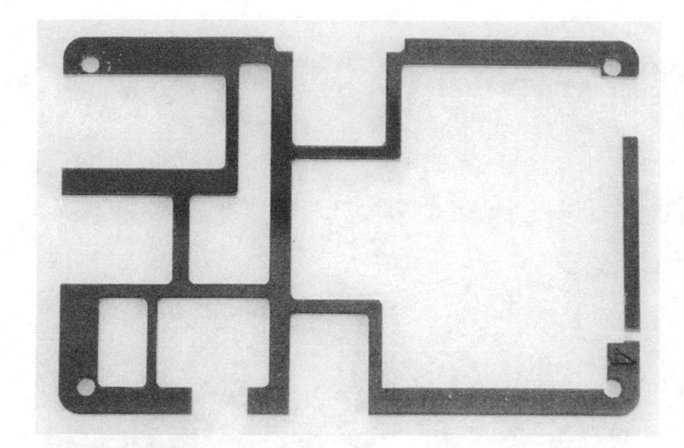

Figure 2-12 Modification to PiBow stack layer four.

carefully align this cable with the socket because serious problems will result if it is not inserted correctly into the socket. Figure 2-14 shows the PiBow stack arranged up to and including stack layer four to help guide you with case installation.

The Pi Cobbler cable should now easily fit in the new opening created by cutting the plastic piece from the PiBow stack layer four. I would suggest that you now complete the PiBow case installation by placing the remaining stacking layers and fastening them with the long nylon screws and nuts. I would also caution you that I had to slightly file the right-hand side tips of the PiTFT printed circuit board (PCB) to provide clearance for the two nylon screws. You may not have to modify them because I only had to file them down about 1/32 of an inch, and manufacturing tolerances are such that yours may fit as built.

Button Installation

This section concerns the optional installation of four 6-mm slim tactile buttons on the PiTFT PCB, one of which is shown in Figure 2-15.

Figure 2-13 Back view of the PiTFT.

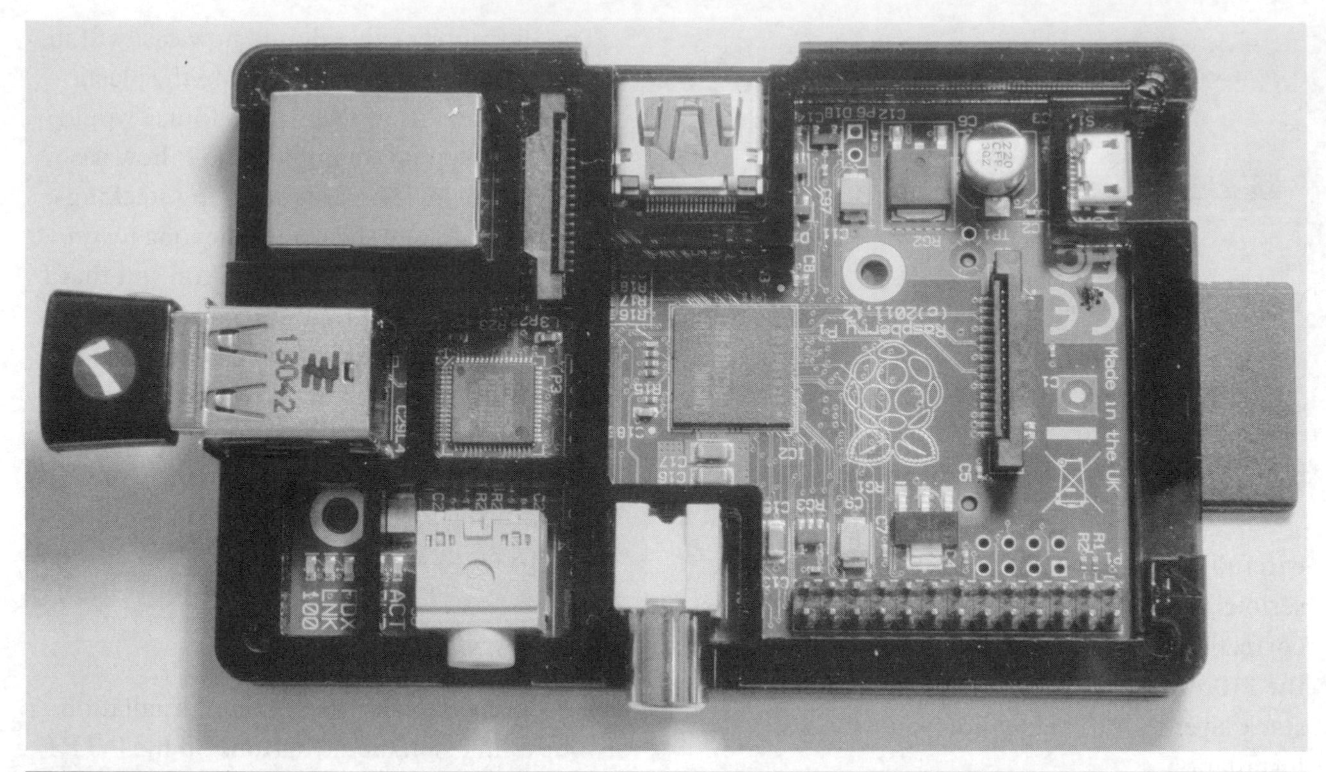

Figure 2-14 Partial PiBow case installation.

A package of 20 of these buttons may be purchased from Adafruit (Part No. 1489).

The four buttons will need to be soldered into the PCB holes located at the bottom of the touchscreen. These tactile buttons, when pressed, will ground RasPi GPIO pins 18, 20, 22, and 23. Their functionality is uncommitted and available for use in a program. The button connected to pin 23 may function as a RasPi reset when the PiTFT software is loaded. You must choose this feature during the PiTFT configuration process.

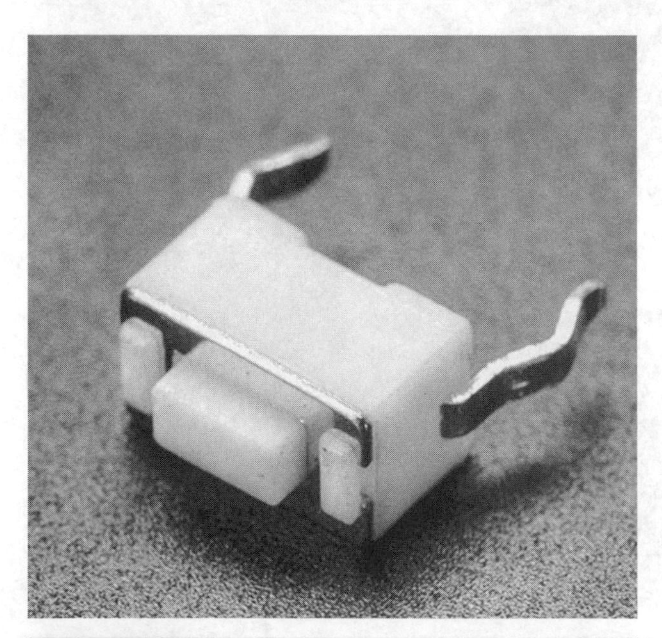

Figure 2-15 Tactile switch.

Completing the optional button installation finishes the physical installation. It is now time to discuss software installation.

Software Installation

There are three approaches you can take when it comes to installing the driver software that supports the PiTFT. The first is to use a prebuilt Wheezy image that already includes all the PiTFT drivers as well as some testing software. This image is available courtesy of Adafruit at http://adafruit-download.s3.amazonaws.com/2015-02-16-raspbian-pitft28r_150312.zip. This is a 3.28-GB image that also would need to be written onto an SD card using the procedure detailed in Chapter 1. The folks at Adafruit refer to this approach as an "easy install," which it is, and I would definitely recommend it to my readers who are not comfortable with running scripts or compiling and building code.

The second approach involves downloading and running a script or helper file from the Adafruit learn website at https://learn.adafruit.com/adafruit-pitft-28-inch-resistive-touchscreen-display-raspberry-pi/easy-install. The helper file found on this site automates many process steps, including downloading files and compiling and installing them onto an existing Wheezy distribution. I would recommend this approach for several reasons. The first is that you may already have a customized Linux distribution set up that you do not want to discard. Another reason is that you will learn a lot about the PiTFT setup and will be able to modify it to suit your specific needs, which is not possible with a prebuilt image.

The third approach involves manually integrating a series of libraries and support code into an existing Wheezy distribution. This involves essentially following a manual series of steps that have been automated by the helper file used in the second approach. I would recommend this approach only to those experienced readers who are very comfortable compiling, building, and installing Linux kernel code.

I will present the second approach because it is a good learning experience yet not too overwhelming. This software installation generally follows the second approach, where a helper file is used to install the necessary software. For readers interested in this process, I will also present an example later on in this chapter on how to compile and build a program.

Helper File Software Installation

This installation generally follows the Adafruit tutorial, as mentioned earlier. I have arranged it in a step-by-step sequence, which you should follow to ensure a successful completion. I will also be using a headless or Secure Shell Protocol (SSH) link to the RasPi because it was easier for me to read the text on a laptop screen rather than attempting to read the smaller text on the PiTFT screen. You certainty may use the PiTFT screen if you're comfortable with the text size. It will not be possible to use both the PiTFT and the HDMI displays simultaneously because of the framebuffer selection, which I will explain later.

1. I will start by assuming that you have a fresh install of the latest Wheezy Linux distribution. The latest is always available from http://www.raspberrypi.org/downloads/.
2. Update and upgrade the installed distribution.
   a. `sudo apt-get update`
   b. `sudo apt-get upgrade`
3. Download and install the helper file. Be patient; the install can take up to 20 minutes.
   a. `curl -SLs https://apt.adafruit.com/add | sudo bash`
   b. `sudo apt-get install -y adafruit-pitft-helper`
4. Run the actual PiTFT configuration using the helper file.
   a. `sudo adafruit-pitft-helper -t 28r`
5. At the end of the configuration process, you will be asked two questions. The first is whether you want the PiTFT to be the console text screen. The second will be

whether you want the button connected to GPIO pin 23 to act as a reset, as I mentioned earlier. Answer the questions to suit your personal preferences. Figure 2-16 is a screenshot showing these questions.

6. All that's left is to reboot the RasPi to have the PiTFT completely set up.
   a. `sudo reboot`

I will show you how to check the PiTFT installation in the following calibration section.

Configuration

The good news for this step is that all the PiTFT configuration has already been accomplished if you used the helper file approach. However, I will cover some details about manually configuring this touchscreen because it will help you to understand the process.

The PiTFT configuration process involves modifying an existing text file, which will hold the device's parameters. I would strongly suggest that you use the already installed nano text editor because I have found it to be easy to use and very convenient. Enter the following to start modifying this configuration file:

```
sudo nano /boot/config
```

Enter the following code after the last line in the existing config file:

```
[pi1]
device_tree=bcm2708-rpi-b-plus.dtb
[pi2]
device_tree=bcm2709-rpi-rpi-2-b.dtb
[all]
dtparam=spi=on
dtparam=i2c1=on
dtparam=i2c_arm=on
dtoverlay=pitft.
rotate=90,speed=32000000,fps=20
```

After you enter all the code, press CTRL-O (^O) to save the modified file. Then press CTRL-X (^X) to exit the nano editor.

There are few things that you should know regarding the PiTFT parameters. In the last line, the parameter `pitft.rotate=90` sets the view for the PiTFT screen. There are four choices for the view prospective:

- 0: Portrait with the screen bottom near the USB jacks
- 90: Landscape with the screen bottom near the headphone jack
- 180: Portrait with the screen top near the USB jacks
- 270: Landscape with the screen top near the headphone jack

The parameter `speed=32000000` sets the touchscreen clock rate. You might need to adjust this downward to `16000000` if the screen starts appearing erratic or jumpy.

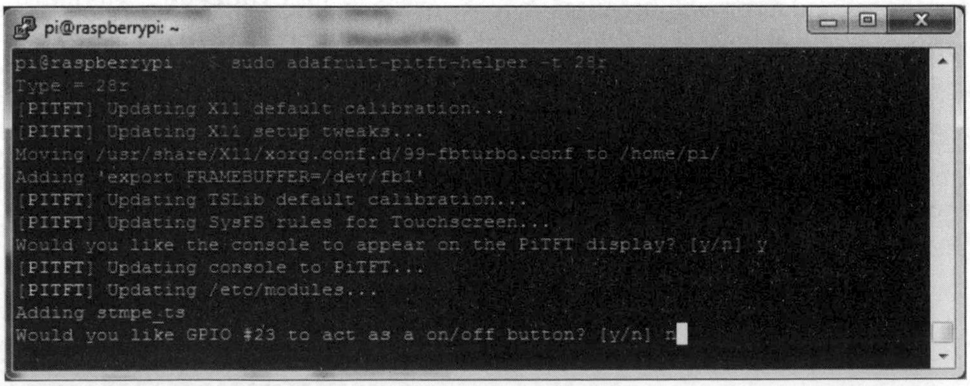

```
pi@raspberrypi ~ $ sudo adafruit-pitft-helper -t 28r
Type = 28r
[PITFT] Updating X11 default calibration...
[PITFT] Updating X11 setup tweaks...
Moving /usr/share/X11/xorg.conf.d/99-fbturbo.conf to /home/pi/
Adding 'export FRAMEBUFFER=/dev/fb1'
[PITFT] Updating TSLib default calibration...
[PITFT] Updating SysFS rules for Touchscreen...
Would you like the console to appear on the PiTFT display? [y/n] y
[PITFT] Updating console to PiTFT...
[PITFT] Updating /etc/modules...
Adding stmpe_ts
Would you like GPIO #23 to act as a on/off button? [y/n] n
```

Figure 2-16 PiTFT configuration questions.

Finally, the parameter `fps=20` sets the number of frames shown per second. The default value should be fine, but if you reset the clock rate, this parameter might have to be lowered by a few frames per second.

The next section concerns calibration. Again, as with the configuration section, all the required calibrations have already been done if you used the helper file for the PiTFT installation.

Calibration

For a touchscreen, calibration is really about matching the logical coordinates with the physical screen, where the LCD origin coordinates (0, 0) are located at the appropriate screen corner and the maximum LCD coordinates (320, 240) are at the diagonally opposite corner. Once these are set, all the logical touchpoints should linearly match the physical ones. This means that if you touch the screen in precisely the center with a stylus, you should return the midpoint LCD coordinates (160, 120). It is also important to realize that the touchscreen controller returns absolute numbers based on the number of ADC bits. The PiTFT uses a multichannel 12-bit ADC, which means that the reported values will range from 0 to 4095. Remember from your high school math that 12 bits can represent 4096 separate whole numbers or, more precisely, $2^{12} = 4096$. Now the *X*-axis must be physically matched to the LCD's 320-pixel range, which means that the absolute value must be scaled or reduced proportionally to match the 0- to 319-pixel range. The same is true for the *Y*-axis, except that the scale factor matches the 0 to 4095 absolute range with a 0- to 239-pixel range. Now let's move on to starting the calibration process.

udev Rule

The first step in this calibration process is to create an udev rule, which will fix or persist a logical name for the PiTFT. The current Linux device manager is named udev, and it uses text-based rules to manage all the devices connected to the Linux computer on which it is installed. Enter the following to use the nano editor to create a text file to hold the rule:

```
sudo nano /etc/udev/rules.d/95-stmpe
.rules
```

Next, enter the following into this file:

```
SUBSYSTEM=="input",
ATTRS(name)=="stmpets",
ENV(DEVNAME)=="*event*",SYMLINK+=$
```

I had to break this statement up to fit this book's format. You should keep it all in one line when you enter it.

After you save and exit the nano editor, you will need to remove and reinstall the touchscreen driver in order for the new udev rule to take effect. Enter the following to accomplish this:

```
sudo rmmod stmpe_ts
sudo modprobe stmpe_ts
```

You should next check to see if the driver installation and rule change were effective. Enter the following:

```
ls -l /dev/input/touchscreen
```

This command results in an `eventx` being returned, where the `x` stands for a number starting at 0. Figure 2-17 shows the result of running this command on my RasPi.

Your number may be different from 0 because the event number is assigned based on the number of keyboards/mice and similar input devices that are actively connected to the RasPi. You are now ready to install the touchscreen library and test the software, which will allow you to both debug and calibrate the touchscreen.

```
pi@raspberrypi ~ $ ls -l /dev/input/touchscreen
lrwxrwxrwx 1 root root 6 Apr 13 19:57 /dev/input/touchscreen -> event0
pi@raspberrypi ~ $ 
```

Figure 2-17 Screenshot for the `ls` command executed on the touchscreen.

evtest and tslib

Enter the following command to install both the test software and the supporting touchscreen library:

```
sudo apt-get install evtest tslib
libts-bin
```

Next, run the `evtest` application by entering

```
sudo evtest /dev/input/touchscreen
```

You should now be viewing the initial `evtest` screen, which shows some configuration information and is awaiting a touch event or input, as shown in Figure 2-18.

Next, use a plastic stylus to carefully and briefly touch the center point of the touchscreen.

You should see a series of events captured by the `evtest` application, as shown in Figure 2-19.

Actually, 19 events were captured, from which I shortened the display to 10. Each event is time-stamped with the number of elapsed seconds from 1/1/1970. Of course, we are only interested in the time frame surrounding the events, which lasted only 0.172861 second or approximately 173 milliseconds. Two principal event types are shown, 1 and 3. Event type 1 is the `BTN_TOUCH` with a value of 1 to start and a value of 0 after the touch is lifted. Event 3 has three subcodes, which report the absolute *X*- and *Y*-coordinates as well as an absolute pressure reading. If you closely examine Figure 2-19, you will see that the *X*-coordinate values hover around 2300, and the *Y*-coordinate values hover around 2050. This

```
donnorris — pi@raspberrypi: ~ — ssh — 74×46
pi@raspberrypi ~ $ evtest /dev/input/touchscreen
Input driver version is 1.0.1
Input device ID: bus 0x18 vendor 0x0 product 0x0 version 0x0
Input device name: "stmpe-ts"
Supported events:
  Event type 0 (EV_SYN)
  Event type 1 (EV_KEY)
    Event code 330 (BTN_TOUCH)
  Event type 3 (EV_ABS)
    Event code 0 (ABS_X)
      Value      0
      Min        0
      Max     4095
    Event code 1 (ABS_Y)
      Value      0
      Min        0
      Max     4095
    Event code 24 (ABS_PRESSURE)
      Value      0
      Min        0
      Max      255
Properties:
Testing ... (interrupt to exit)
```

Figure 2-18 Initial `evtest` screenshot.

```
Testing ... (interrupt to exit)
Event: time 1429106466.299075, type 3 (EV_ABS), code 0 (ABS_X), value 2301
Event: time 1429106466.299075, type 3 (EV_ABS), code 1 (ABS_Y), value 2037
Event: time 1429106466.299075, type 3 (EV_ABS), code 24 (ABS_PRESSURE), value 108
Event: time 1429106466.299075, type 1 (EV_KEY), code 330 (BTN_TOUCH), value 1
Event: time 1429106466.299075, -------------- SYN_REPORT ------------
Event: time 1429106466.308426, type 3 (EV_ABS), code 0 (ABS_X), value 2299
Event: time 1429106466.308426, type 3 (EV_ABS), code 24 (ABS_PRESSURE), value 119
Event: time 1429106466.308426, -------------- SYN_REPORT ------------
Event: time 1429106466.316767, type 3 (EV_ABS), code 0 (ABS_X), value 2300
Event: time 1429106466.316767, type 3 (EV_ABS), code 1 (ABS_Y), value 2000
Event: time 1429106466.316767, -------------- SYN_REPORT ------------
Event: time 1429106466.325114, type 3 (EV_ABS), code 0 (ABS_X), value 2301
Event: time 1429106466.325114, type 3 (EV_ABS), code 1 (ABS_Y), value 2031
Event: time 1429106466.325114, type 3 (EV_ABS), code 24 (ABS_PRESSURE), value 125
Event: time 1429106466.325114, -------------- SYN_REPORT ------------
Event: time 1429106466.333517, type 3 (EV_ABS), code 0 (ABS_X), value 2298
Event: time 1429106466.333517, type 3 (EV_ABS), code 1 (ABS_Y), value 2044
Event: time 1429106466.333517, type 3 (EV_ABS), code 24 (ABS_PRESSURE), value 130
Event: time 1429106466.333517, -------------- SYN_REPORT ------------
                                         •
                                         •
                                         •
Event: time 1429106466.429402, type 3 (EV_ABS), code 0 (ABS_X), value 2293
Event: time 1429106466.429402, type 3 (EV_ABS), code 1 (ABS_Y), value 2083
Event: time 1429106466.429402, type 3 (EV_ABS), code 24 (ABS_PRESSURE), value 126
Event: time 1429106466.429402, -------------- SYN_REPORT ------------
Event: time 1429106466.437948, type 3 (EV_ABS), code 0 (ABS_X), value 2297
Event: time 1429106466.437948, type 3 (EV_ABS), code 1 (ABS_Y), value 2064
Event: time 1429106466.437948, type 3 (EV_ABS), code 24 (ABS_PRESSURE), value 129
Event: time 1429106466.437948, -------------- SYN_REPORT ------------
Event: time 1429106466.446265, type 3 (EV_ABS), code 0 (ABS_X), value 2302
Event: time 1429106466.446265, type 3 (EV_ABS), code 1 (ABS_Y), value 2058
Event: time 1429106466.446265, type 3 (EV_ABS), code 24 (ABS_PRESSURE), value 124
Event: time 1429106466.446265, -------------- SYN_REPORT ------------
Event: time 1429106466.471936, type 3 (EV_ABS), code 24 (ABS_PRESSURE), value 0
Event: time 1429106466.471936, type 1 (EV_KEY), code 330 (BTN_TOUCH), value 0
Event: time 1429106466.471936, -------------- SYN_REPORT ------------
```

Figure 2-19 Touchpoint event screenshot.

makes sense when you consider that the absolute *X* and *Y* ranges are 0 to 4095. I touched the stylus near the screen center, which resulted in having approximate center value readings.

The origin point or coordinate values (0, 0) are at the upper left-hand corner, and the maximum coordinate values of (4095, 4095) are at the lower right-hand corner because the screen has been logically rotated 90°, as was shown in the configuration section. Also recall that these absolute coordinate values must be scaled to match the LCD pixel range, as I explained earlier.

The measurement of absolute pressure is a bit more complex because it involves measuring the cross-layer resistance between the *X*-ITO and *Y*-ITO layers. Suffice it to say that a higher reported number corresponds to higher applied touchpoint pressure. It would depend on the specific application using the touchscreen whether or not the pressure measurement was actually used. However, I do believe that pressure measurements are used in the normal internal operation of the touchscreen controller in order to achieve specific functionalities.

Automagic Calibration Script

Adafruit has provided the following Python script to automatically run the touchscreen calibration. You will need to enter the following:

```
sudo wget https://raw.githubusercontent
.com/adafruit/PiTFT_Extras/master/pitft
_touch_cal.py
```

NOTE **The preceding must be entered on the same line. It was split to match this book's format.**

After you have downloaded the Python script, enter the following to run it:

```
sudo adafruit-pitft-touch-cal
```

Figure 2-20 shows the result of running this script on the touchscreen I previously set up with my RasPi using the helper file.

You will be presented with a question at the end of the script results asking whether or not to update the current configuration. I answered yes because the current configuration exactly matched the old configuration.

Manual Calibration

I will now discuss a manual calibration procedure that might come in handy for readers who believe that the automatic calibration is not quite precise enough for their use. You will need to have already downloaded and installed both the `tslib` and `libts-bin` packages for the manual calibration to work. I discussed how this was done in the `evtest` and `tslib` section.

Enter the following to start the manual calibration:

```
sudo TSLIB_FBDEVICE=/dev/fb1 TSLIB_
TSDEVICE=/dev/input/touchscreen ts_
calibrate
```

NOTE **The preceding must be entered on the same line. It was split to match this book's format.**

Figure 2-21 shows the result of entering the preceding command.

You need to touch the center of the touchpoint shown in the figure. Four more touchpoints will appear on the screen, each one at a time to be touched. Figure 2-22 shows the result of the five touches and the set of new calibration constants that will be used as a result of this manual calibration.

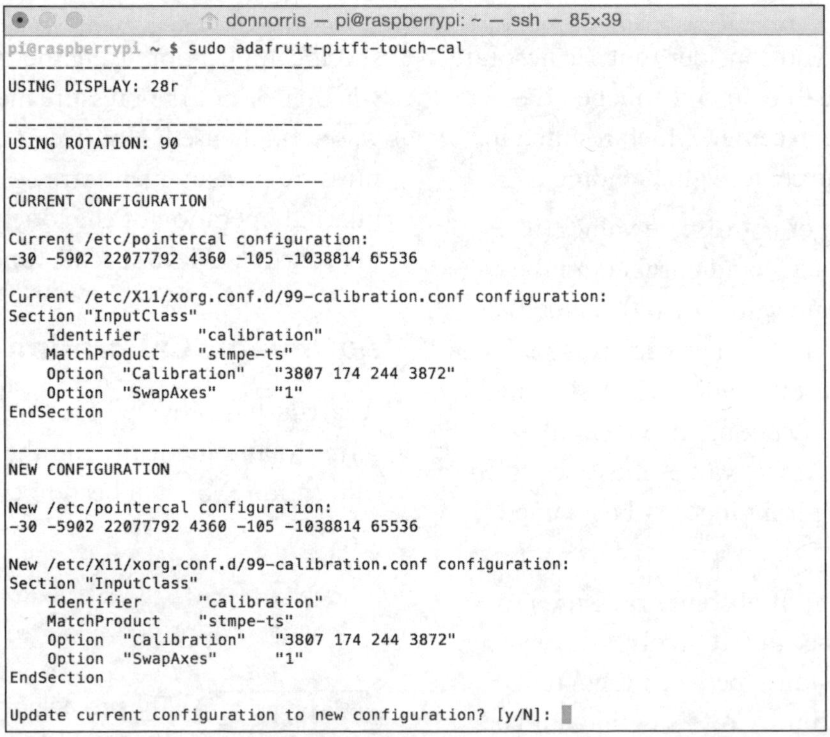

```
donnorris — pi@raspberrypi: ~ — ssh — 85×39
pi@raspberrypi ~ $ sudo adafruit-pitft-touch-cal
----------------------------------
USING DISPLAY: 28r

----------------------------------
USING ROTATION: 90

----------------------------------
CURRENT CONFIGURATION

Current /etc/pointercal configuration:
-30 -5902 22077792 4360 -105 -1038814 65536

Current /etc/X11/xorg.conf.d/99-calibration.conf configuration:
Section "InputClass"
    Identifier      "calibration"
    MatchProduct    "stmpe-ts"
    Option  "Calibration"   "3807 174 244 3872"
    Option  "SwapAxes"      "1"
EndSection

----------------------------------
NEW CONFIGURATION

New /etc/pointercal configuration:
-30 -5902 22077792 4360 -105 -1038814 65536

New /etc/X11/xorg.conf.d/99-calibration.conf configuration:
Section "InputClass"
    Identifier      "calibration"
    MatchProduct    "stmpe-ts"
    Option  "Calibration"   "3807 174 244 3872"
    Option  "SwapAxes"      "1"
EndSection

Update current configuration to new configuration? [y/N]:
```

Figure 2-20 Screenshot of the automatic calibration script results.

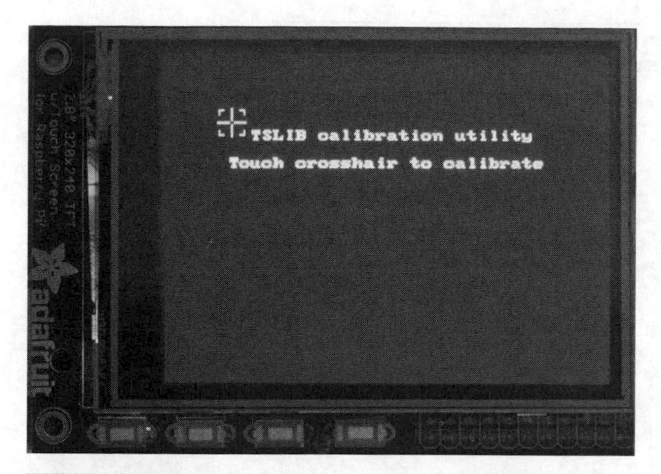

Figure 2-21 Manual calibration screenshot.

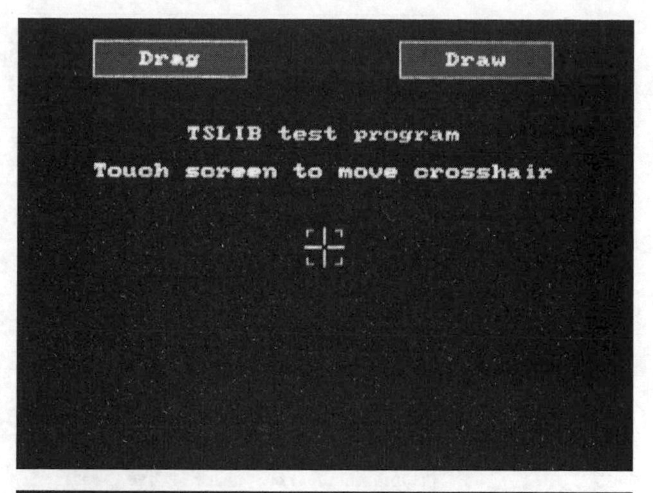

Figure 2-23 `ts_test` touchscreen display.

You may run the following test to check on how well the new calibration constants are working. Enter the following to initiate the test:

```
sudo TSLIB_FBDEVICE=/dev/fb1 TSLIB_
TSDEVICE=/dev/input/touchscreen ts_test
```

The preceding must be entered on the same line. It was split to match this book's format.

Figure 2-23 shows what will appear on the touchscreen as a result of running this command.

A corresponding coordinate display will also appear on the console screen. However, the coordinates shown will be in terms of the LCD pixels and not the absolute coordinates shown in previous figures. Apparently, the `ts_test` program automatically scales the absolute coordinates before displaying them.

I next drew a short line from the displayed touchpoint to the left in order to generate some sample coordinates. Figure 2-24 is a console screenshot, which has been shortened, to show the results of the line drawn on the touchscreen.

You may see from the figure that the *X* values increase from 160 to about 290, while the *Y* values remain reasonably constant around 120. These numbers translate into a horizontal line drawn on the touchscreen from the center to

```
donnorris — pi@raspberrypi: ~ — ssh — 85×39
pi@raspberrypi ~ $ sudo TSLIB_FBDEVICE=/dev/fb1 TSLIB_TSDEVICE=/dev/input/touchscreen
 ts_calibrate
xres = 320, yres = 240
Took 73 samples...
Top left : X = 1012 Y = 3154
Took 96 samples...
Top right : X =  973 Y =  721
Took 79 samples...
Bot right : X = 3141 Y =  713
Took 70 samples...
Bot left : X = 3178 Y = 3156
Took 63 samples...
Center : X = 2085 Y = 1923
334.720978 -0.000127 -0.090233
-12.287567 0.064603 -0.001006
Calibration constants: 21936274 -8 -5913 -805278 4233 -65 65536
pi@raspberrypi ~ $
```

Figure 2-22 Manual calibration results.

```
donnorris — pi@raspberrypi: ~ — ssh — 94×39
pi@raspberrypi ~ $ sudo TSLIB_FBDEVICE=/dev/fb1 TSLIB_TSDEVICE=/dev/input/touchscreen ts_test
1429152108.254094:     161    112     50
1429152108.261655:     160    112     65
1429152108.278557:     163    112     80
1429152108.278557:     164    112     83
1429152108.294299:     163    112     97
1429152108.294299:     162    112    105
1429152108.311244:     162    113    106
1429152108.311244:     161    113    107
1429152108.319071:     161    113    104
1429152108.327427:     161    113    100
1429152108.343146:     161    113    101
1429152108.351519:     161    113    104
1429152108.359707:     161    113    108
1429152108.368566:     161    113    114
1429152108.376880:     160    113    118
1429152108.400199:     160    113    121
1429152108.400199:     160    113    123
                        •
                        •
                        •
1429152108.696454:     244    120    133
1429152108.704734:     249    120    130
1429152108.704734:     253    120    127
1429152108.721693:     257    120    126
1429152108.729856:     260    120    126
1429152108.738352:     263    120    125
1429152108.746814:     267    120    124
1429152108.755248:     270    120    124
1429152108.763584:     273    120    124
1429152108.772263:     276    120    123
1429152108.772263:     278    120    123
1429152108.789965:     282    120    121
1429152108.789965:     284    120    120
1429152108.797381:     285    120    120
1429152108.813986:     287    120    120
1429152108.813986:     287    120    120
1429152108.830086:     287    120    120
1429152108.838599:     288    120    120
1429152108.846973:     289    120    119
1429152108.855239:     289    120    119
1429152108.863547:     290    120    119
1429152108.871686:     290    120    120
1429152108.880628:     289    120    121
1429152108.889132:     289    120    121
1429152108.897525:     289    120    121
1429152108.905860:     289    120    120
1429152108.914055:     289    120    121
1429152108.934068:     288    120    120
1429152108.934068:     288    120    120
1429152108.950084:     288    120    120
1429152108.958384:     288    120    120
1429152108.966526:     288    120    120
1429152108.978477:     288    121    119
1429152108.987133:     287    121    116
1429152109.013194:     287    121      0
```

Figure 2-24 Console display for a straight horizontal line drawn on the touchscreen.

the right edge, which is exactly what I did. The pressure readings average about 120, which corresponds to medium force applied to the stylus used to draw the line. I also later tried drawing another line using more force on the stylus and recorded pressure readings up to about 139. I later determined that simply using a fingertip on the touchscreen could generate readings up around 220. I can only conclude that the area of applied pressure plays some role in determining the ultimate pressure reading. Based on my findings, I would be a bit leery of relying on pressure readings to derive any conclusions about any specific touchscreen event other than when it started and stopped.

I did another check using the `ts_test` application in which I tapped both the origin and maximum touchpoints on the touchscreen. Figure 2-25 shows these two points as tiny white dots. I touched the Draw button on the screen to show where I placed these two dots.

The console screen displayed the minimum point as (5, 5) and the maximum point as (319,

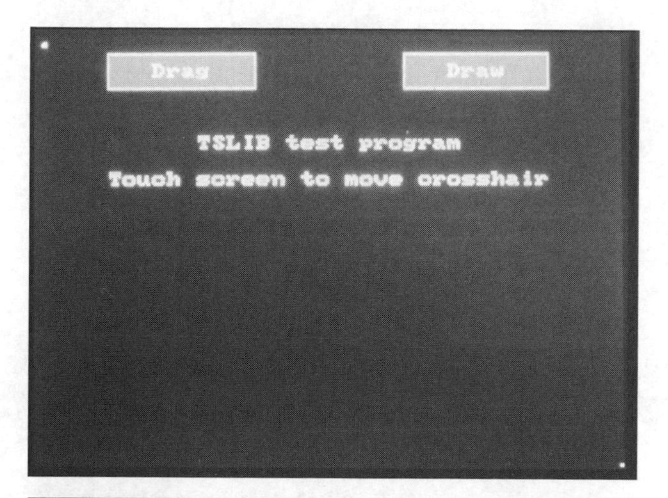

Figure 2-25 Max and min touchpoint test.

237). These are very close to the expected values of (0, 0) and (320, 240), respectively. If you examine the figure very closely, you will see that the origin touchpoint was slightly down and away from the true origin, which is the upper right-hand corner of the slightly lighter-colored touchscreen region. The maximum touchpoint is almost dead centered on the lower right-hand corner, which is reflected in the reported coordinate position. Also notice that the touchscreen region precisely fills the complete LCD display area. This may be due to the physical area needed by the sense lines, or it could be an artifact of the manual calibration that I did just prior to this test.

This last test completes my calibration discussion. I do want to make you aware that the Adafruit PiTFT tutorial I referenced earlier contains several more calibration techniques that you can do, especially if you are using the touchscreen with X-Windows. I will next discuss the framebuffer, a term I used earlier but deferred explaining what it is and how it is used in conjunction with a touchscreen.

Framebuffer

It is important for you to understand the framebuffer concept because it affects how the PiTFT functions with regard to the regular HDMI display. A framebuffer (FB) may simply be defined as a place in dynamic memory that contains one complete video frame to be displayed. Each location in an FB contains 3 bytes whose values represent the RGB intensities for each displayed pixel. Video frame sizes vary depending on the display device. The underlying touchscreen LCD used in this project is 320 × 240 pixels in size for a total of 76,800 pixels. This makes the total memory required for an FB to be 230,400 bytes or 225 kB because each pixel uses 3 bytes.

The actual hardware implementation used in the RasPi is fairly complex because it involves the central processing unit (CPU), a graphical processing unit (GPU), and a series of mailbox registers to control the flow of video data between the two processors. I elected not to include a detailed discussion of the hardware implementation because it would significantly add complexity to this chapter with little reader benefit because you have no programmatic control of internal FP operations. You can, however, use a nice little Adafruit application to view the FB contents.

Framebuffer Image (FBI) Viewer

Enter the following to download and install the FBI application:

```
sudo apt-get install fbi
```

You will need a test image to check out this viewer application. Once again, Adafruit comes to the rescue. Enter the following to download a test image precisely sized for the touchscreen LCD:

```
wget http://adafruit-download.s3
.amazonaws.com/adapiluv320x240.jpg
```

NOTE **The preceding must be entered on the same line. It was split to match this book's format.**

View it by entering

```
sudo fbi -T 2 -d /dev/fb1 -noverbose -a
adapiluv320x240.jpg
```

Figure 2-26 shows the image displayed on the touchscreen.

Touchscreen Logical Name

Sharp-eyed readers may have noticed that I used the identifier `fb1` in the preceding command to display the downloaded image. This is the PiTFT *logical name*, which is the identifier associated with the PiTFT touchscreen. For all intents and purposes, it is the driver name, which the operating system (OS) uses to send and receive data to/from the device. I had also used `fp1` previously in the manual calibration command sequence. There is also another FB device installed on the OS, and that is `fb0`, which is the HDMI display. You can only have one FB device active at a time, which is the reason I mentioned much earlier in this chapter that the touchscreen and the HDMI display could not be used simultaneously. You should also realize that the composite output is the analog conversion of the HDMI display, so you cannot "cheat" the system and try to run the touchscreen and get a composite output. This is all the more reason to use a headless development station, from which you get the best of both worlds, a working touchscreen and a large development display.

Video Player

In this brief section I will show how to set up and run a video on the touchscreen. This video player application nicely demonstrates the LCD screen video capabilities, even though running a video player has nothing to do with typical touchscreen operations.

The first step is downloading a video, which will be the source of the demonstration. I would recommend the classic *Big Buck Bunny* video, which has been sized for the PiTFT and may be

Figure 2-26 Adapiluv320x240.jpg displayed on the PiTFT.

downloaded from Adafruit using the following command:

```
wget http://adafruit-download.s3
.amazonaws.com/bigbuckbunny320p.mp4
```

NOTE The preceding must be entered on the same line. It was split to match this book's format.

Once the video is downloaded, you will need to download and install the player application. Enter the following to install the `mplayer` application:

```
sudo apt-get install mplayer
```

Then enter the following to run this video:

```
mplayer -vo fbdev2:/dev/fb1 -x 240 -y
320 -framedrop bigbuckbunny320p.mp4
```

NOTE The preceding must be entered on the same line. It was split to match this book's format.

NOTE I ran this without audio, but I am fairly certain that a music track accompanies the video. You might try plugging an amplified speaker into the RasPi stereo jack to see whether the music plays.

At this point, I will conclude my touchscreen background discussion and proceed to explain this chapter's project.

Touchscreen Project

The demonstration screen project is quite simple: display a button on the touchscreen that when clicked by either a stylus or fingertip will toggle a light-emitting diode (LED) connected to one of the RasPi's GPIO pins. While the project goals are quite easy, getting to an actual implementation will involve a bit of work.

The code for this project is written in the C language, which has been around for a comparatively long time. Do not fear if you are skittish about using C because I have provided step-by-step instructions on how to compile and build the source code, and I have made the source code available on this book's companion website. Before jumping into the project code, I make a brief diversion into the concept of Linux file input-output (I/O).

Linux File I/O

Some readers may already know that the Linux OS treats almost everything as a file if you also consider a directory as a special file type. This means that I/O commands can be directed to a file where it effectively interfaces to the outside world. This is great news and can simplify programming to a great extent. I will use a practical example to demonstrate how this file I/O functions.

The PiTFT has four LEDs built in, which provide backlighting for the display. As most of you know, a liquid crystal display (LCD) is essentially a dark display without backlighting. I will demonstrate how to control the touchscreen's backlighting using file I/O.

You first need to ensure that the PiTFT is plugged into the 26-pin connector. Next, enter the following command:

```
sudo sh -c "echo 508 > /sys/class/gpio/
export"
```

This command created a series of files and directories that will allow us to control the backlights. Enter the next command to confirm that the master directory was created:

```
ls -l /sys/class/gpio
```

You should see a display similar to Figure 2-27.

Just ensure that you have a directory with the number 508 in the title. Given that the 508 directory exists, enter the following to create an output type direction for this newly create GPIO

```
donnorris — pi@raspberrypi: ~ — ssh — 140×23
pi@raspberrypi ~ $ sudo sh -c "echo 508 > /sys/class/gpio/export"
pi@raspberrypi ~ $ ls -l /sys/class/gpio
total 0
-rwxrwx--- 1 root gpio 4096 Apr 16 17:35 export
lrwxrwxrwx 1 root gpio    0 Apr 16 08:13 gpio4 -> ../../devices/soc/20200000.gpio/gpio/gpio4
lrwxrwxrwx 1 root gpio    0 Apr 16 17:35 gpio508 -> ../../devices/soc/20204000.spi/spi_master/spi0/spi0.1/stmpe-gpio/gpio/gpio508
lrwxrwxrwx 1 root gpio    0 Dec 31  1969 gpiochip0 -> ../../devices/soc/20200000.gpio/gpio/gpiochip0
lrwxrwxrwx 1 root gpio    0 Dec 31  1969 gpiochip506 -> ../../devices/soc/20204000.spi/spi_master/spi0/spi0.1/stmpe-gpio/gpio/gpiochip506
-rwxrwx--- 1 root gpio 4096 Dec 31  1969 unexport
pi@raspberrypi ~ $
```

Figure 2-27 `ls` command run on the `/sys/class/gpio` directory.

control file (also be looking at the touchscreen when you enter this command):

```
sudo sh -c "echo 'out' > /sys/class/
gpio/gpio508/direction"
```

If all went as it should, the PiTFT should go dark because the backlighting turned off. It turned off because the preceding command set the file direction as an output, and a 0 was automatically sent to the corresponding GPIO pin. Enter the following to turn the backlighting on again:

```
sudo sh -c "echo '1' > /sys/class/gpio/
gpio508/value"
```

You will now see the touchscreen display because the backlighting is turned on again. Note that nothing in these commands changed the content of what is displayed; they only toggled the backlighting LEDs to allow you to see or not see the touchscreen. You change the 1 to a 0 in the preceding command to turn off the backlighting. Try experimenting with these commands to convince yourself of their utility.

buttonExample

This project sets up a rectangular button on the touchscreen that will be used to control a LED. The LED will simply be turned on or off, which is known as *toggling*. The button color and text will also change reflecting the state of the LED.

Hardware Setup

You will need a little hardware setup to run this project. I used a LED in series with a 220-Ω resistor connected on a solderless breadboard. I also used a Pi Cobbler to extend all the GPIO lines from the touchscreen. The complete physical setup was shown earlier in Figure 2-10. The wiring schematic is shown in Figure 2-28.

Software Setup

I will now show you the C program that creates a button on the touchscreen that controls a LED. This program is an adaptation of an existing program named buttonExample.c created by Mark Williams. The original, unaltered version is available for download from GitHub at https://github.com/mwilliams03/Pi-Touchscreen-basic/blob/master/buttonExample.c. My modified version is as follows:

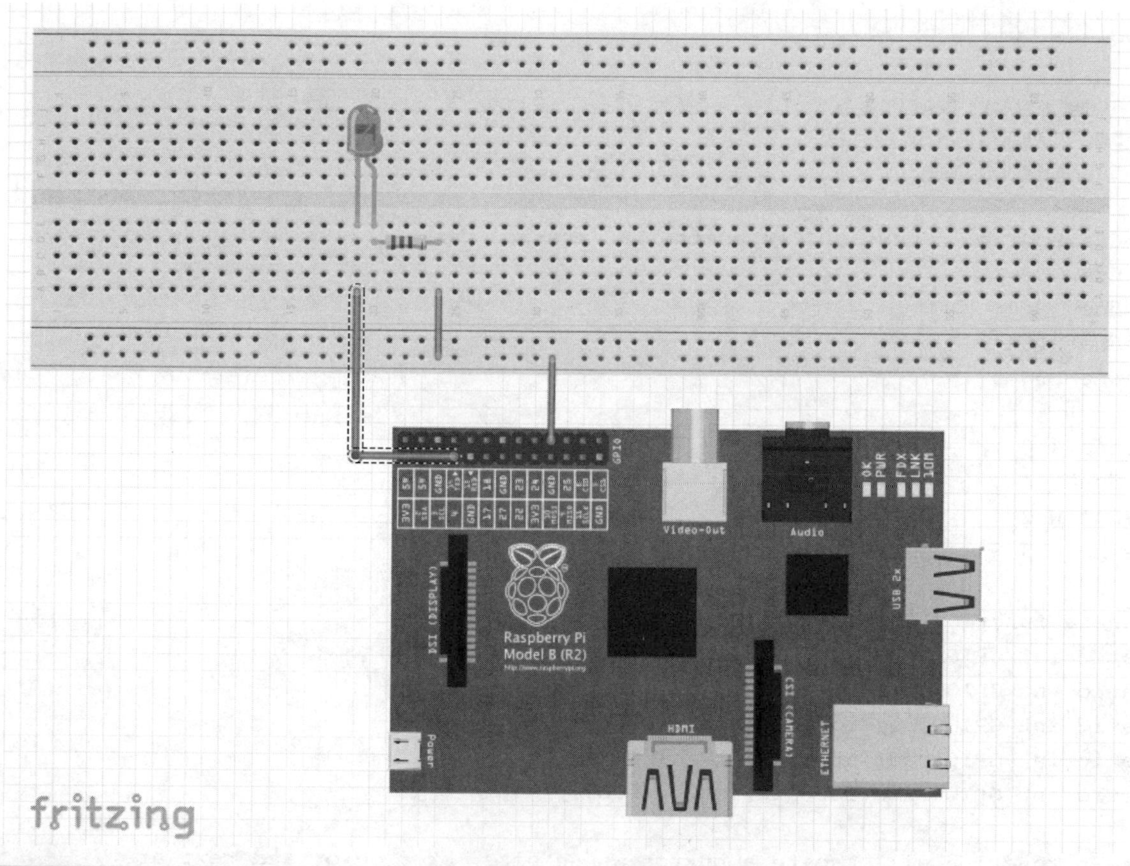

Figure 2-28 buttonExample schematic.

```
/*
original by Mark Williams, entitled buttonExample.c
heavily modified by D. J. Norris. Kept the original title

A program that demonstrates how to code for a touchscreen. Specifically for a PiTFT
attached to a Model B Raspberry Pi.
*/

#include <linux/input.h>
#include <sys/stat.h>
#include <sys/types.h>
#include <unistd.h>
#include <string.h>
#include <fcntl.h>
#include <stdio.h>
#include "touch.h"
#include "touch.c"
#include "framebuffer.c"
#include <signal.h>
#include <time.h>
#include <stdlib.h>
```

```
#include "wiringPi.h"

// some handy constants
#define BUTTON_ON 1
#define BUTTON_OFF 0

#define POUT 4 // uses GPIO04 as the LED control pin
#define OUT 1
#define IN 0
#define HIGH 1
#define LOW 0

#define X 0
#define Y 1
#define W 2
#define H 3

//these functions handle all the Linux file I/O
static int GPIOExport(int pin);
static int GPIOUnexport(int pin);
static int GPIODirection(int pin, int dir);
static int GPIORead(int pin);
static int GPIOWrite(int pin, int value);

//this function will create a button whose size, bkgd color and text are
//controllable
void createButton(int x, int y, int w, int h, char *text, int backgroundColor, int
foregroundColor)

{
    int size = sizeof(text);
    char *p = text;
    int length = 0;

    //get length of the string *text
    while(*(p+length))
       length++;

    if((length*8)> (w-2)){
       printf("####error,button too small for text####\n");
       exit(1);
    }

    //Draw button outline
    drawSquare(x-2,y-2,w+4,h+4,backgroundColor);

    //Draw button foreground
    drawSquare(x,y,w,h,foregroundColor);
    //Place text on the button. Try and center it in a primitive way
```

```
    put_string(x+((w-(length*8))/2), y+((h-8)/2),text,4);
    }

//this function configures the desired GPIO pin
int SetPinsOut()
{
GPIOExport(POUT);
GPIODirection(POUT, OUT);

}
//a function to calculate the elapsed time in milliseconds
int mymillis()
{
struct timeval tv;
gettimeofday(&tv, NULL);
return (tv.tv_sec) * 1000 + (tv.tv_usec)/1000;
}

//this is an interrupt handler, which is required for event based processing
void INThandler(int sig)
{

    signal(sig, SIG_IGN);
    closeFramebuffer();
exit(0);
}

//the starting point. Every C program has one
int main()
{
    signal(SIGINT, INThandler);
int xres,yres,x;
int screenXmax, screenXmin;
    int screenYmax, screenYmin;
float scaleXvalue, scaleYvalue;
int rawX, rawY, rawPressure, scaledX, scaledY;
//Used to monitor the timer for the button
    int buttonTimer3 = mymillis();

    //To keep track of the button state
    int button3 = BUTTON_OFF;

    //The button. X, Y, Width, Height
    int buttonCords3[4] = {135,75,60,80};
    if (openTouchScreen() == 1)
        perror("error opening touch screen");
getTouchScreenDetails(&screenXmin,&screenXmax,&screenYmin,&screenYmax);

    //initialize the touchscreen frame buffer now
```

```
    framebufferInitialize(&xres,&yres);

    scaleXvalue = ((float)screenXmax-screenXmin) / xres;
    printf ("X Scale Factor = %f\n", scaleXvalue);
    scaleYvalue = ((float)screenYmax-screenYmin) / yres;
    printf ("Y Scale Factor = %f\n", scaleYvalue);

    createButton(buttonCords3[X],buttonCords3[Y],buttonCords3[W],buttonCords3[H],"Off"
,GREY,GREEN);

    //Setup pin
    SetPinsOut();

    while(1){

  getTouchSample(&rawX, &rawY, &rawPressure);

  scaledX = rawX/scaleXvalue;
  scaledY = rawY/scaleYvalue;

//See if the results retuned by the touch event fall within the coordinates of the
//button
if((scaledX > buttonCords3[X] && scaledX < (buttonCords3[X]+buttonCords3[W])) &&
(scaledY > buttonCords3[Y] && scaledY < (buttonCords3[Y]+buttonCords3[H])))
    if (mymillis() - buttonTimer3 > 500)
    //Has 500ms passed since the last time this button was pressed?
    if(button3){
    createButton(buttonCords3[X],buttonCords3[Y],buttonCords3[W],buttonCords3[H],
"Off",GREY,GREEN);
    button3= BUTTON_OFF;
    buttonTimer3 = mymillis();
    GPIOWrite(POUT,LOW);
    }
  else{
  createButton(buttonCords3[X],buttonCords3[Y],buttonCords3[W],buttonCords3[H],"On",
WHITE,LIGHT_GREEN);

    button3= BUTTON_ON;
    buttonTimer3 = mymillis();
    GPIOWrite(POUT,HIGH);
    }

    }
}

int GPIOExport(int pin)
{
#define BUFFER_MAX 3
    char buffer[BUFFER_MAX];
```

```
    ssize_t bytes_written;
    int fd;

    fd = open("/sys/class/gpio/export", O_WRONLY);
    if (-1 == fd) {
        fprintf(stderr, "Failed to open export for
        writing!\n");
        return(-1);
    }

    bytes_written = snprintf(buffer, BUFFER_MAX, "%d", pin);
    write(fd, buffer, bytes_written);
    close(fd);
    return(0);
}

int GPIOUnexport(int pin)
{
    char buffer[BUFFER_MAX];
    ssize_t bytes_written;
    int fd;

    fd = open("/sys/class/gpio/unexport", O_WRONLY);
    if (-1 == fd) {
        fprintf(stderr, "Failed to open unexport for
        writing!\n");
        return(-1);
  }

    bytes_written = snprintf(buffer, BUFFER_MAX, "%d", pin);
    write(fd, buffer, bytes_written);
    close(fd);
    return(0);
}

int GPIODirection(int pin, int dir)
{
    static const char s_directions_str[] = "in\0out";

#define DIRECTION_MAX 35
    char path[DIRECTION_MAX];
    int fd;

    snprintf(path, DIRECTION_MAX, "/sys/class/gpio/gpio%d/direction", pin);
    fd = open(path, O_WRONLY);
    if (-1 == fd) {
  //if you get this error you probably failed to use sudo in
  //the command
        fprintf(stderr, "Failed to open gpio direction for
```

```
        writing!\n");
        return(-1);
    }

    if (-1 == write(fd, &s_directions_str[IN == dir ? 0 : 3],
    IN == dir ? 2 : 3)) {
        fprintf(stderr, "Failed to set direction!\n");
        return(-1);
}
    close(fd);
    return(0);
}

int GPIORead(int pin)
{
#define VALUE_MAX 30
    char path[VALUE_MAX];
    char value_str[3];
    int fd;

    snprintf(path, VALUE_MAX, "/sys/class/gpio/gpio%d/value",
    pin);
    fd = open(path, O_RDONLY);
    if (-1 == fd) {
        fprintf(stderr, "Failed to open gpio value for
        reading!\n");
        return(-1);
    }

    if (-1 == read(fd, value_str, 3)) {
        fprintf(stderr, "Failed to read value!\n");
        return(-1);
    }

    close(fd);

    return(atoi(value_str));
}

int GPIOWrite(int pin, int value)
{
    static const char s_values_str[] = "01";

    char path[VALUE_MAX];
    int fd;

    snprintf(path, VALUE_MAX, "/sys/class/gpio/gpio%d/value",
    pin);
    fd = open(path, O_WRONLY);
```

```
    if (-1 == fd) {
        fprintf(stderr, "Failed to open gpio value for
        writing!\n");
        return(-1);
    }

    if (1 != write(fd, &s_values_str[LOW == value ? 0 : 1],
    1)) {
        fprintf(stderr, "Failed to write value!\n");
        return(-1);
    }

    close(fd);
    return(0);
}
```

I will not attempt to go through all the code because it is fairly extensive, but I will point out that I generously sprinkled comments throughout the code to help you understand what was happening in a particular code section or module.

Here are the steps to compile, build, and run this program:

1. Download and install the wiringPi package found at wiringpi.com. Do not use the GitHub version of wiringPi because it is not the version required for this installation.
2. Download and install the zip file from Mark Williams' GitHub site at https://github.com/mwilliams03/Pi-Touchscreen-basic.
3. Extract all the files into a single directory. My directory was named Pi-Touchscreen-basic-master. The following files should be in this directory:
   a. README
   b. buttonExample.c
   c. font_8x8.c
   d. framebuffer.c
   e. main.c
   f. touch.c
   g. touch.h
4. Download the source code from the book's companion website. It is entitled buttonExample.c.
5. Replace the existing buttonExample.c with this version.
6. Compile it by running this command:

   ```
   sudo gcc -g -o buttonExample
   buttonExample.c -l wiringPi
   ```

7. Build it by entering:

   ```
   make
   ```

8. Install it by entering:

   ```
   make install
   ```

9. Run it by entering:

   ```
   sudo ./buttonExample
   ```

If all goes well, you should see the button displayed on the touchscreen, as shown in Figure 2-29.

Now press the button on the touchscreen, and two things should happen. First, the button should change to what is shown in Figure 2-30.

Second, the LED should be on. If not, check that you used `sudo` in the command; otherwise, the OS will not permit the file to be written.

Figure 2-29 Initial touchscreen with the buttonExample program running.

Figure 2-30 Touchscreen after pressing the button.

This is a permissions issue, which I discussed in Chapter 1. The other reason could be that the LED is connected backwards.

A connected console screen also will show all the events as they are happening, just as it was with earlier demonstrations. Figure 2-31 is a snapshot of some button-pressing events for this application.

This concludes my simple demonstration project. It obviously may be extended to include additional buttons that can trigger other I/O events as well as start or stop other processes. You can really dive deep into this area because most readers will realize that most smart phones have a myriad of touchscreen activities that trigger all sorts of events, e.g., “Hello Siri.”

```
donnorris — pi@raspberrypi: ~/Pi-Touchscreen-basi
Event type is Sync = Start of New Event
Event type is Absolute & Event code is X(0) & Event value is 2103
Event type is Absolute & Event code is Y(1) & Event value is 2039
Event type is Absolute & Event code is Pressure(24) & Event value is 101
Event type is Sync = Start of New Event
Event type is Absolute & Event code is X(0) & Event value is 2104
Event type is Absolute & Event code is Y(1) & Event value is 2020
Event type is Absolute & Event code is Pressure(24) & Event value is 93
Event type is Sync = Start of New Event
Event type is Absolute & Event code is Y(1) & Event value is 2027
Event type is Absolute & Event code is Pressure(24) & Event value is 79
Event type is Sync = Start of New Event
Event type is Absolute & Event code is X(0) & Event value is 2100
Event type is Absolute & Event code is Y(1) & Event value is 2032
Event type is Absolute & Event code is Pressure(24) & Event value is 75
Event type is Sync = Start of New Event
Event type is Absolute & Event code is X(0) & Event value is 2092
Event type is Absolute & Event code is Y(1) & Event value is 2036
Event type is Absolute & Event code is Pressure(24) & Event value is 73
Event type is Sync = Start of New Event
Event type is Key & Event code is TOUCH(330) & Event value is 0 = Touch Finished
Event type is Sync = Start of New Event
```

Figure 2-31 Event log for button presses.

One last item: I strongly recommend that you do not disassemble your touchscreen project if you intend to do the software-defined radio (SDR) project in Chapter 7. The touchscreen PiTFT is used as is in that project, and it would be a shame to have to rebuild it all if you want to explore the wonderful world of SDR.

Summary

This was a chapter about touchscreens, and it concluded with a simple demonstration of a single button that toggled a LED. I also dug deeply into what makes a touchscreen function, pointing out the different types and their pros and cons. Linux file I/O also was covered, and I demonstrated how to use simple file I/O to control the touchscreen LCD.

CHAPTER 3

Arduino Coprocessor

IN THIS CHAPTER, I WILL SHOW YOU how to efficiently use an Arduino coprocessor with the RasPi. I will also address the common question of why a coprocessor is even necessary for a RasPi application. There is also a demonstration project that amply illustrates how a coprocessor can allow a project to run successfully on a RasPi, where it could not previously run the project without incorporating it.

What Is a Coprocessor?

In the simplest sense, a coprocessor is just another computer that runs concurrently with a RasPi and is somewhat controlled by the RasPi. The Arduino coprocessor I will be using in this project essentially runs autonomously from the RasPi and interacts with the RasPi only as necessary to fulfill its project responsibilities. The Arduino has its own separate dynamic and flash memory, GPIO, communications links, power regulator, and processor clock source. The RasPi would be considered the main processor, and the Arduino is the coprocessor, as shown in the block diagram in Figure 3-1.

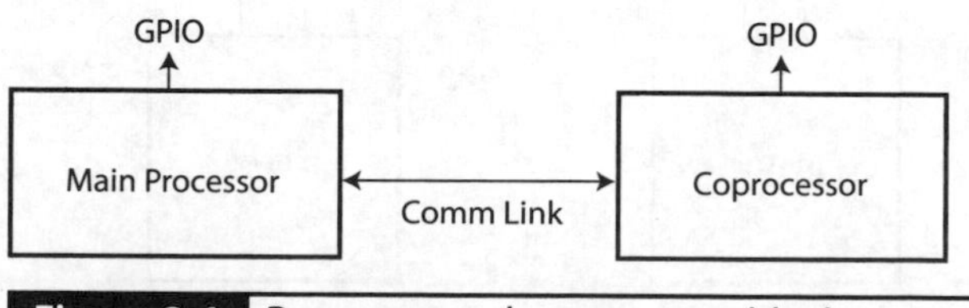

Figure 3-1 Processor and coprocessor block diagram.

The key element that I want to point out in the figure is the comm link shown connecting the two processors. This is how they interact with the main processor, either initiating an action or responding to an event from the coprocessor.

Chapter 3 Parts List

Item	Model	Quantity	Source
Arduino Uno	rev 3	1	amazon.com
RasPi	B+	1	adafruit.com
XBee module	Pro, wire antenna	2	digi.com
XBee Arduino shield	20-011-902	1	sainsmart.com
XBee SIP adapter	32402	1	parallax.com
Lidar module	LidarLite	1	pulsedlight.com

Communication Implementation Techniques

A number of implementation techniques can be used for the comm link, as listed next and discussed separately:

- UART
- I2C
- SPI

- GPIO poll/respond
- GPIO interrupt
- Specialized with external hardware

UART Serial Protocol

Universal asynchronous receive transmit (UART) is also commonly referred to as *serial data transmission*. It is a bit serial protocol that uses ASCII characters for duplex or simultaneous two-way communication between two devices. Figure 3-2 is a simplified block diagram of a generic UART comm link.

The RasPi transmits data on pin TXD0 and receives on pin RXD0. There is also no concept of a master or slave in this protocol because it is used primarily for data communications as opposed to control, which is the focus of both the SPI and I2C interfaces (discussed next).

There is an existing protocol specification (RS232) that controls the physical, electrical, and logical configurations for this type of data-transmission link. The RasPi and Arduino both use a modified approach to UART transmissions in which two voltage levels, 0 and 3.3 V, are used, which represent high and low bits or 1s or 0s, respectively. The RasPi then sends a bit on the TX line, which is received on the RX line by the Arduino. The opposite is true when the Arduino is sending to the RasPi.

UART's are *asynchronous*, meaning that no clock pulses are required to either initiate a data transmission or decode a received bit stream. The synchronization happens when the transmitting side changes the normally idle state voltage, which then creates a start pulse. Normally, eight pulses follow and represent the ASCII character being sent, and finally, a stop pulse is sent. All these pulses are collectively known as a *frame*. The number of frames sent per second is termed the *baud rate*, which is a legacy name that comes from the old days of teletypewriters. Common baud rates are 4800, 9600, and 115,200. There are many variations on how a frame is constituted per the specifications, but what I have stated is today's common practice when interconnecting processors such as the RasPi and Arduino.

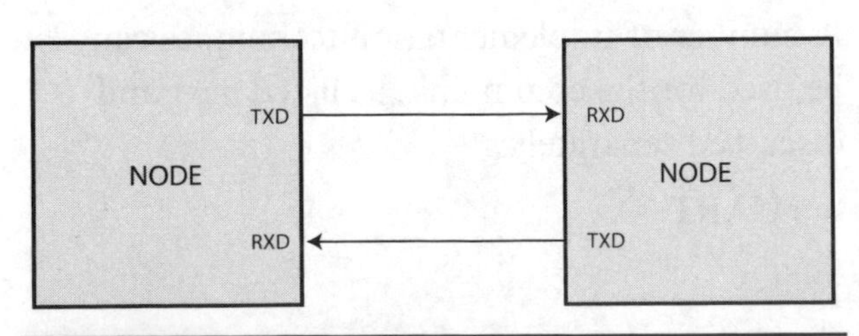

Figure 3-2 UART block diagram.

The ASCII characters sent and received then are used to create messages between the two processors, which now constitute the formal communications channel. The exact nature of the messages and how each processor interacts are determined by the nature of the application.

I2C Serial Protocol

The next serial protocol is the *interintegrated circuit interface* or *I2C* (pronounced "eye-two-cee" or "eye-squared-cee), which is a *synchronous* serial data link. Figure 3-3 shows a block diagram of the I2C interface with one master and one slave. This configuration is also known as a *multidrop* or *bus network*.

I2C supports more than one master as well as multiple slaves. This protocol was created by the Philips Company in 1982 and is a very mature technology, meaning that it is extremely reliable.

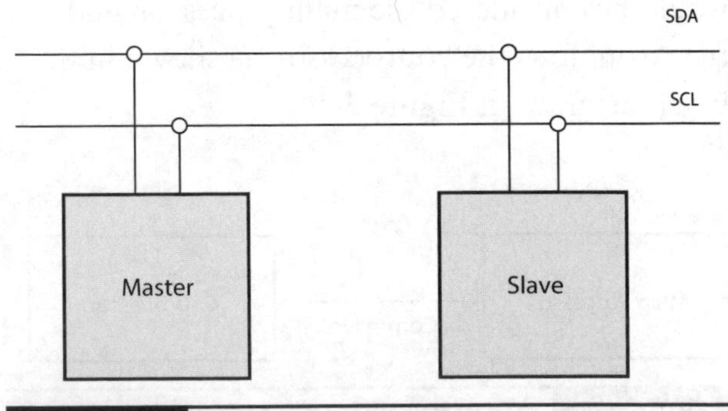

Figure 3-3 I2C block diagram.

Only two lines are used: SCLK for the serial clock and SDA for serial data. Table 1-1 shows the RasPi names for both the clock and data lines.

TABLE 3-1 I2C Signal Lines

Signal Name	Description	RasPi Name
SCL	Clock	SCL0
SDA	Data	SDA0

SPI Serial Protocol

A block diagram of the *serial peripheral interface* (SPI) is shown in Figure 3-4. The SPI interface (pronounced "spy" or "ess-pee-eye") is a *synchronous* serial data link. A clock signal is needed because it is synchronous. It is also a full-duplex protocol, meaning that data can be sent and received simultaneously from the host and slave. SPI is also referred to as a *synchronous serial interface* (SSI) or *four-wire serial bus.*

The four interconnecting signal lines between the SPI host and SPI slave shown in Figure 3-4 are explained in Table 3-2.

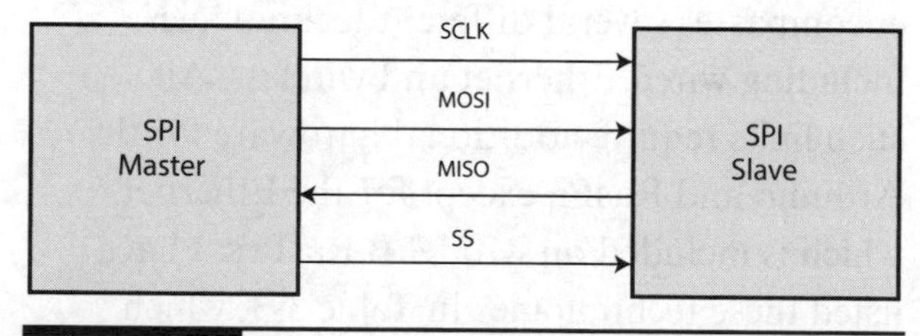

Figure 3-4 SPI block diagram.

TABLE 3-2 SPI Signal Lines

Signal Name	Description	RasPi Name
SCLK	Clock	SPI_SCLK
MOSI	Master out, slave in	SPI_MOSI
MISO	Master in, slave out	SPI_MISO
SS	Slave select	SPI_CEn*

*There are two pin connections on the RasPi named SPI_CE0 and SPI_CE1. These connections allow two separate SPI slave devices to be individually selected, or *enabled.* In fact, the "CE" in the RasPi name is short for "chip enable."

GPIO Poll/Respond

This type of communication link uses two GPIO pins to transfer a signal bit between the processors. I will use the hypothetical oven control system shown in Figure 3-5 to help clarify this communications link example.

This system controls the oven temperature for the world-famous French pastry chef Jacque M., who demands an accuracy of ±0.5°F for his oven temperature preset. Measuring oven temperatures to this degree of precision normally requires the use of an industrial resistance temperature detector (RTD). RTDs use an absolute resistance measurement in ohms that relates to a real temperature. The

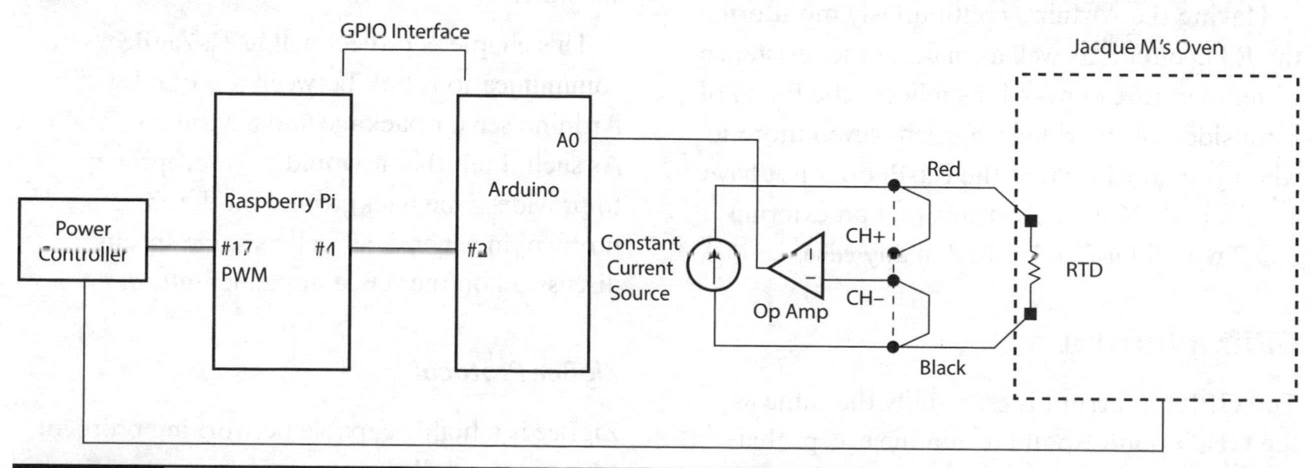

Figure 3-5 GPIO communications link.

Arduino coprocessor shown in the figure is set up to measure the RTD, which is part of a standardized circuit. It does this by measuring the voltage across the RTD and then comparing this voltage with a voltage drop across a precision resistor, which is also part of the standardized circuit containing the RTD. These two voltage readings are then used in a calculation made by the Arduino to derive a compensated voltage that will be directly proportional to the RTD resistance. The compensated voltage is then converted into a 10-bit binary number using the Arduino's ADC. However, this number now needs to be converted into an actual temperature, which the Arduino does by using lookup table calibration data, which are supplied by the RTD's manufacturer.

The Arduino can be easily set to turn on a GPIO pin, which is set as an output when the oven reaches the preset temperature that Jacque demands for baking his exquisite pastries. This Arduino GPIO pin is also connected to a RasPi GPIO pin, which is set as an input. The RasPi is programmed to periodically check the status of the selected Arduino GPIO pin to see if it has changed states. This periodic checking is also known as *polling*. When a state change is detected by the RasPi, it could then branch to a different part of its program, which could be Jacque's kitchen automation application.

Having the Arduino continuously monitoring the RTD circuit as well as making the resistance-to-temperature conversions relieves the RasPi of a considerable workload burden—even more so when you consider that the RasPi does not have a built-in ADC, which means that an external ADC would have to be used in any case.

GPIO Interrupt

The GPIO interrupt is essentially the same as the GPIO poll/respond technique except that the RasPi does not poll the selected GPIO input pin looking for a state change. Instead, any state change, as described next, will trigger a hardware interrupt. Every RasPi GPIO pin can accommodate interrupts. An *interrupt* is an event that stops or "interrupts" the normal programming flow and directs the RasPi to execute some special handler program or code based upon the source of the interrupt. Interrupts maybe triggered by the following state changes:

- HIGH level detected
- LOW level detected
- HIGH to LOW transition detected
- LOW to HIGH transition detected

Using these state changes will improve RasPi performance over polling, but at the expense of adding a certain level of complexity to the software.

Specialized with External Hardware

This last communications link approach actually encompasses several different technologies, including wired Ethernet and wireless. All such links require additional hardware for the Arduino and RasPi, except for the Ethernet, which is included on Model B RasPi's. I have listed these technologies in Table 3-3, which also has entries for the required hardware and software.

This chapter's project will use a ZigBee communications link between a portable Arduino sensor package and a Model B RasPi. As such, I felt that it would be appropriate to provide some background on the ZigBee protocol in general, as well as some specific discussion on the XBee implementation.

ZigBee Protocol

ZigBee is a highly capable networking protocol, which is also called a *personal area network*

TABLE 3-3 Additional Communications Link Hardware

Technology	Wired	Wireless	RasPi Hardware	Arduino Hardware	Software
Ethernet	X		None for Model B's	Ethernet shield	Socket
RS485	X		RS485/GPIO shield	RS485 shield	UART code
ZigBee		X	Wireless shield with RGB LED XBee module	XBee shield XBee Module	UART code
Bluetooth		X	Inateck Bluetooth 4.0 adaptor	Bluefruit EZ-Link	BT V4 drivers UART code

(PAN) because its intended radiofrequency (RF) range is about 100 m. ZigBee was designed to be compliant with the ISO seven-layer network model. As such, its inherent design is based on proven computer network concepts that are robust, efficient, and well understood by most system designers. Figure 3-6 shows the ZigBee logical network stack with the corresponding ISO layer number. All subsequent network software developed for the ZigBee network follows this model.

Data sent through the ZigBee network is in *packets*, similar to the Ethernet format. Figure 3-7 shows how these packets are constituted initially at Layer 2, or MAC, as it is referred to in the figure. These packets may be modified subsequently at higher layers, as needed, to suit the real-time network communication needs.

There are four packet types that exist in ZigBee:

1. Beacon
2. Data
3. MAC command
4. ACK

Actual data packets are formed at the MAC, or Layer 2, level, where the data are prepended with both the source and destination addresses. A sequence number is also assigned to allow the receiver to determine the correct sequence of received packets. It is relatively easy to receive out-of-sequence packets in this type of network. Frame control bytes are also appended for error checking, which is the reason why ACK packets are required. ZigBee is a type of connection network, similar to Ethernet, that

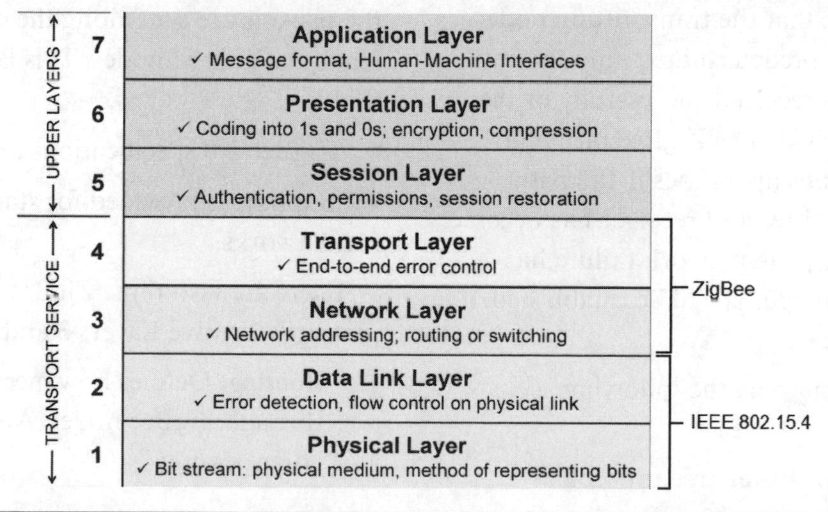

Figure 3-6 ZigBee and the ISO network layers.

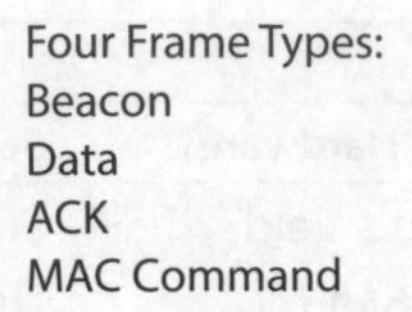

Four Frame Types:
Beacon
Data
ACK
MAC Command

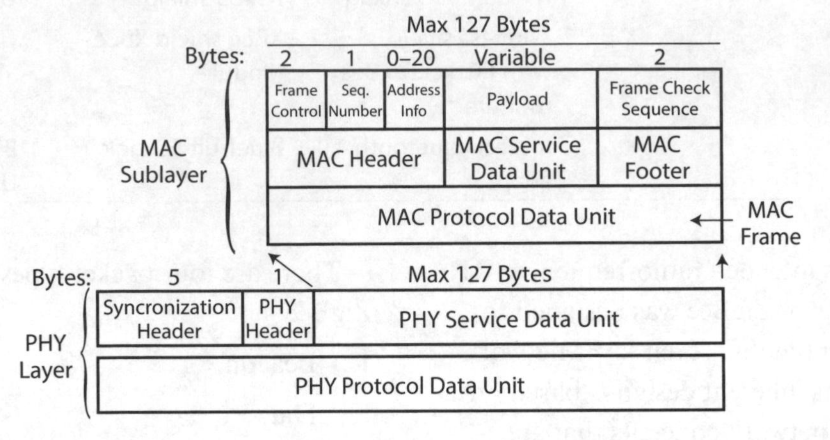

Figure 3-7 ZigBee packet formation.

has a very robust way of ensuring that packets get where they need to go. ZigBee Layer 3 uses an *acknowledgment packet* (ACK).

The receiver performs a 16-bit *cyclic redundancy check* (CRC) to verify that the packet was not corrupted during transmission. If a good CRC is determined, the receiver then will transmit an ACK; this action allows the transmitting XBee node to know that the data were received properly. The packet is discarded if the CRC indicates that the packet was corrupted, and no ACK is transmitted. The network should be configured such that the transmitting node will resend up to a predetermined amount until either the packet is received successfully or the resend limit is reached. The ZigBee protocol provides self-healing capabilities if the path between the transmitter and receiver has become unreliable or a complete network failure has happened. Alternate paths will be established, if physically possible.

Layers 1 and 2 support the following standards:

- Star, mesh, and cluster-tree topologies
- Beaconed networks
- GTS for low latency
- Multiple power-saving modes (idle, doze, hibernate)

Layers 3 and 4 further refine the packets by identifying what the packet type is, where it is going, and where it has been. They also set the data payload and support the following:

- Point-to-point and star network configurations
- Proprietary networks

Layer 4 sets up the routing, thus ensuring that the packets are sent along the correct paths to reach the desired nodes. This layer also ensures that:

- ZigBee 1.0 specifications are met.
- Support is provided for star, mesh, and tree networks.

There are also three ZigBee standards that primarily involve Layers 3 and 4:

1. **Routing.** Defines how messages are sent through ZigBee nodes. Also referred to as "digi-peating."

2. **Ad hoc network.** Creates a network automatically without any operator involvement.
3. **Self-healing mesh.** Determines automatically whether a malfunctioning node exists and reroutes messages, if physically possible.

Layer 5 is responsible for security, which is enforced by using the Advanced Encryption Standard (AES) 128-bit security key.

XBee Implementation

XBee transceivers were selected to implement the ZigBee communications link because they are small, lightweight, inexpensive, and totally compatible with the Uno and RasPi boards. XBee is the brand name for a series of digital RF transceivers manufactured by Digi International. Figure 3-8 shows one of the XBee Pro transceivers that I used.

There are two rows of 10 pins on each side of the module. These pins are spaced at 2 mm between each one, which is incompatible with the standard 0.1-inch spacing used on solderless breadboards. This means that a special connector socket must be used with the XBee module to interconnect it with the Uno. This special socket is part of an XBee Arduino shield, which is shown in Figure 3-9.

Figure 3-8 XBee Pro transceiver.

Figure 3-9 XBee Arduino shield.

This shield contains all the functionality needed to effectively interface the Arduino Uno with an XBee module. The shield and accompanying software make it very easy to create a useful RF communications link with very little effort.

The XBee used with the RasPi is connected using a different XBee interface board known as a SIP, which I purchased from the Parallax Corp. Figure 3-10 shows one of these SIP boards mounted on a solderless breadboard. Note that there are two rows of pins at the bottom of the board, which are connected together. These parallel pin sets provide additional physical support to the board when it is plugged into the breadboard.

The RasPi connects to the SIP board using both UART signal lines as well as power and ground leads. Figure 3-11 shows a RasPi connected to a SIP adapter board by a Pi Cobbler. The SIP adapter also has an XBee Pro module mounted on it.

Table 3-4 details the interconnections between the RasPi and SIP adapter.

TABLE 3-4 Interconnections Between the RasPi and XBee SIP Adapter

RasPi	RasPi Pin Number	SIP Adapter
5 V	2	+5 V
GND	6	GND
TXD	8	Din
RXD	10	Dout

Next, I will examine the XBee hardware to show how this clever design makes wireless transmission so easy.

XBee Hardware

All the electronics in the XBee hardware, except for the antenna, are contained in a slim metal case located on the bottom of the module, as may be seen in Figure 3-12.

If you look closely at the figure, you should see the bottom of the antenna wire, which is located near the top-left corner of the case.

Figure 3-10 Parallax XBee SIP adapter board.

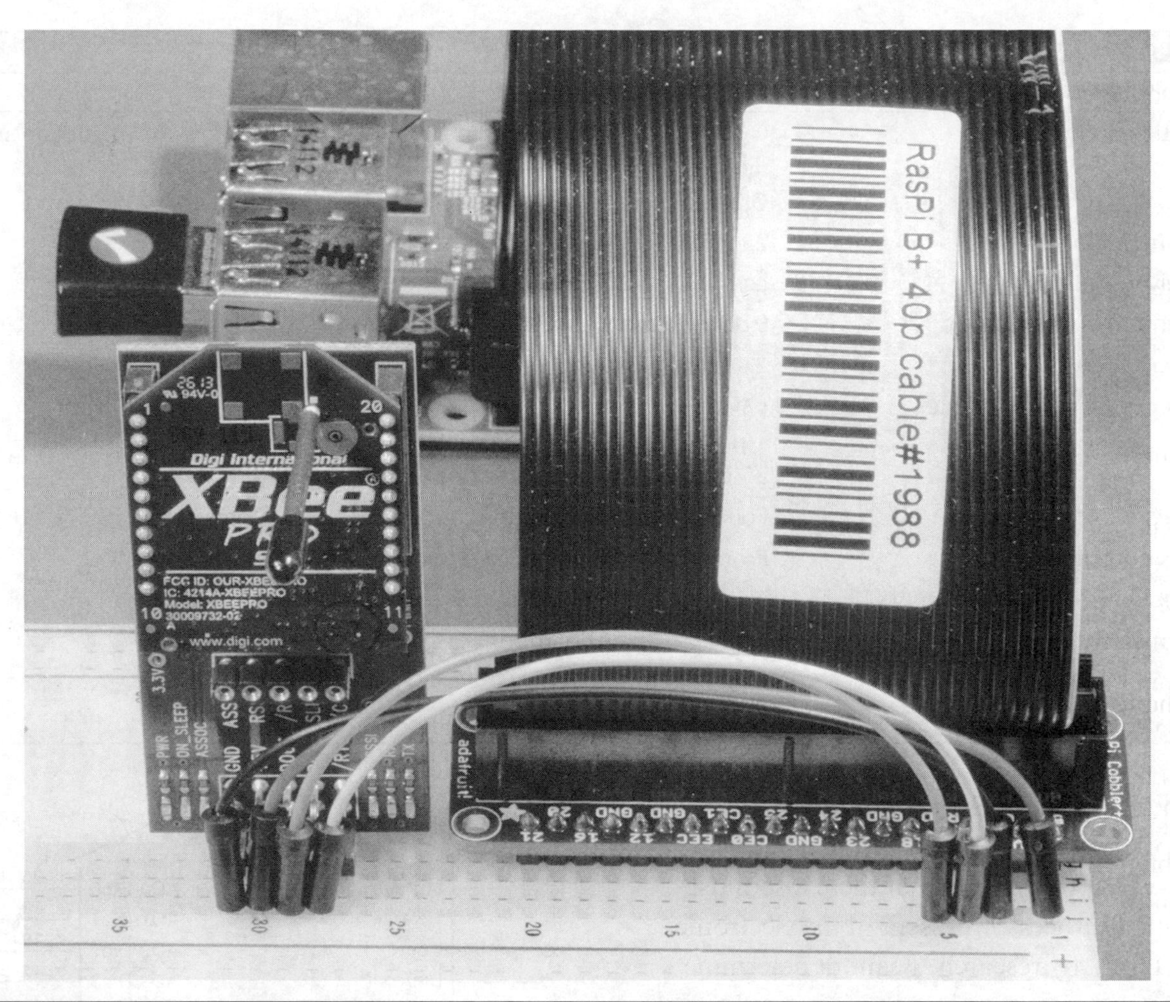

Figure 3-11 RasPi connected to an XBee module on a SIP adapter board.

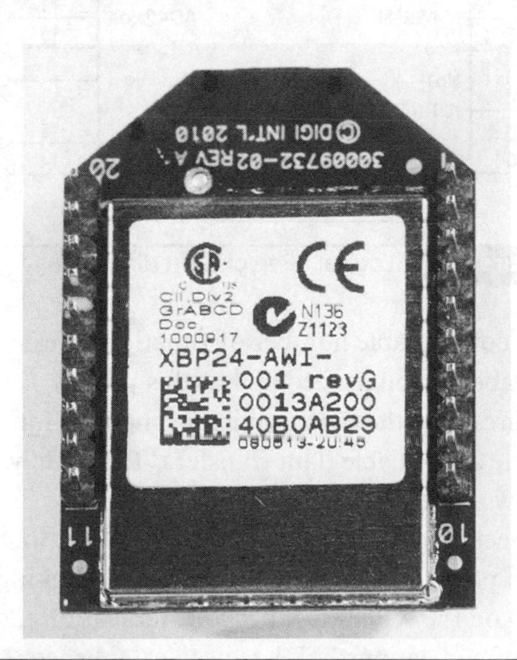

Figure 3-12 Close-up of the XBee electronics case.

While Digi International is not too forthcoming regarding what makes up the electronic contents of the case, I did determine that the earlier versions of the XBee Pro transceivers used the Freescale Model MC13192 RF transceiver. This chip is a hybrid type, meaning that it is made up of both analog and digital components. The analog components make up the RF transmit-and-receive circuits, while the digital components implement all the other chip functions. It is a complex chip, which is the reason why the XBee module is so versatile and able to automatically perform a remarkable number of networking functions. Table 3-5 shows a select number of features and specifications for the MC13192.

TABLE 3-5 Freescale MC13192 Features and Specifications

Features/Specifications	Description
Frequency/modulation	O-QPSK data in 5.0-MHz channels and full spread-spectrum encode and decode (modified DSSS) Operates on one of 16 selectable channels in the 2.4-GHz ISM band
Maximum bandwidth	250 kbps (compatible with the 802.15.4 Standard)
Receiver sensitivity	Less than –92 dBm (typical) at 1.0% packet error rate
Maximum output power	0 dBm nominal, programmable from –27 to 4 dBm
Power supply	2.0 to 3.4 V
Power conservation modes	<1-µA off current 1-µA typical hibernate current 35-µA typical doze current (no CLKO)
Timers/comparators	Four internal timer comparators available to supplement MCU resource
Clock outputs	Programmable frequency clock output (CLKO) for use by MCU
Number of GPIO pins	7
Internal oscillator	16 MHz with onboard trim capability
Operating temperature range	–40 to 85°C
Package size	QFN-32 small form factor (SFF)

The XBee module implements a full network protocol suite, but from a hardware perspective, this means that there also must be a microprocessor present in the electronics case. From my research, I cannot determine which type of microprocessor it is, but I am willing to make an educated guess that it would be a Freescale chip, based on the reasonable assumption that the MC13192 would be designed to be highly compatible with the company's own line of microprocessors. One other factor supporting my guess is that Digi International has recently introduced a line of programmable XBee modules named XBee Pro SB that use the 8-bit Freescale S08 microprocessor.

The XBee pins are detailed in a logical arrangement in Figure 3-13 for your information. Just be aware that only four of the pins are needed for this project, and they are shown with an asterisk next to the pin number.

All the pin and function descriptions are shown in Table 3-6.

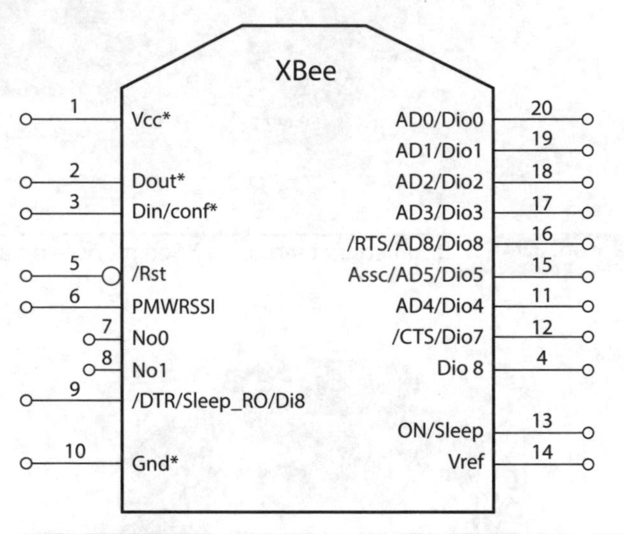

Figure 3-13 Logical XBee pin-out diagram.

A considerable number of functions are available to you if needed, but this project requires only the most minimal functions for simple and reliable data transfers. Thankfully, the two XBee modules will automatically connect and establish reliable communications when power is applied to them. A red blinking LED on the XBee shield is your indication that a communications link has been established. I

TABLE 3-6 XBee Pin Descriptions and Functions

Pin Number	Name(s)	Description
1*	Vcc	Power supply, 3.3 V
2*	Dout	Data out (TXD)
3*	Din	Data in (RXD)
4	DIO12	GPIO pin 12
5	Reset	XBee module reset, pin low
6	PWM0/RSSI/DIO10	Pulse-width modulation (PWM Analog 0), received signal strength indicator (RSSI), GPIO pin 10
7	DIO7	GPIO pin 7
8	Reserved	Do not connect (DNC)
9	DTR/SLEEP_RQ/DIO8	Data terminal ready (DTR), GPIO sleep assertion (pin low), GPIO pin 8
10*	GND	Ground or common
11	DIO4	GPIO pin 4
12	CTS/DIO7	Clear to send (CTS), GPIO pin 7
13	ON/SLEEP	Pin high when *not* sleeping
14	V_{REF}	Voltage reference level (used with analog-to-digital conversion)
15	ASSOC/DIO5	Pulse signal when connected to a network, GPIO pin 5
16	RTS/DIO6	Request to send (RTS), GPIO pin 6
17	AD3/DIO3	Analog input 3, GPIO pin 3
18	AD2/DIO2	Analog input 2, GPIO pin 2
19	AD1/DIO1	Analog input 1, GPIO pin 1
20	AD0/DIO0/COMMIS	Analog input 0, GPIO pin 0, commissioning button

will next discuss the Arduino IDE because it is a prerequisite to running the Lidar software on the Arduino.

Arduino Board and Arduino IDE

I will start this section with a brief overview of the Arduino Uno development board because I suspect that most readers will already be familiar with it. If not, I would highly recommend reading Simon Monk's excellent book on programming Arduino boards, *Programming Arduino: Getting Started with Sketches* (McGraw-Hill, ISBN 978-0071784221). The word *sketches* in the book's title refers to the name the Arduino Project gives to programs written for Arduino development boards. I will discuss sketches and other related programming elements in the software section, but first I want to provide a brief tour of the Arduino hardware I will be using in this project.

Arduino Uno Development Board

The Arduino Uno board I used is shown in Figure 3-14. It is a revision (rev) 3 board, which is important to note because the pin sockets changed slightly between the board revisions.

You can quickly identify rev 3 boards because the reset button was relocated from the center right-hand side on earlier versions to the upper left-hand side on rev 3 boards. The key

Figure 3-14 Arduino Uno rev 3 board.

specifications of the Uno board are listed in Table 3-7.

Don't be concerned if you do not understand some of the specification abbreviations in this table because I will explain them if they are needed for the project. I would recommend looking at Atmel's ATMEGA328P datasheet if you want to learn more about the detailed microprocessor specifications.

The single most important item to be mindful of regarding the Uno is that it is a microcontroller board and not a fully operational computer such as a RasPi. The significant difference is that the Uno has no

TABLE 3-7 Arduino Uno Key Specifications

Item	Value	Remarks
Microcontroller	8-bit Atmel ATMEGA328P	28-pin DIP socket
Operational voltage	5 V	Input range: 7–12 V
Digital GPIO	14	6 capable of PWM
Analog I/O	6	10 bits
Program memory	Flash 32 kB, EEPROM 1 kB	SRAM 2 kB
Clock speed	16 MHz	
USB	Type B socket	
Programmer	In-system firmware	USB-based
Serial communications	SPI, I2C	Software UART
Other	RTC, watchdog, interrupts	

capability of hosting an operating system and cannot support any programming development using only the board. It must be connected to an external computer to be programmed. This does not make the Uno inferior to the RasPi; it is just designed for a different approach for controlling embedded projects compared to the RasPi.

The open-source Arduino Project may be accessed at http://arduino.cc, which is the homepage and contains many links to other pages that I know you will find very informative. In fact, I would highly recommend that you stop reading this book for a while and go to this site and become acquainted with the Arduino concept because it will help you to comprehend the software underlying Arduino boards.

Arduino Software

The key software that you need to program the Uno is an *integrated development environment* (IDE). The IDE is available as a free download from the Arduino website (provided earlier). The current IDE that I will be using is the 1.05, which will likely change in the future because improvements and upgrades are constantly being added by the very smart folks who run and maintain the Arduino Project. One nice feature is that the existing Arduino hardware will always run on the latest version of the IDE. There is no planned or unplanned obsolescence in this arena.

I would recommend that you power-on your Uno and connect it to the computer running the IDE using a standard USB cable. Almost any "wall wart" power supply that uses a 2.1-mm outer barrel with a positive center connector will work. Remember that the supply must be between 7 and 12 VDC. I used a surplus power supply that provides 7.5 V at 1.5 A, which is more than ample for this project. Your computer should show a dialog box indicating that a driver is being installed after the Uno is plugged into

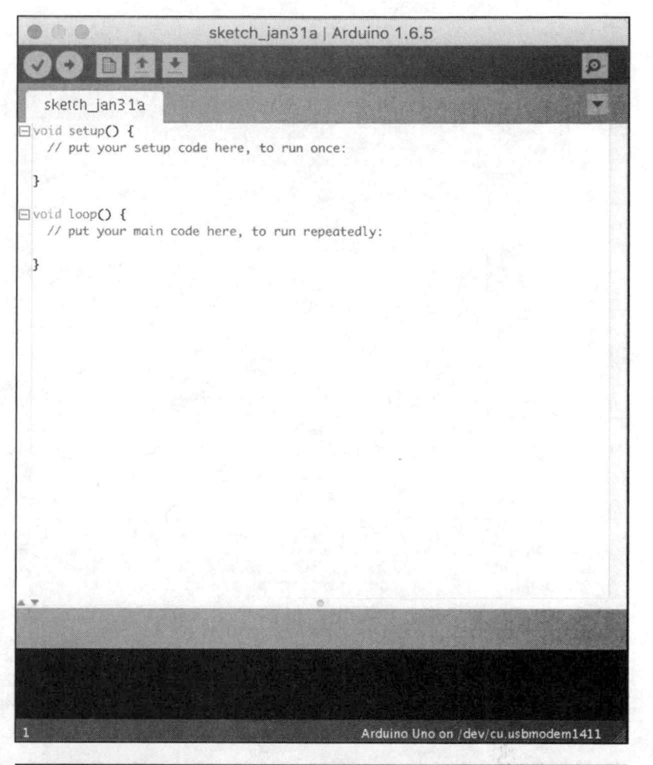

Figure 3-15 Arduino IDE startup screen.

the computer. Wait until the driver has been installed before starting the IDE. Figure 3-15 is a screen shot of the Arduino v1.05 start screen on a Windows laptop.

The IDE automatically created a default sketch entitled sketch_jan31a, which obviously contains the date that I ran the IDE program. You would normally use this blank sketch to create a program and then rename it to whatever suits your application. I will not be creating a sketch for this demonstration but will instead load a prestored example to demonstrate the classic LED blink program. There are many example programs that are automatically loaded into the computer during the IDE download. The program I opened was aptly named "Blink" and was loaded by following this sequence:

Click progressively on File → Open → Examples → 01.Basics → Blink → Blink

Figure 3-16 is a screen shot of the loaded Blink program, which appears in its own

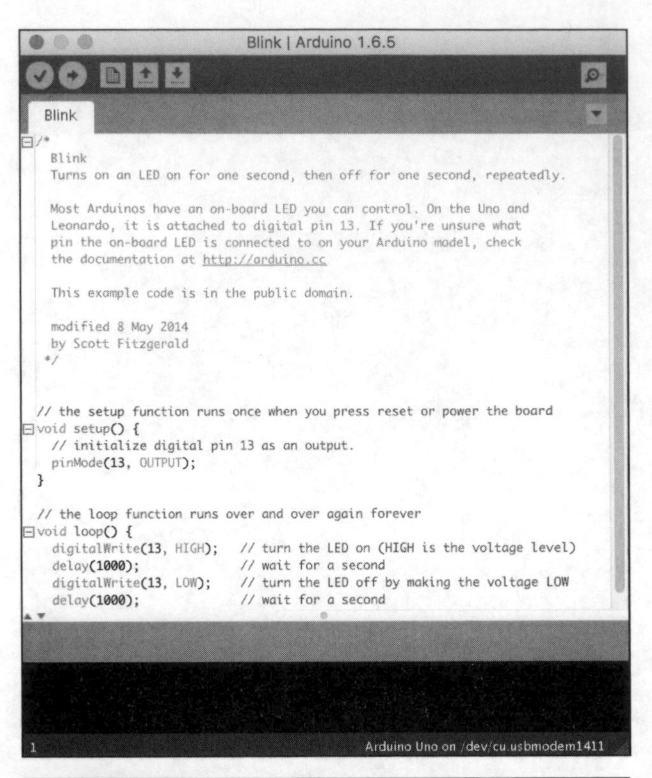

Figure 3-16 Blink code screen shot.

window. Note that the original window for the sketch_jan31a is still open in the background. This makes working on multiple programs very easy and convenient because all you need to do is click in the desired window to resume development in that program.

I show the Blink code next in order to point out some key program parts for this introductory example. I will not normally list example program code because such codes are easy to load and examine by yourself.

```
/*
  Blink
  Turns on an LED on for one second,
then off for one
  second, repeatedly.
  /This example code is in the public
domain.
  */
  // Pin 13 has an LED connected on most
Arduino boards.
// give it a name:
int led = 13;
// the setup routine runs once when you
press reset:
void setup() {
  // initialize the digital pin as an
output.
  pinMode(led, OUTPUT);
}
// the loop routine runs over and over
again forever:
void loop() {
  digitalWrite(led, HIGH); // turn the
LED on (HIGH is the
  voltage level)
  delay(1000); // wait for a second
  digitalWrite(led, LOW); // turn the
LED off by making the
  voltage LOW
  delay(1000); // wait for a second
}
```

This sketch has two methods named `setup` and `loop`. The `setup` method is always run first, followed by the `loop` method. The `setup` method provides the logical name `led` to the LED attached to the Uno's pin 13. It also makes GPIO pin 13 an output.

The `loop` method is a forever loop that alternately turns on the LED for 1 second and then turns it off for 1 second. The `digitalWrite` method is the means by which the Uno controls pin 13 and, ultimately, the LED.

You should note that I didn't mention that any physical wiring was required for this demonstration because the Uno board already has a yellow LED permanently connected to pin 13. You can easily see this LED in Figure 3-14 because it is labeled with an "L" and is located just to the left and above the ARDUINO silkscreen name.

Clicking on the right-facing arrow in the toolbar shown in Figure 3-15 will cause the program to be compiled and uploaded to the Uno. The Blink program will start immediately and continue indefinitely. You also might be a bit confused because the LED probably was already blinking before you uploaded the Blink program. That blinking was due to the default "heartbeat" that runs when no program had been previously loaded. You can prove to yourself that the Blink program functions as expected by changing the delay time and observing that the new blink rate matches whatever you entered. Simply enter new values for the delay time, say, 2000, which will make the LED blink at a 2-second rate. Compile and upload the changed program by pressing the right-facing arrow, and watch the LED slowly blink every 2 seconds. This completes my brief introduction to the Arduino hardware and software, which should provide you with sufficient knowledge to continue building this chapter's project.

The next part of this chapter starts the demonstration project. What's unique about this project is that it will be a large component in a later chapter project concerning robotics. It is my intention to try to reuse chapter projects in order to reduce component costs as well as improve the learning process by providing some level of continuity from chapter to chapter.

Lidar Demonstration Project

In this chapter project, I will demonstrate how to use a RasPi with a *light-detection and ranging system* (Lidar), which is also commonly referred to as *laser radar*. Just as a point of interest, the term *Lidar* was coined in the early 1960s as a portmanteau of the words *light* and *radar*. I will simply refer to it as Lidar from this point on.

The Lidar sensor system in this will project will be directly controlled by an Arduino. The Arduino also will be acting as a coprocessor for the RasPi. In addition, the Arduino and RasPi will use two XBee modules as a wireless communications link. The block diagram for this system is shown in Figure 3-17.

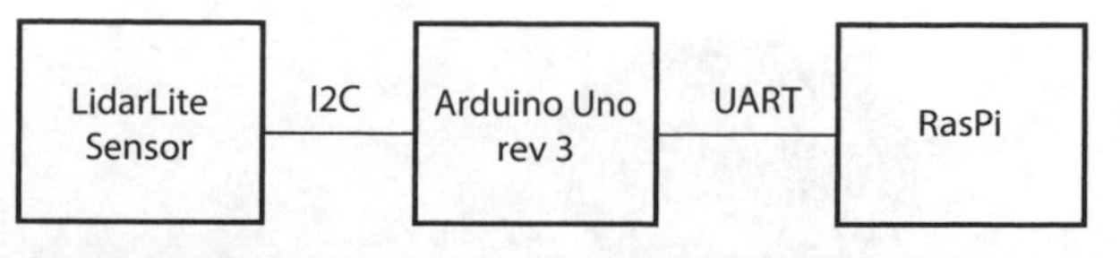

Figure 3-17 Lidar project block diagram.

The main objective of this project is to demonstrate the effectiveness of an Arduino coprocessor. There is also a side objective of showing that the Lidar sensor itself also contains a signal-processing system, which means that the Lidar is acting as coprocessor to the Arduino, while the Arduino is simultaneously acting as a coprocessor to the RasPi. Two levels of coprocessing are at play in this system, which is really not unusual when so-called smart sensors are used in a system. There are at least two additional coprocessors involved when you really analyze all the components that make up the project. These two additional coprocessors are the XBee controllers in the XBee modules, which constitute the wireless communications link. These XBee controllers should be considered "transparent" to the user because they are not specifically programmed by the user, and their operation is normally considered autonomous and automatic.

I will discuss the Lidar sensor next to provide you with a good background for this interesting technology.

Lidar Technology

The Lidar sensor system used in this project is the LIDAR-Lite model manufactured by PulsedLight LLC and is shown in Figure 3-18.

Lidar has been in existence for quite a while, with the first systems deployed in the early 1960s. These systems are used in a wide variety of applications, including but not limited to:

- High-resolution map making
- Archeology
- Geology

Figure 3-18 PulsedLight LIDAR-Lite sensor system.

- Seismology
- Forestry
- Atmospheric sensing
- Airborne altimeters
- Contour mapping

One of the more interesting Lidar deployments was Apollo 15's lunar surface mapping exercise in the early 1970s.

Main Lidar Components

Three principal components go into a Lidar system such as the one I will be using for this project:

1. **Laser emitter.** This is the device that emits, or sends out, a light pulse, which subsequently is reflected from a target and detected by a photo detector (discussed later). Lidar lasers may operate in ultraviolet (UV), visible, and near-infrared (near-IR) wavelengths depending on the specific use of the sensor. The LIDAR-Lite model uses a near-IR light wavelength on the order of 905 nm, which is not in the visible spectrum but still could be harmful if viewed head-on. The laser is Class 1, meaning that it must not be operated outside its approved enclosure and should not be viewed directly while operating. Lidar lasers also may be high, medium, or low power, again depending on their intended use. This project laser is classified as low power, even though it has over 1 W of peak power. The brief pulse time is the reason it is in the low-power category.
2. **Scanning and optics.** Scanning for a Lidar is a combination of panning and tilting. By far the most common scanning technique is to mount the Lidar emitter and receiver on a platform, which can be swept or panned about in a specific angular region. The angle of the panning servo can be measured so that the sensor controller "knows" the angular displacement from the center, or 0° forward bearing. Tilting is where another servo tilts or displaces the sensor platform from the normal or level plane plus or minus a certain number of degrees, usually within the range of ±20°. The Lidar used in this initial project is in a fixed position and will neither pan nor tilt. The optics refers to using a lens with the laser and photo detector to alter the focus and field of view of both devices, thus improving overall effectiveness. Typically, optics are quite simple, consisting of a tube with a lens attached near the end of the tube. This project Lidar has these tubes, as may be seen in Figure 3-18. The optics help to increase the distance-measurement range of this system to about 40 m with a remarkable resolution of about 2.5 cm or about 1 inch.
3. **Photo detector and signal-processing electronics.** The receiver uses a sensitive photo detector, which normally is a photodiode tuned to the same wavelength

as the laser emitter. The photo detector is critical in any Lidar system. Not being able to reliably detect reflected laser light pulses would render any Lidar sensor useless. The laser pulses are also encoded in such a manner as to minimize to the maximum extent possible any effects of ambient sunlight. The controller within the Lidar does all the encoding and decoding of the light pulse as well as some sophisticated signal processing, which I discuss in further detail later.

NOTE **The LIDAR-Lite photo detector is located 2.5 cm from the lens at the top of the light receiver tube. This means that you will need to add 2.5 cm to any calibrated distance measurements with respect to the tube end.**

Figure 3-19 shows a LIDAR-Lite system block diagram to help explain some of the sophisticated signal processing, which continually operates to allow this Lidar to function with remarkable range and accuracy. This diagram is excerpted from an earlier user's PDF manual, which is no longer readily available because the PulsedLight website now uses a hyperlinked reference manual.

You will see two transmitters in the figure, with one connected to the laser emitter and the other to a LED that is located next to the photo detector or optical receiver, as labeled in the figure. The LED creates a reference signal, which is required for processing, along with the actual received signal from the target. These two signals are both conditioned by filters and then digitally

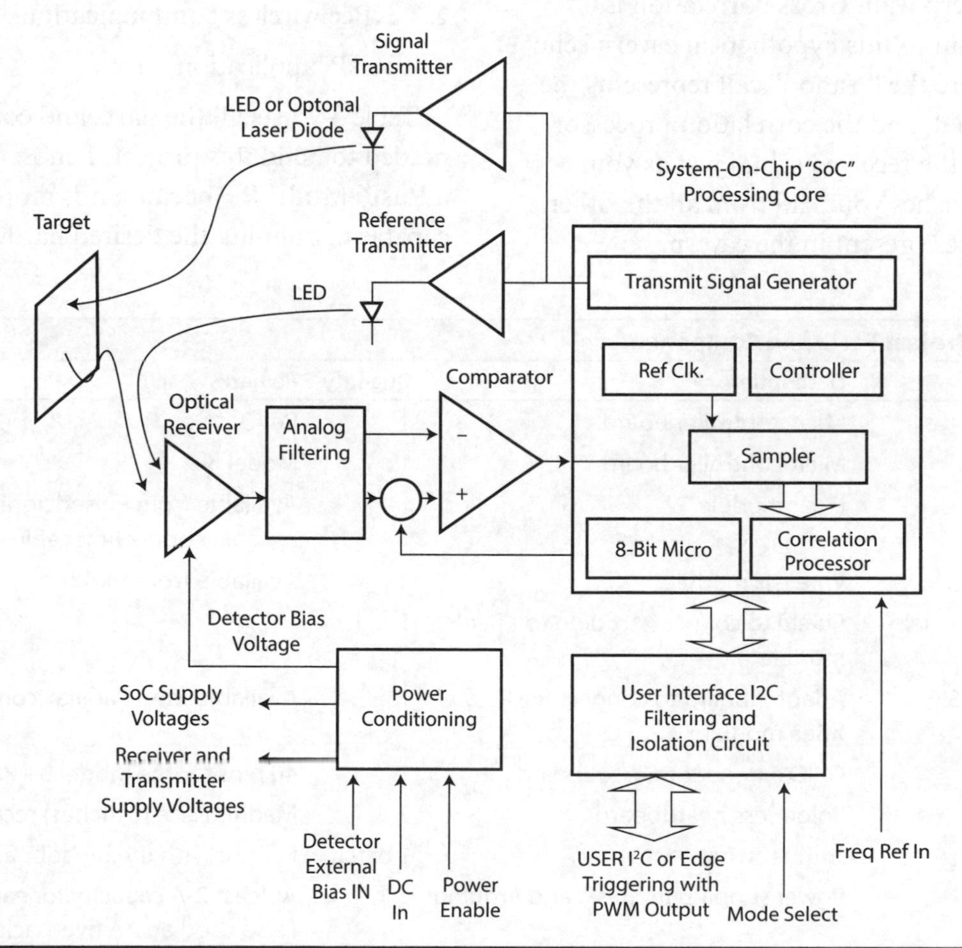

Figure 3-19 LIDAR-Lite system block diagram.

sampled before being sent into a correlation processor. Without getting too technical, I will explain that this specialized processor performs a cross-correlation between the reference and received signals to extract any time delay detected when the reference signal was applied to the receiver and a reflected version of it that might be present in the actual signal. One way to think of this process is to imagine a natural cavern that produces great echoes. Now, it would get very confusing if there were a bunch of people in the cavern with you, each one yelling something to test the echoes. But you are pretty clever, so you shout "Yahoo" and wait for that to return, ignoring every other audible signal. The time delay in seconds between you shouting and hearing your specific "Yahoo" call is related to the twice the distance between you and the reflecting cavern wall. Cross-correlation is something akin to this hypothetical cavern echo example, where the "Yahoo" call represents the reference signal, and the correlation processor distinguishes the received signal just as your brain distinguishes your call from all the other calls and noises present in the cavern.

There is also a block in the figure labeled, in part, "User Interface I2C," which is the interface logic between the Lidar and the Arduino. The Arduino supports the I2C bit-serial protocol, which I discussed earlier in this chapter. It takes only a few wires to interconnect the Lidar with the Arduino, which I will shortly demonstrate. At this point, I have completed the Lidar background discussion, so it's time to proceed to building the real project.

Building the Lidar Project

I have elected to present the project in three parts to ensure that it is clear how to interconnect all the components as well as set up and test the software:

1. Lidar and Arduino
2. XBee wireless communications link
3. RasPi application

Table 3-7 lists all the parts and components needed to build this project. I chose to use a RasPi Model B+ because it is more than capable of running the desired hardware and

TABLE 3-7 Project Parts and Components

Model Number	Description	Quantity	Remarks
Arduino Uno	Microcontroller board	1	Rev 3
Raspberry Pi	Microcontroller board	1	Model B+
LIDAR-Lite	Lidar module	1	Available from PulsedLight; comes with module connection cable
XBee Pro	Wireless module	2	Available from Digi.com
Arduino XBee shield	Shield to connect Arduino to XBee module	1	
Parallax XBee SIP	Adapter board to connect RasPi to XBee module	1	Available from Parallax.com
Pi Cobbler	GPIO extender to breadboard	1	40 pins for the Model B+ RasPi
Various	Solderless beardboard	2	Medium size (6 inches) recommended
	Jumper wires	1 batch	For use with the breadboards
	Power supplies for RasPi and Arduino	2	At least 2-A capacity for each supply; I also used an AA five-pack with the LIDAR-Lite to make it portable

software. The RasPi is also set up as a standalone workstation just to ease the development process. You can optionally use a headless configuration if that meets your needs.

An Arduino Uno rev 3 was selected because it is an extremely popular and low-cost board. You can easily change the boards to whichever ones you happen to have with few modifications to the build instructions and likely no modifications to the software.

Lidar and Arduino

Connecting the Arduino to the Lidar module will be the first item that needs to be done. Figure 3-20 is a Fritzing diagram showing how to interconnect these two devices using a solderless breadboard.

The interconnection is relatively simple, involving only four wires, as shown in Figure 3-20. The LIDAR-Lite uses the I2C bus to communicate with the Arduino, which is implemented using two wires, one for data (SDA) and the other for the clock (SCL). Just be careful about how you connect the two wires to the Arduino because they are not labeled on the Arduino board. You won't damage anything if you transpose the connections, but the LIDAR-Lite will not be able to communicate with the Arduino. Figure 3-21 shows the physical setup of the LIADAR-Lite with the Arduino, with a solderless breadboard used for the interconnections.

NOTE **The pretinned wires that are part of the LIDAR-Lite cable assembly are too narrow to fit snuggly into the Arduino pin sockets. Using a solderless breadboard along with jumper wires eliminates any potential connection issues.**

You will now need to install a required software library to enable the I2C protocol.

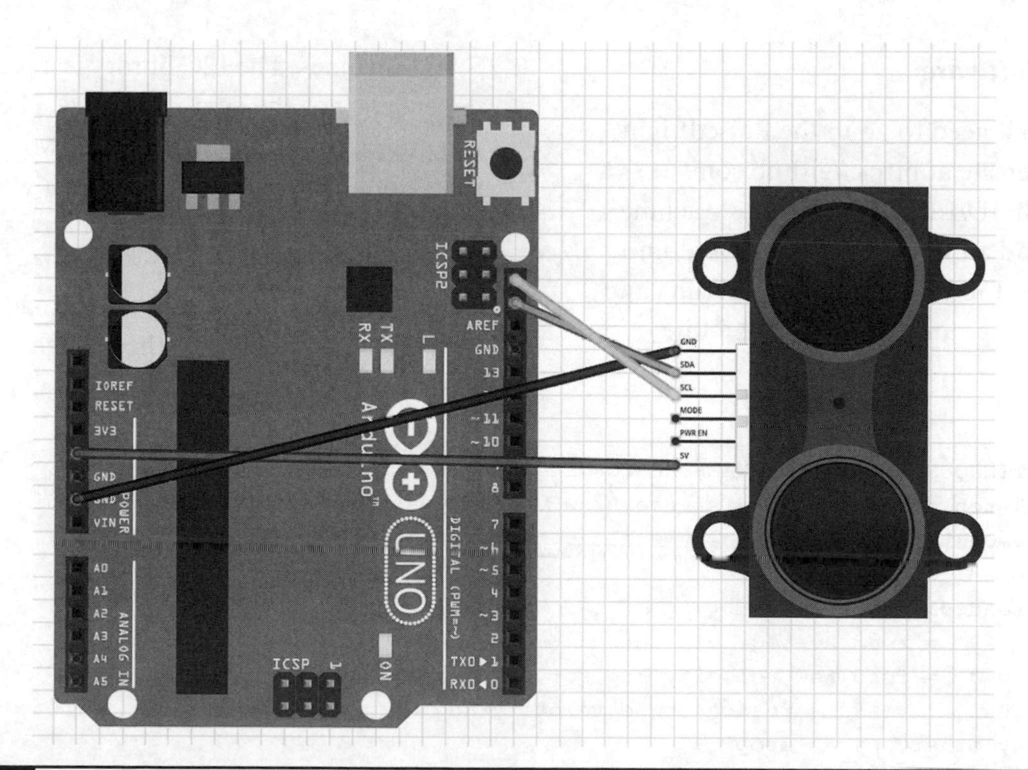

Figure 3-20 Interconnection diagram for the Arduino and Lidar module.

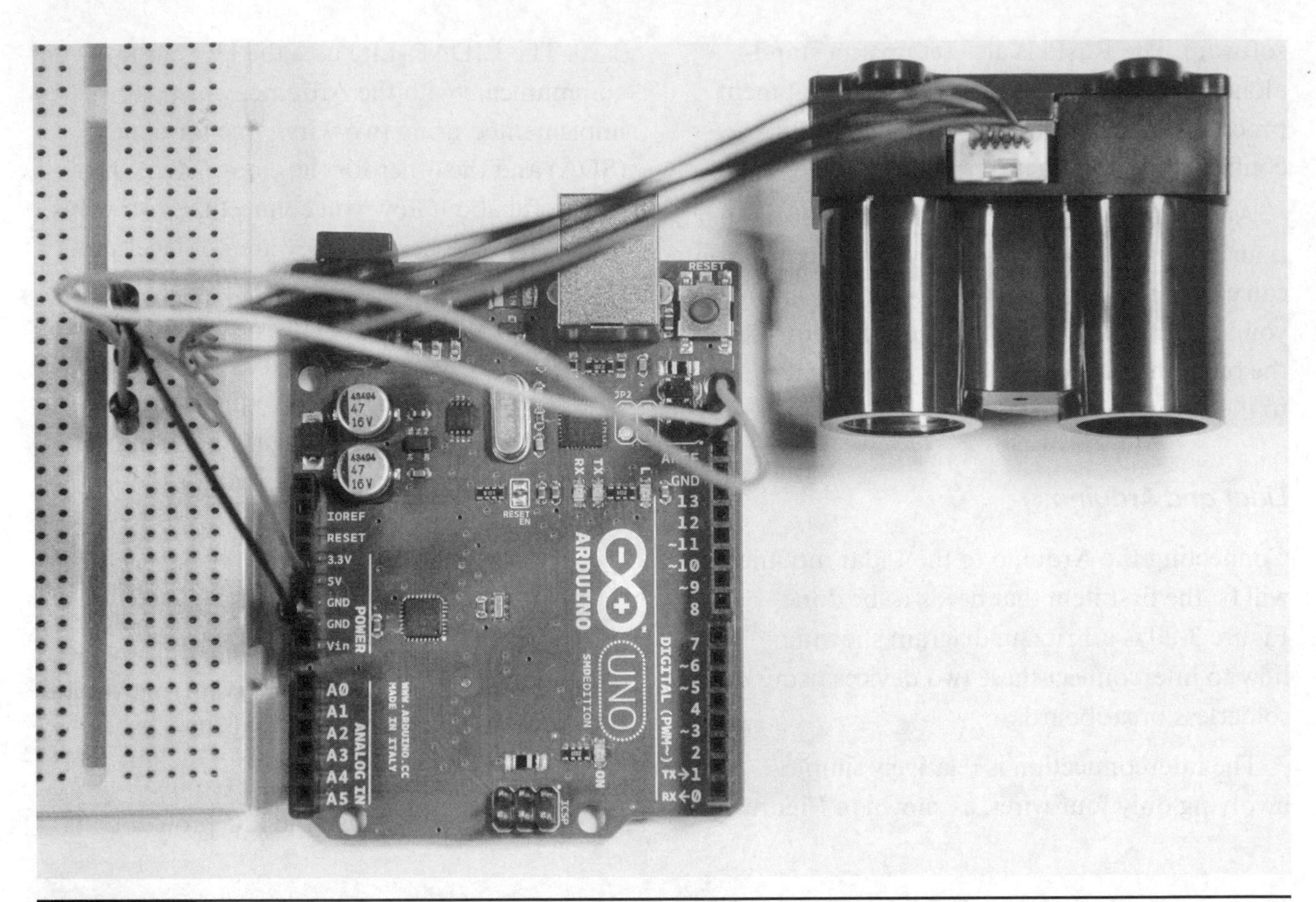

Figure 3-21 Physical setup of the LIDAR-Lite with an Arduino.

Arduino Software

You will first need to go to the PulsedLight GitHub website at https://github.com/PulsedLight3D/LIDARLite_StateMachine to download and install both the "Arduino I2C Master Library" from DSS Circuits and the demo program named "LIDARLite_StateMachine." The I2C library is a required dependency of the demo program. The demo program is listed next along with two changes that I found were needed to make it work properly with my setup. These changes are near the end of the listing and are indicated with a bold arrow (**←**) pointing to the change followed by a brief explanation of the change.

```
/* CONTINUOUS READ STATE MACHINE FOR LIDAR-LITE
This sketch demonstrates distance, raw velocity and smoothed velocity readings. It
also demonstrates the velocity scaling feature.
USAGE:
'm' = read distance
'n' + '2-9' = read average distance
'v' + '0-3' = read raw velocity and set scaling register
'w' + '0-3' + '2-9' = read average velocity and set scaling register
Type any other key to stop
Examples:
```

```
- 'm' will read distance
- 'n2' will read average distance for two readings
- "v0" will read velocity at 0.1m/s
- 'w34' will read velocity at 1m/s and average 4 readings

More information about velocity scaling is available in the LIDAR-Lite Operating
Manual on page 22:
http://pulsedlight3d.com/pl3d/wp-content/uploads/2014/10/LIDAR-Lite-Operating-
Manual-PRELIM.pdf
It utilizes the 'Arduino I2C Master Library' from DSS Circuits:
http://www.dsscircuits.com/index.php/articles/66-arduino-i2c-master-library
You can find more information about installing libraries here:
http://arduino.cc/en/Guide/Libraries */

#include <I2C.h>
// Global Variables
char LIDARLite_ADDRESS = 0x62; // LIDAR-Lite I2C Address

void setup(){
  Serial.begin(9600); //Opens serial conn at 9600 baud.
  I2c.begin(); // Opens & joins the irc bus as master
  delay(100); // Waits to make sure everything is powered up
  I2c.timeOut(50); // Sets a timeout to ensure no lock up
}

void loop(){
  smRunStateMachine(); // Run the State Machine that controls actions based on user
input
}
/* Basic read and write functions for LIDAR-Lite, waits for success message (0 or
ACK) before proceeding */

// Write a register and wait until it responds with success

void llWriteAndWait(char myAddress, char myValue){
  uint8_t nackack = 100; //variable to hold ACK/NACK
  while (nackack != 0){ // Wait until ACK is received
    nackack = I2c.write(LIDARLite_ADDRESS,myAddress,
    myValue); // Write to LIDAR-Lite Address with Value
    delay(2); // Wait 2 ms to prevent overpolling
 }
}

// Read 1-2 bytes from a register and wait for ACK
byte llReadAndWait(char myAddress, int numOfBytes, byte arrayToSave[2]){
  uint8_t nackack = 100; // Setup variable to hold ACK/NACK
  while (nackack != 0){ // Wait until ACK is received
    nackack = I2c.read(LIDARLite_ADDRESS,myAddress,
    numOfBytes, arrayToSave); // Read 1-2 Bytes from Lidar
```

```
    delay(2); // Wait 2 ms to prevent overpolling
  }
  return arrayToSave[2]; // Return array
}

/* Get 2-byte distance from sensor and combine into single 16-bit int */

int llGetDistance(){
  llWriteAndWait(0x00,0x04); // Write 0x04 to register 0x00
  byte myArray[2]; // array to store bytes from read
  llReadAndWait(0x8f,2,myArray); // Read 2 bytes from 0x8f
  int distance = (myArray[0] << 8) + myArray[1]; /* Shift high byte [0] 8 to the left
and add low byte [1] to create 16-bit int */
  return(distance);
}

/* Get raw velocity readings from sensor and convert to signed int */

int llGetVelocity(){
  llWriteAndWait(0x00,0x04); // Write 0x04 to register 0x00
  llWriteAndWait(0x04,0x80); // Write 0x80 to 0x04 to switch on velocity mode
  byte myArray[1]; // Array to store bytes from read
  llReadAndWait(0x09,1,myArray); // Read 1 byte from register 0x09 to get velocity
measurement
  return((int)((char)myArray[0])); /* Convert 1 byte to char   and then to int to get
signed int value for velocity   measurement */
}

/* Average readings from velocity and distance
int numberOfReadings - the number of readings you want to average (0-9 are possible,
2-9 are reccomended)*/

int llGetDistanceAverage(int numberOfReadings){
  if(numberOfReadings < 2){
    numberOfReadings = 2; /* If the number of readings to be    taken is less than 2,
default to 2 readings */
  }
  int sum = 0; // Variable to store sum
  for(int i = 0; i < numberOfReadings; i++){
    sum = sum + llGetDistance(); // Sum readings
  }
  sum = sum/numberOfReadings; // Calc average
  return(sum);
}

int llGetVelocityAverage(int numberOfReadings){
  int sum = 0; // Variable to store sum
  for(int i = 0; i < numberOfReadings; i++){
    sum = sum + llGetVelocity(); // Sum readings
```

```
  }
  sum = sum/numberOfReadings; // Calc average
  return(sum);
}

/* SET VELOCITY SCALING VALUES AND RESET REGISTERS BEFORE READING DISTANCE*/

void llConfigureRegisters(char myFunction, int velocityScaling, int numberOfReadings)
{
  unsigned char setScalingValue[] = {0xC8,0x50,0x28,0x14};   /* Array of velocity
scaling values, see "Velocity  Measurment" in operating manual for details */
  if(myFunction == 'v' || myFunction == 'w'){
    if(velocityScaling < 4){
      llWriteAndWait(0x68,setScalingValue[velocityScaling]);      // Set scaling
value based on scaling choice
  }else{
    llWriteAndWait(0x68,setScalingValue[0]); /* If scaling choice is out of array
range, use default scaling */
    }
  }else if(myFunction == 'm' || myFunction == 'n'){
    llWriteAndWait(0x00,0x00); /* reset device to defaults  for distance measurement
*/
  }
  smConfigureRegistersPrintStatements(myFunction, velocityScaling, numberOfReadings);
}

/* SERIAL INTERACTION AND DISPLAY FUNCTIONS */

void smPrintFunctionInit(String myString){
 Serial.println("\n=========================================================");
  Serial.println(myString);
 Serial.println("=========================================================\n");
}

String smPrintUserManaual(){
  String myString = "Continuous read state machine for  LIDAR-Lite User Manual\n\
nUSAGE: \n'm' = read  distance\n'n' + '2-9' = read average distance\n'v' + '0-3'  =
read raw velocity and set scaling register\n'w' + '0-3'  + '2-9' = read average
velocity and set scaling register  \nType any other key to stop\n\nExamples:\n- 'm'
will read  distance\n- 'n2' will read average distance for two readings\n- 'v0'
will read velocity at 0.1m/s\n- 'w34'  will read velocity at 1m/s and average 4
readings";
  return myString;
}

// SET VELOCITY SCALING PRINT OUTPUT

void smConfigureRegistersPrintStatements(char myFunction, int velocityScaling, int
numberOfReadings){
```

```
  String myString;
  if (myFunction == 'v'){myString = "Reading Raw Velocity";}
  else if (myFunction == 'w'){myString = "Reading Velocity
  Smooth";}
  else if(myFunction == 'm'){myString = "Reading Raw Distance";}
  else if(myFunction == 'n'){myString = "Reading Smooth Distance";}
  if(myFunction == 'v' || myFunction == 'w'){
    if (velocityScaling == 1){
      myString += " @ 0.25 m/s";
    }else if (velocityScaling == 2){
      myString += " @ 0.5 m/s";
    }else if (velocityScaling == 3){
      myString += " @ 1 m/s";
    }else{
      myString += " @ 0.1 m/s";
    }
  }
  if(numberOfReadings != 0){
    myString += " averaging ";
    myString += int(numberOfReadings);
    myString += " readings";
  }
  smPrintFunctionInit(myString);
}

// GLOBAL VARIABLES FOR STATE MACHINE
int configureFlag = 0;
int i = 0;
char serialArray[3];
char serialRead;

// The state machine
void smRunStateMachine(){
  if(Serial.available() > 0){
    while(Serial.available() > 0){
      if(i==0){
        configureFlag = 0;
      }
      serialRead = Serial.read();
      if(serialRead == 10 || serialRead == 32 ){
      }else{
        serialArray[i] = serialRead;
        i++;
      }
    }
  }else{
    i = 0;
  }
```

```
switch(serialArray[0]){
  case 'v': case 'V':
    if (configureFlag == 0){
     llConfigureRegisters(serialArray[0],int(serialArray[1])-48,0);
     configureFlag++;
    }
    Serial.println(llGetVelocity());
  break;
  case 'w':case 'W':
    if (configureFlag == 0){
     char scaleChar = Serial.read();
     llConfigureRegisters(serialArray[0],int(serialArray[1])-
48,int(serialArray[2])-48);
     configureFlag++;
  }
  Serial.println(llGetVelocityAverage(int(serialArray[2])-48));
break;
case 'm':case 'M':
  if (configureFlag == 0){
    llConfigureRegisters(serialArray[0],0,0);
    configureFlag++;
  }
  Serial.println(llGetDistance());
break;
case 'n':case 'N':
  if (configureFlag == 0){
    llConfigureRegisters(serialArray[0],0,int(serialArray[1])
-48);
    configureFlag++;
  }
  Serial.println(llGetDistanceAverage(int(serialArray[1])-48));
break; ← insert, missing in original version
default:
  if (configureFlag == 0){
     Serial.println("Did not recognize command"); ← replacement
     configureFlag++;
    }
  }
}
```

This program is heavily commented, including a discussion on all the operating commands, which I will not repeat. I will simply say that I found it very easy to use this program with both the Arduino and the RasPi.

All you need to do to run this program is open it in the Arduino IDE and click on the right-facing arrow icon located on the menu bar. This will cause the program to be compiled and uploaded into the Arduino. The initial program load is shown in Figure 3-22.

Next, click on the magnifying glass icon in the upper right-hand corner of the IDE to open the serial monitor screen. You will need to check that the baud rate is set to 9600 in the lower right-hand corner text box. Next enter **'m'** in the send text box, and you should start seeing distance readings in centimeters on the monitor. They will continually scroll by at a fast pace. Figure 3-23 shows a snapshot of these distance readings on the serial monitor screen.

It is now time to install the RasPi program, which will also be used to display LIDAR-Lite measurements.

RasPi Software

I used a GUI terminal program named CuteCom to display the LIDAR-Lite measurements. Enter the following to install CuteCom:

```
sudo apt-get install CuteCom
```

You will need to run the X-Windows server once CuteCom is installed. Simply enter **startx** to see the desktop. The CuteCom application

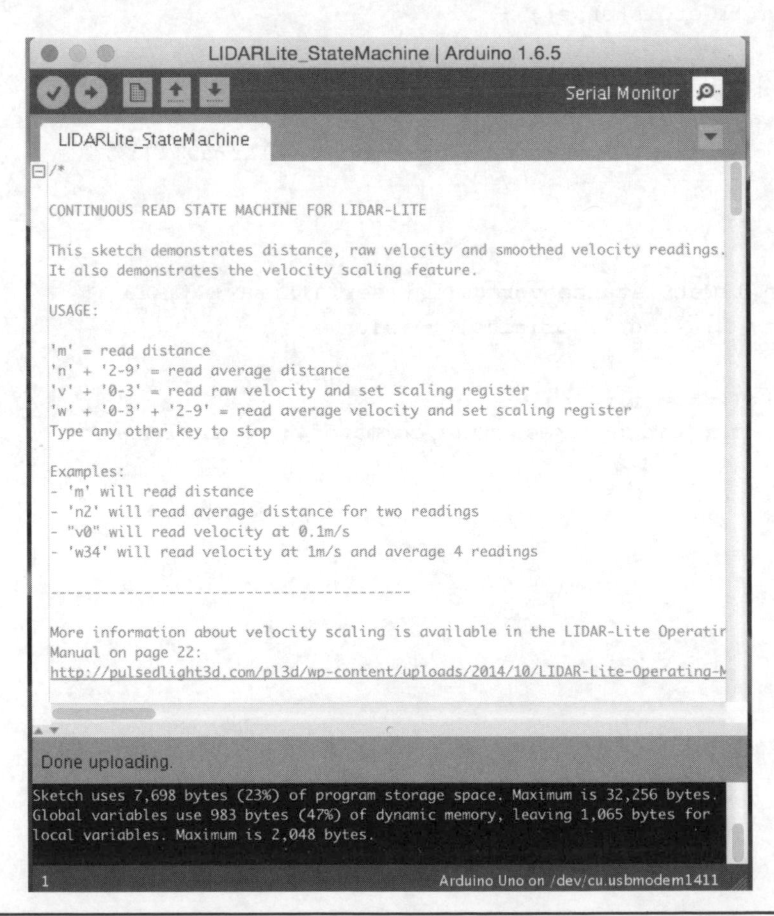

Figure 3-22 Initial load for the `LIDARLite_StateMachine` program.

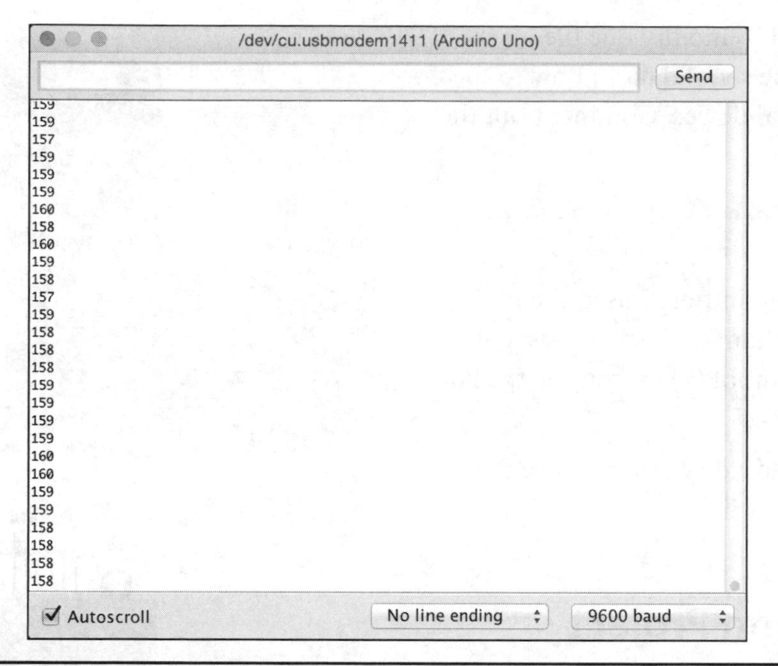

Figure 3-23 Distance readings on the Arduino serial monitor.

is located in the "Other" folder, which becomes visible when you click on the icon in the lower left-hand corner of the desktop. Figure 3-24 shows the CuteCom opening screen with the following settings:

Device	/dev/ttyAMA0
Baud rate	9600
Data bits	8
Stop bits	1
Parity	None

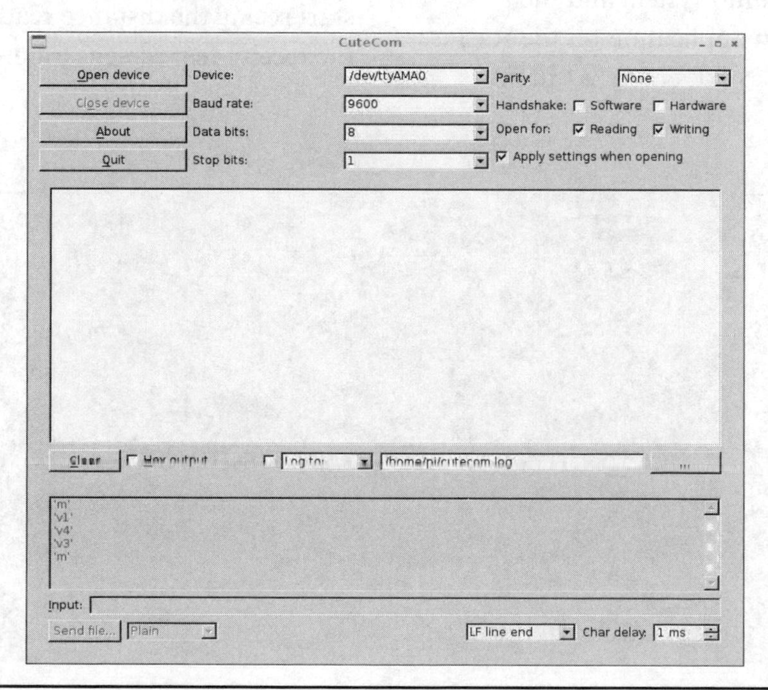

Figure 3-24 CuteCom opening screen.

You will also need to modify one file on the RasPi in order for the serial data to flow to the /dev/ttyAMA0 logical device. Comment out the line:

```
TO:23:respawn:/sbin/getty -L ttyAMA0
115200 vt100
```

which is in the file /etc/inittab. I used the nano editor to make this change, which consists of placing a number symbol (#) in front of the line, as shown in Figure 3-25.

The RasPi now should be ready to use the XBee communications link.

Demonstration Project

In this demonstration project, the RasPi is connected to an XBee module using the SIP adapter, as discussed earlier in this chapter. The Arduino board has an XBee shield attached with an Xbee module mounted on the shield. The LIDAR-Lite is connected to the Arduino board and Xbee shield board as shown in Figure 3-26.

Running the demonstration consists of powering on the Arduino system and then running the CuteCom application on the RasPi, as I earlier discussed. Next, enter `'m'` in the

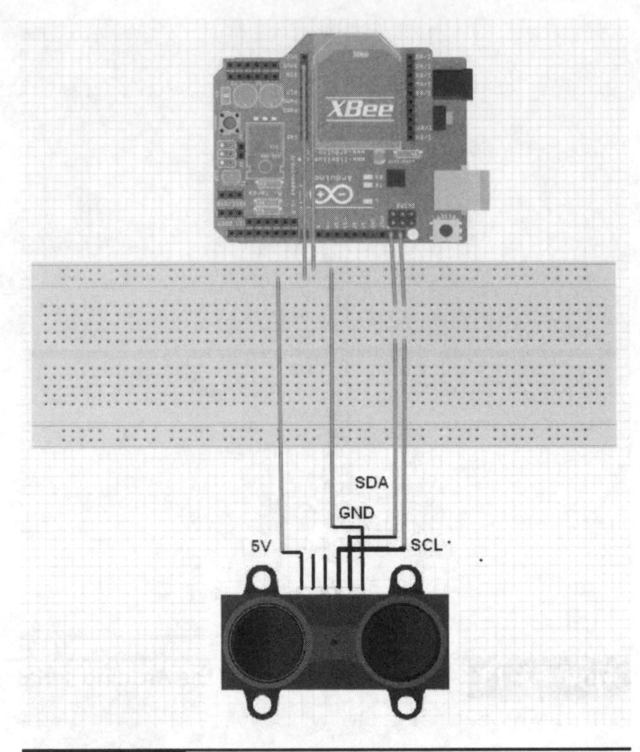

Figure 3-26 Arduino and Xbee shield connected to the LIDAR-Lite.

CuteCom input text box. You will likely see the following response: `"Did not recognize the command"` in the CuteCom receive text box. Just reenter the `'m'` again, and you should start seeing the distance readings scroll through the receive text screen. Figure 3-27 shows these

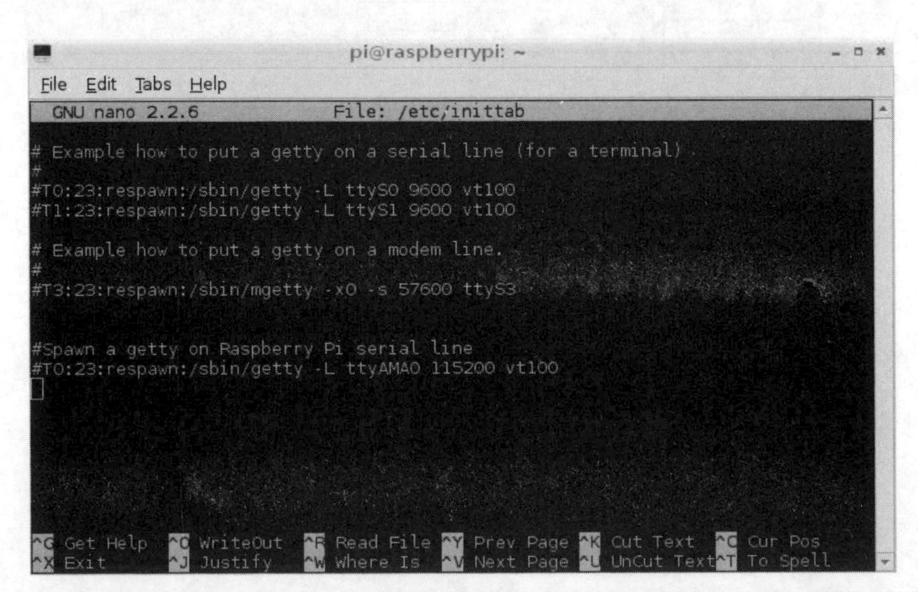

Figure 3-25 Comment out the ttyAMA0 preset configuration.

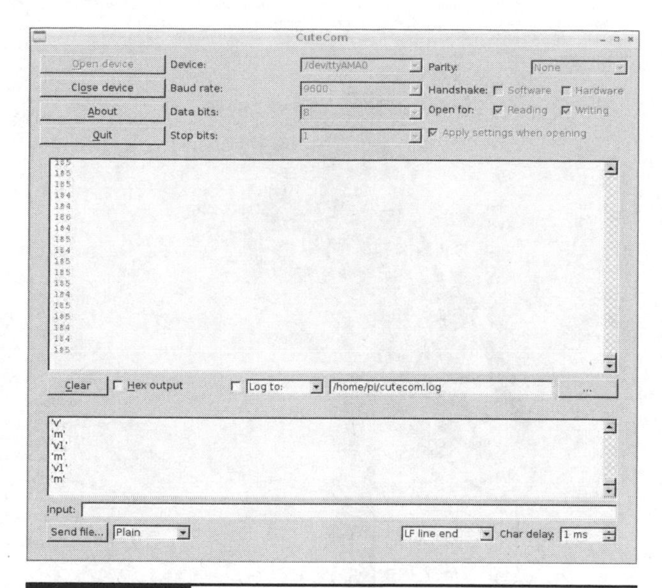

Figure 3-27 Distance readings scrolling on the RasPi CuteCom application.

measurements as they are streamed from the Lidar/Arduino subsystem via the XBee link.

You should note that no additional setup or configuration commands are needed to start the XBee communications link. The two XBee modules will automatically discover each other and set up a "transparent" serial data link, as I discussed during the earlier discussion regarding the XBee modules.

This last demonstration completes the main purpose of this chapter's project, which was to show how an Arduino coprocessor could be used effectively with a RasPi. It is also entirely possible to use a specialized application to process the Lidar data in lieu of using the CuteCom program to display the received measurements. Before completing this project, I want to show how to make the Arduino-Lidar system a self-contained, portable package.

Portable Arduino–LIDAR-Lite System

I put the Arduino board with the XBee shield attached into a small plastic case. The LIDAR-Lite was mounted on the case face in an effort to make the system highly portable and easy to deploy. This system was also powered by a five-AA battery pack. This portable system is shown in Figure 3-28.

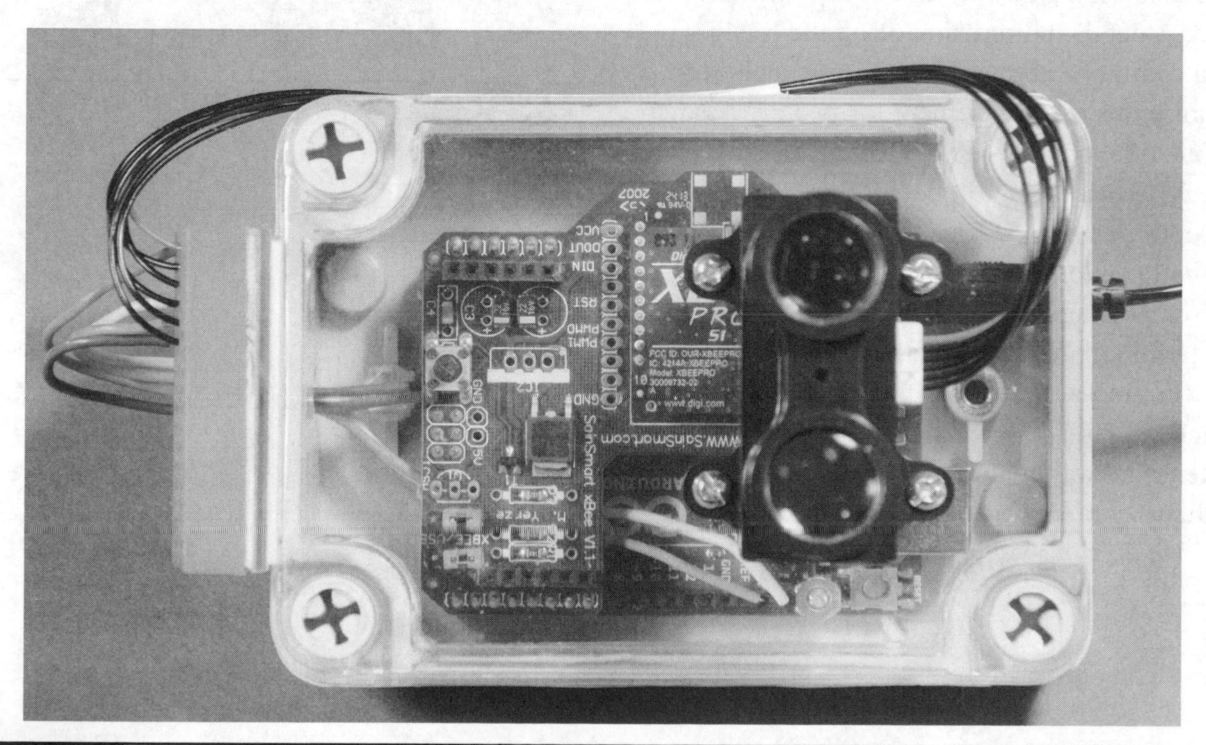

Figure 3-28 Portable Arduino–LIDAR-Lite system.

I also mounted this package on a camera tripod, which made it very easy to place it anywhere that remote distance measurements are required. This tripod arrangement is shown in Figure 3-29.

The system has a RF range in excess of 100 m, which means that the RasPi workstation can be located quite a distance from the portable, unattended Lidar unit. This feature could be used to trigger the RasPi to take some action if an object came within a preset distance from the LIDAR-Lite.

This concludes the RasPi coprocessor demonstration project. This is but one of many similar applications that can take advantage of using a coprocessor along with the main RasPi processor.

Figure 3-29 Portable package mounted on a tripod.

Summary

This chapter demonstrated how a coprocessor can effectively increase a RasPi's usefulness as well as lessening its real-time computational load. A detailed discussion of all the RasPi serial data communications protocols was included because this is a key element for two processors to function together in an effective manner. The XBee wireless protocol also was discussed in depth because this was the selected wireless communications link between the Arduino coprocessor and the RasPi, which was running the terminal program. Finally, a good in-depth presentation was done for the LIDAR-Lite, which is a laser-based range-determination system that was connected directly to the Arduino coprocessor.

CHAPTER 4

RGB LED Matrix Display

In this chapter, I will show you how to build a project that is certain to wow your family, friends, and coworkers. I will be using RGB LED matrix displays driven by a RasPi 2 to display both text and graphics. The displays also will scroll, which will add some animation to enhance the interests of people viewing the display.

32 × 64 RGB LED Matrix

The first matrix I will demonstrate is a 32 × 64 LED unit, which is shown in Figure 4-1.

There are 2048 separate RGB LED modules shown in the figure. I used the word *module* to highlight the fact that each one contains a red, green, and blue LED. Consequently, there are really 6144 separate LEDs mounted in the matrix. Each of these LEDs must be individually controlled, which means that there is a large computational task facing the RasPi to control such an array. Also note that I have used both the words *array* and *matrix* to describe the display. There is no real linguistic distinction between the two words, and most developers use both when describing these displays.

I would like to point out several terms that are also used in describing this matrix. A *row* is a single horizontal line and a *column* is a single vertical column of LEDs. Therefore, the matrix shown in the figure has 32 rows and 64 columns.

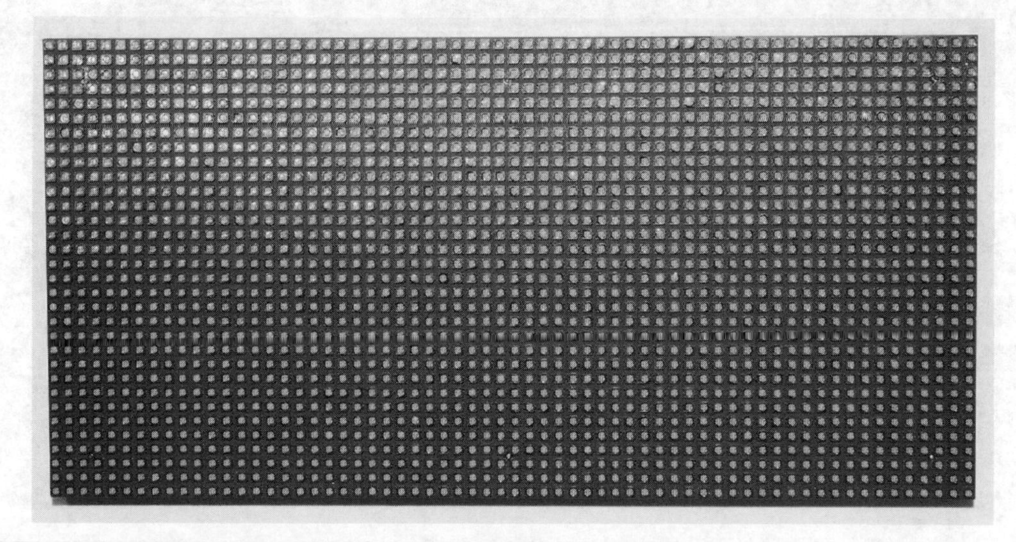

Figure 4-1 32 × 64 RGB LED matrix.

Typically, the number of rows is stated first in the matrix description, followed by the number of columns. The rows are numbered starting with 0 going to 31, while the column numbering ranges from 0 to 63. This numbering scheme is very consistent with the way conventional numeric matrices are labeled, making it easier for developers to write programs using standard notation.

Chapter 4 Parts List

Item	Model	Quantity	Source
32 × 64 RGB LED matrix	2279	1	adafruit.com
RGB Matrix HAT + RTC	2345	1	adafruit.com
Raspberry Pi 2 Model B	2358	1	adafruit.com
5-V, 4-A power supply with 2.1-mm connector	1466	1	adafruit.com

The reverse side of the 32 × 64 matrix is shown in Figure 4-2.

There are 46 integrated circuits (ICs) in the figure. The 24 driver ICs are dedicated to directly controlling the LEDs, and the other 22 ICs are used to drive the addressing lines that determine which LEDs are actually displayed for a specific display instruction. The drivers and the address logic must function together in a very efficient manner to be able to handle all 6144 LEDs that I mentioned earlier.

You should also notice a power connector located in the board middle; this is the means by which a 5-V power supply is connected to the matrix. This board came with a preassembled power cable, which is shown in Figure 4-3.

The power requirements for this board depend on the content being displayed at any moment. If it were even possible to turn on all 6144 LEDs simultaneously, it would require about 122 A, which assumes that each LED uses 20 mA. This is over 610 W of power. Fortunately, in reality, only a fraction of the LEDs is activated at any

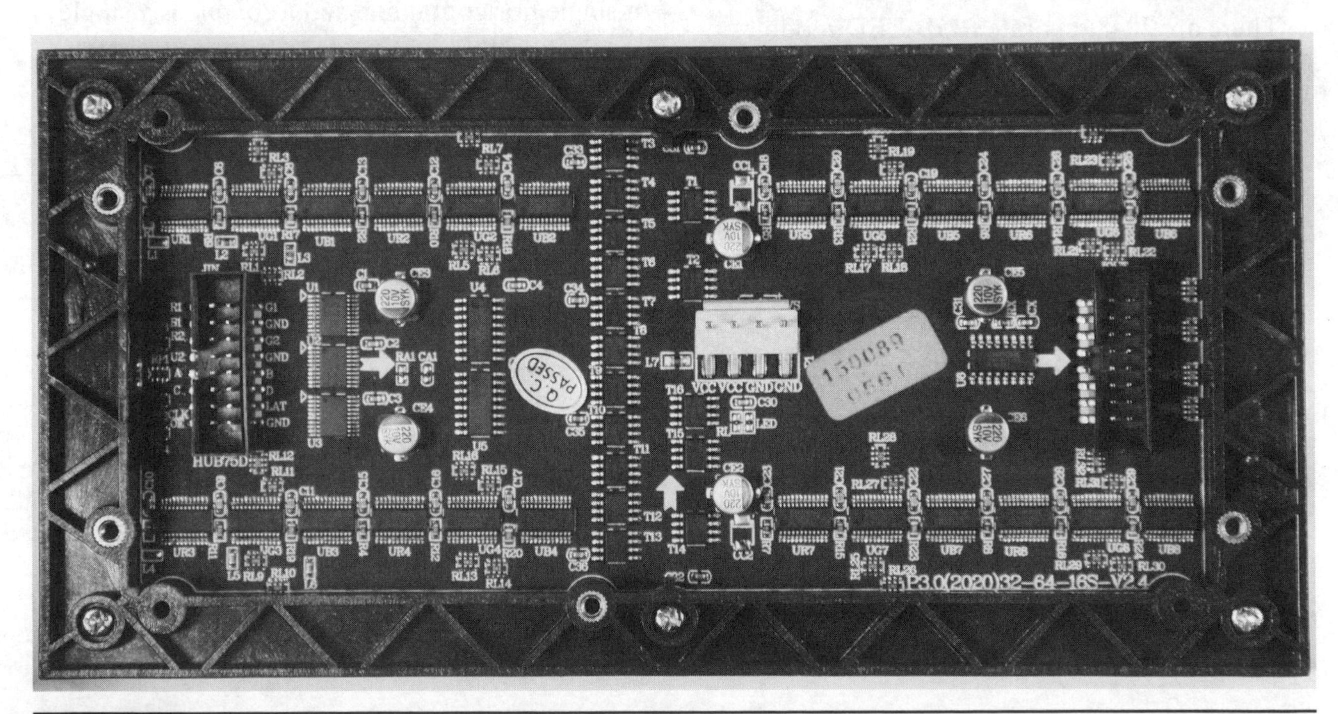

Figure 4-2 Reverse side of the 32 × 64 RGB LED matrix.

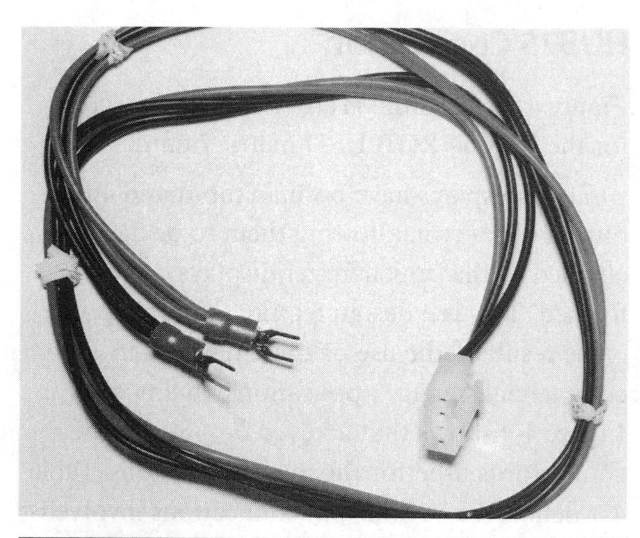

Figure 4-3 Preassembled RGB LED matrix power cable.

given moment, which means that the current draw is more in the range of 2 to 4 A. I will explain in detail which LEDs are active in the following section on how the matrix works.

How the RGB LED Matrix Works

Let me start this discussion by stating there really isn't a lot of technical information readily available regarding this type of display. As far as I can determine, these displays are actually sourced from digital signage manufacturers, who primarily make large-scale commercial displays from these modules. The modules made available through maker community distributors such as Adafruit or MCM Electronics are likely production overages from signage manufacturers looking to optimize sales for their production runs. All this means is that there is no strong incentive for the original equipment manufacturers to provide technical support for these modules because they are considered a secondary market. But this didn't deter the smart folks at Adafruit, who reverse engineered these modules to help makers understand how they function and how to program microcontrollers to drive them effectively. The following discussion is based largely on this Adafruit effort.

First, consider that a 32-row display is split into two sections, with the first section consisting of rows 0 to 15 and the second consisting of rows 16 to 31. Next, pair the first line in each section to be displayed simultaneously. This means that rows 0 and 16 are followed by rows 1 and 17 and so forth until rows 15 and 31 are activated. Then the process restarts with rows 0 and 16. Using row pairs in this fashion is referred to as *interleaving* and is very effective in minimizing power requirements and at the same time producing a pleasing display with little or no flicker or visual artifacts.

Now each row has 64 RGB LEDs, meaning that there are 192 separate LEDs to be turned on or off in each row. Two rows being active at a time translates to a maximum of 384 LEDs that could be on at any instant. This would mean that a maximum current draw of 7.680 A would be needed for this unlikely situation, assuming 20 mA per LED. Of course, the peak current flow is only for a very short time because the row pairs are rapidly scanned during the course of a full-matrix display output. The real current draw is typically 2 to 4 A, which is provided by a robust 5-V power supply with a rating of at least 6 to 10 A. Of course, the RasPi by itself cannot power this display and must use an external power supply. A bit later I will show how the external power supply will power both the display and the RasPi.

Each driver IC mentioned earlier has 16 constant-current outputs, with 12 ICs controlling one row. This means that 192 digital outputs are available to control a single row, which exactly matches the LED count previously specified. This board has 24 driver ICs, which means that two rows can be output simultaneously, which also meets the interleave display requirement.

There are four row address lines labeled A, B, C, and D, which select one of the 16 row pairs to be displayed. Once the row address is set, 192 bits are clocked out from the RasPi representing each of the column LED positions in the array. The row data are then latched, and the driver ICs are enabled, turning on or off the entire row of LEDs connected to the driver IC. The address is then incremented, and the process is repeated until all the row pairs have been output. This addressing and outputting repeat rapidly at least 20 times per second to produce a flicker-free display.

The only significant downside to this simple display interface is that the LEDs are either fully on or off, which prevents full-color presentation. The display must be rewritten very quickly to allow for more than the default eight colors to be displayed. This approach establishes what I would classify as a quasi–pulse-width-modulation (PWM) scheme, but it will allow for much fuller color rendition as well as for smoother animations on the display. The PWM approach requires a very fast processing timing, which will push the RasPi to its operational limits.

The next item I want to discuss is the interface connection that is used with these displays. This is called the HUB75.

Figure 4-4 HUB75 input connector.

HUB75 Connection

Figure 4-4 shows the HUB75 input connector for the 32 × 64 RGB LED matrix board.

These displays have both an input and an output connector, allowing them to be daisy-chained so that much larger displays can be formed. This is a design feature that I suppose is the result of the use of these displays in commercial signage applications. In any case, Figure 4-4 shows that a keyed 2- × 8-pin IDC connector is used for the interconnections. Table 4-1 details the various pin connections involved with the HUB75 connection scheme. Refer to Figure 4-4 to see the physical placement for each pin.

TABLE 4-1 HUB75 Pin Connections

Pin Label	Use	Remarks
R1	First row pair—red	
G1	First row pair—green	
B1	First row pair—blue	
GND	Ground	
R2	Second row pair—red	
G2	Second row pair—green	
B2	Second row pair—blue	Shown as U2 in Figure 4-4
GND	Ground	
A	Row-pair address line—A	
B	Row-pair address line—B	
C	Row-pair address line—C	
D	Row-pair address line—D	
CLK	Clock	
LAT	Latch	
OE	Output enable	Active low
GND	Ground	

Note that some of the available matrix displays use 0-based labeling, meaning that the RGB lines are labeled R0, G0, B0, R1, G1, and B1. It makes no difference regarding programming or actual operations but simply adds a bit of nonstandardization for the HUB75 specification.

You also should be aware that the HUB75 input and/or output connectors are not guaranteed to be identified as such on these boards. I have found that the output connector is consistently identified with both the word *output* and a right-facing arrow next to the connector. Also, the output connector seems to be located on the right side of the PCB, but I don't think this is a standard.

This concludes my brief discussion on how a RGB matrix works, and it is time to show you the RasPi interface board.

RasPi Interface Board

I used an Adafruit RGB Matrix HAT + RTC for Raspberry Pi Minikit, product ID 2345, as the interface between a RasPi 2 Model B and the 32 × 64 RGB LED matrix. This kit made it very easy to interface the RasPi to the display, especially when you consider that the driver software was designed specifically to be compatible with this kit. Figure 4-5 shows a front view of a completed kit.

The Adafruit website contains comprehensive instructions on how to assemble the board. It really is quite easy because only a few components need to be soldered, although I did find soldering the 40-pin RasPi GPIO socket a bit tedious. Just take your time, and you will easily be able to build the kit.

Figure 4-5 Front view of the Adafruit interface board.

The board also has a real-time clock (RTC) onboard, which the company added because it added the PCB space according to its write-up. I didn't use this RTC initially, but it does offer some advantages for keeping track of time, especially if you do not have the RasPi connected to the Internet. Figure 4-6 shows the back of the board, which is clean except for the 40-pin GPIO connector that plugs into the RasPi.

I do want to mention that this board complies with the Raspberry Pi Foundation's HAT specification. HAT is short for "Hardware attached on top." It is a recent approach taken by the folks at the Raspberry Pi Foundation to help standardize the form factor and software interface for current and future RasPi add-on boards, which use the 40-pin GPIO connector.

The following is taken from the foundation's website regarding the HAT specification:

> In a nutshell a HAT is a rectangular board (65 × 56 mm) that has four mounting holes in the (nicely rounded) corners that align with the mounting holes on the B+, has a 40-W GPIO header and supports the special autoconfiguration system that allows automatic GPIO setup and driver setup. The automatic configuration is achieved using two dedicated pins (ID_SD and ID_SC) on the 40-W B+ GPIO header that are reserved for an I2C EEPROM. The EEPROM holds the board manufacturer information, GPIO setup and a thing called a "device tree" fragment—basically a description of the attached hardware that allows Linux to automatically load the required drivers.
>
> What we are not doing with HATs is forcing people to adopt our specification. But you can only call something a HAT if it follows the spec.

Figure 4-6 Back view of the Adafruit interface board.

The RasPi 2 Model B used in this project is also completely compatible with the HAT specification. Table 4-2 details the interconnections between the RasPi GPIO pins and the HUB75 connector mounted on the interface board.

TABLE 4-2 RasPi GPIO to HUB75 Connections

RasPi	HUB75	Remarks
5	R1	First row pair—red
13	G1	First row pair—green
6	B1	First row pair—blue
12	R2	Second row pair—red
16	G2	Second row pair—green
23	B2	Second row pair—blue
4	OE	Output enable (active low)
17	CLK	Clock
21	LAT	Latch
22	A	Row-pair address line—A
26	B	Row-pair address line—B
27	C	Row-pair address line—C
20	D	Row-pair address line—D

Of course, the ground leads are also connected between the RasPi and HUB75, but these do not have to be identified separately. Simply use any one of the multiple ground points on the HAT board as a reference. Additionally, if you closely inspect Figure 4-5, you will see that all the connections specified in Table 4-2 have been brought out to labeled PCB holes. In fact, all 40 pins from the GPIO connector have been brought out, even ones that are not involved with the HUB75 interconnection. This makes it quite convenient to add additional connections if a project needs access to these additional pins.

The HAT board also contains level-shifter ICs, which adapt the 3.3-V GPIO RasPi outputs to the 5-V inputs required by the RGB LED display. There is also a 2.1-mm barrel connector on the board, where you may plug in an external 5-V power supply as well as a pair of screw terminals where the matrix power cable assembly can be attached. Figure 4-7 shows the interface

Figure 4-7 Interface board with power and data cables connected.

board with the various power and data cables connected to it.

Note that the interface board will power the RasPi, eliminating the need for a separate power supply dedicated to the RasPi. Just ensure that you use a supply with a sufficient current rating to handle both the display and the RasPi, as I discussed earlier.

This section concludes the hardware discussion. It is now time to take on the software.

Software to Drive the RGB LED Matrix

The maker community is quite fortunate to have very smart individuals involved who create open-source software to make all sorts of projects feasible. We all are indebted to Henner Zeller, who created a comprehensive C language library that runs on the RasPi and will drive the RGB LED matrix. I strongly recommend that you set up the RasPi with the HAT interface board before loading any software. You can use either the stand-alone or headless configuration, as I mentioned in Chapter 1. I chose to use a stand-alone configuration because it fit my workstation, and all the components were readily available. Just connect the HAT interface board to the RasPi and the RGB LED matrix to the HAT using the IDC cable that came with the display. Don't forget to attach the power leads from the display to the HAT or you will not see anything. I also used a 5-V, 10-A external power supply that I purchased from Adafruit.

I would also strongly suggest that you update and upgrade your RasPi before attempting to run the matrix software. You must be connected to the Internet before starting the updating and upgrading. Enter these commands in the order shown:

```
sudo apt-get update
sudo apt-get upgrade
```

Mr. Zeller has put the C/C++ RGB LED matrix library on GitHub, where it may be downloaded by entering:

```
wget https://github.com/hzeller/rpi-rgb
-led-matrix/archive/master.zip
```

You will next need to unzip the archive by entering:

```
unzip master.zip
```

Next, change to the subdirectory containing the code by entering:

```
cd rpi-rgb-led-matrix-master
```

You will now need to compile the code by entering:

```
make
```

This is all that's needed to try testing the RasPi with the RGB LED matrix.

If you are anxious to see whether the display works and you have a 32 × 64 matrix attached, enter the following command:

```
sudo ./led-matrix -d -c 2 -t 60 -D 1
runtext.ppm
```

If everything has been set up and configured properly, you should see the runtext.ppm image scroll across the screen for 1 minute. Figure 4-8 shows a snapshot of a portion of this scrolling message, which contains both text and graphics.

Incidentally, I will explain what a ppm file is and how to use it with this software in a later section.

Entering `sudo ./led-matrix` by itself will show a help screen to inform you of the available options that may be used with this program. I have duplicated this help information here with some additional explanations to assist you in understanding this program and getting the most utility from it:

Figure 4-8 Runtext.ppm scrolling message.

```
Usage: ./led-matrix <options> -D
<demo-nr> <optional parameters>
```

Options:

- `-r <rows>`. Display rows: 16 for 16 × 32, 32 for 32 × 32; default is 32.
- `-c <chained>`. Daisy-chained boards; default is 1.
- `-g`. Do a gamma correction (experimental).
- `-l`. Don't do luminance correction (CIE1931).
- `-L`. "Large" display composed of 4 times 32 × 32.
- `-p <pwm-bits>`. Bits used for PWM; values between 1 and 11.
- `-P <parallel>`. For Plus-models or RasPi 2: parallel chains: 1–3; default is 1.
- `-d`. Run as daemon. Use this when starting /etc/init.d but also when running without a terminal (e.g., cron).
- -D <demo-number>. Always needs to be set with a number as follows:
  0. A rotating square
  1. Forward scrolling an image (`-m <scroll-ms>`)
  2. Backward scrolling an image (`-m <scroll-ms>`)
  3. Test image: a square
  4. Pulsing color
  5. Grayscale block
  6. Abelian sandpile model (`-m <time-step-ms>`)
  7. Conway's game of life (-m <time-step-ms>)
  8. Langton's ant (`-m <time-step-ms>`)
  9. Volume bars (`-m <time-step-ms>`)
- `-t <seconds>`. Run for the specified number of seconds and then exit. If neither `-d` nor `-t` is supplied, the program will wait for the ENTER key to be pressed.

To run the actual demos, you need to run this as root so that the GPIO pins can be accessed.

You should now be able to interpret the initial test command:

```
sudo ./led-matrix -d -c 2 -t 60 -D 1
runtext.ppm
```

as follows:

- `sudo`. Run the program at the root level, that is, maximum administrative privileges.

- `./`. Program is in the current directory.
- `led-matrix`. Program name.
- `-d`. Run as a daemon; allows for other processes to be started, that is, nonblocking.
- `-c 2`. A daisy-chain of two 32 × 32 matrices.
- `-t 60`. Run for 60 seconds.
- `-D 1`. Forward scroll the image specified by the following parameter.
- `runtext.ppm`. The image to be displayed; must be in the same directory as the program.

There is another ppm file in the master directory named runtext16.ppm, which, as the name implies, has 16-pixel-high characters and figures in lieu of the 32-pixel-high characters contained in runtext.ppm. I displayed this file by entering the following command:

```
sudo ./led-matrix -d -c 2 -t 60 -D 1
runtext16.ppm
```

Figure 4-9 shows the result of this command.

As you can readily see, the text and figures are half the size of the previous message generated by the runtext.ppm file. Also, the message only scrolls on the top half of the display, leaving the bottom half blank.

The next sections explore the images resulting from running some of the preprogrammed demo programs. I will first show you the command followed by the resulting display.

Rotating Square

Type in the following command to run demo number 1. Figure 4-10 shows the resulting image.

```
sudo ./led-matrix -d -c 2 -t 60 -D 0
```

Square Test

Type in the following command to run demo number 3. Figure 4-11 shows the resulting image.

```
sudo ./led-matrix -d -c 2 -t 60 -D 3
```

Pulsing Color

Type in the following command to run demo number 4. Figure 4-12 shows the resulting image.

```
sudo ./led-matrix -d -c 2 -t 60 -D 4
```

Conway's Game of Life

Type in the following command to run demo number 7. Figure 4-13 is a composite image with the left-hand side showing the cellular

Figure 4-9 Runtext16.ppm scrolling message.

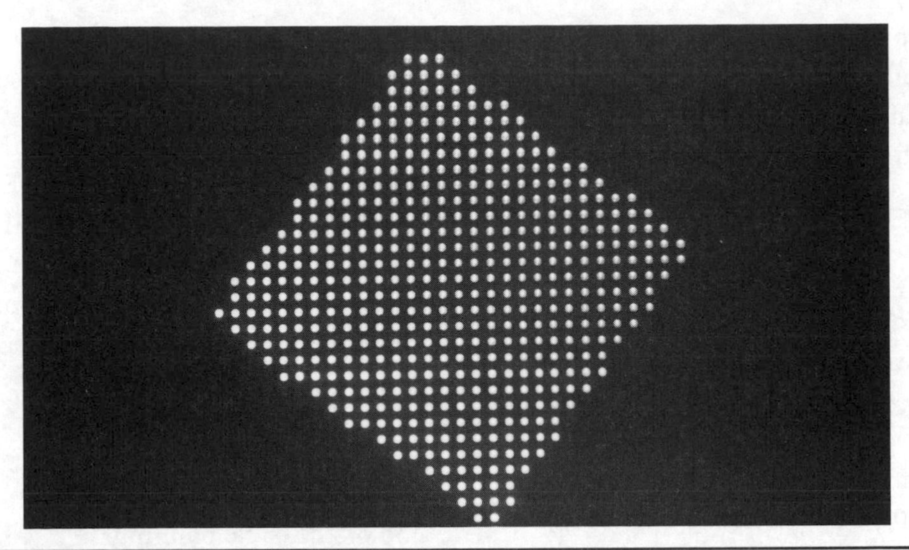

Figure 4-10 Rotating square.

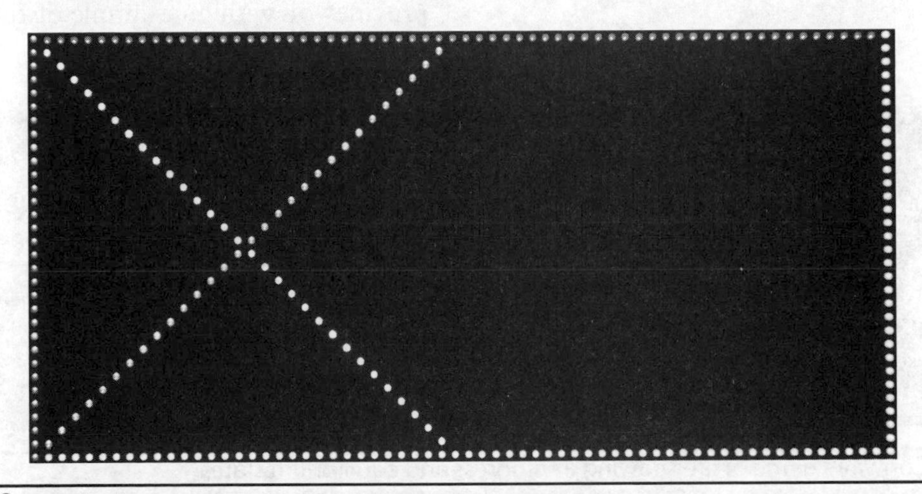

Figure 4-11 Square test.

Figure 4-12 Pulsing color.

automatons in progress and the right-hand side showing the final or equilibrium state. The entire life cycle takes about 25 seconds to progress from an initial to an equilibrium state.

```
sudo ./led-matrix -d -c 2 -t 60 -D 7
```

Volume Bars

Type in the following command to run demo number 9. Figure 4-14 shows the resulting image.

```
sudo ./led-matrix -d -c 2 -t 60 -D 9
```

I will now show you two other interesting demo programs that you might want to try included within Zeller's package.

Minimal-Example Demonstration Program

There are two executable files in the master directory that I want to demonstrate. The first is named minimal-example, and it "paints a filled-in circle in a 32 × 32 matrix using a spiral pattern." When I ran the program, I saw two circles because I used a 32 × 64 matrix. You can run the program by entering:

```
sudo ./minimal-example
```

Figure 4-15 shows a screen capture near the end of the circle painting.

I have also included the code listing for this program to show how compact it is and to provide you with an example of the underlying C/C++ code.

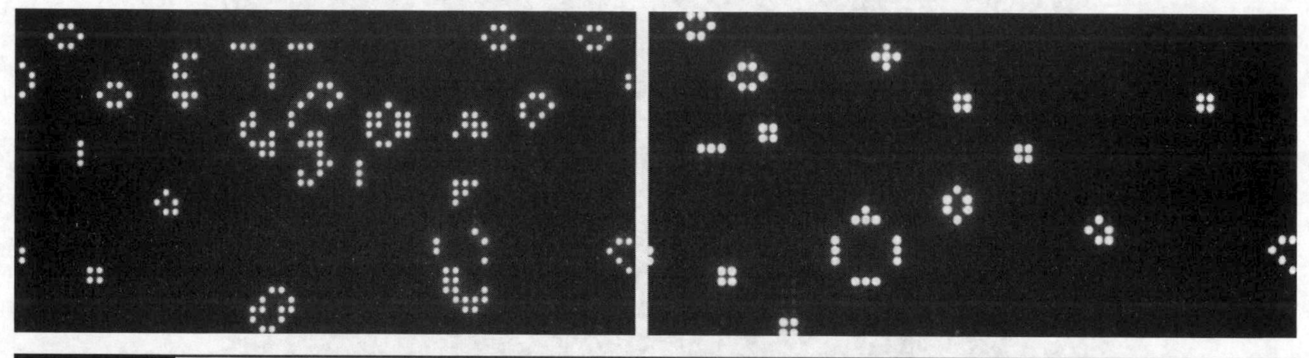

Figure 4-13 Conway's game of life showing in-progress and equilibrium states.

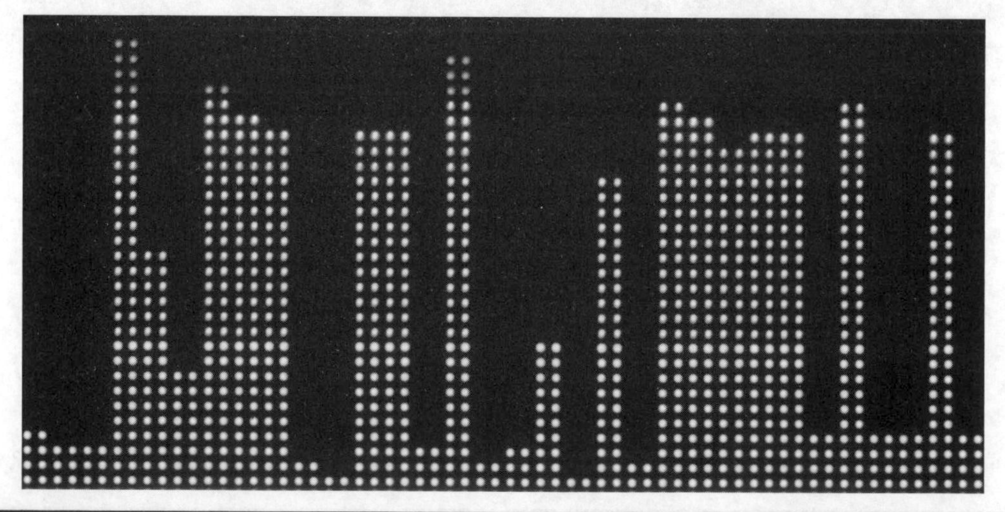

Figure 4-14 Volume bars.

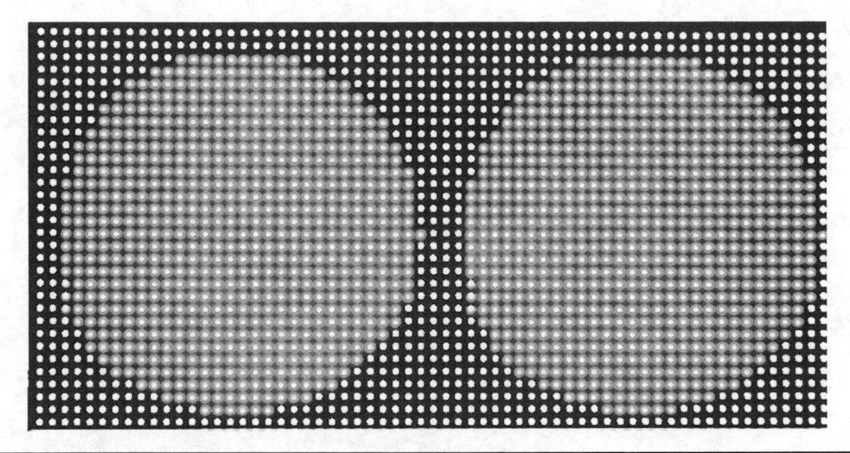

Figure 4-15 Minimal-example program display.

```
// -*- mode: c++; c-basic-offset: 2; indent-tabs-mode: nil; -*-
// Small example how to use the library.
// For more examples, look at demo-main.cc
//
// This code is public domain
// (but note, that the led-matrix library this depends on is GPL v2)

#include "led-matrix.h"

#include <unistd.h>
#include <math.h>
#include <stdio.h>

using rgb_matrix::GPIO;
using rgb_matrix::RGBMatrix;
using rgb_matrix::Canvas;

static void DrawOnCanvas(Canvas *canvas) {
  /*
   * Let's create a simple animation. We use the canvas to draw
   * pixels. We wait between each step to have a slower animation.
   */
  canvas->Fill(0, 0, 255);

  int center_x = canvas->width() / 2;
  int center_y = canvas->height() / 2;
  float radius_max = canvas->width() / 2;
  float angle_step = 1.0 / 360;
  for (float a = 0, r = 0; r < radius_max; a += angle_step, r += angle_step) {
    float dot_x = cos(a * 2 * M_PI) * r;
    float dot_y = sin(a * 2 * M_PI) * r;
    canvas->SetPixel(center_x + dot_x, center_y + dot_y,
 255, 0, 0);
    usleep(1 * 1000); // wait a little to slow down things.
```

```
    }
  }

  int main(int argc, char *argv[]) {
    /*
     * Set up GPIO pins. This fails when not running as root.
     */
    GPIO io;
    if (!io.Init())
      return 1;

    /*
     * Set up the RGBMatrix. It implements a 'Canvas' interface.
     */
    int rows = 32; // A 32x32 display. Use 16 when this is a 16x32 display.
    int chain = 1; // Number of boards chained together.
    Canvas *canvas = new RGBMatrix(&io, rows, chain);


    DrawOnCanvas(canvas); // Using the canvas.

    // Animation finished. Shut down the RGB matrix.
    canvas->Clear();
    delete canvas;

    return 0;
  }
```

For readers who are not familiar with C/C++ programming, I will simply point out that this program relies on three libraries—GPIO, RGBMatrix, and Canvas—that are contained in the name-space called rgb_matrix. The program makes calls to a variety of methods within these libraries to draw the pixel pattern according to the DrawOnCanvas(Canvas * canvas) method, which is defined in this program. Note that a key object called a *canvas* is the logical container for all the pixels displayed.

Text-Example Demonstration Program

The second executable program in the master directory is named text-example. You can use this program to directly display any text that you desire as long as it remains within the pixel dimensions of the array, that is, 32 × 64 in my case. The text-example program also uses similar options as the `led-matrix` program, so it is fairly simple to run. The only new option to be aware of is that you need to specify a font file. All the fonts used by this program are stored in a master subdirectory named *fonts*. The following text was entered manually and displayed using a 4 × 6 font. This meant that 16 characters could be displayed in one row and 5 rows could be displayed on one screen. The 16 characters are the result of each character's width of 4 pixels divided into the row length of 64 pixels. The number of text rows is the result of each character's height of 6 pixels divided into the total matrix height of 32 pixels. Obviously, the remaining 2 pixels are discarded.

Hello World!

0123456789012345

Some sample text

Symbols #@$%&+

Fifth and last row

Entering the following runs the text-example program:

```
sudo ./text-example -c 2 -f ./fonts/4x6
.bdf
```

Figure 4-16 shows the result of executing the command and typing in all the text. You should note that the characters are very bright because the default character color is brilliant yellow.

All the available font files are listed in Table 4-3.

You will rapidly run out of matrix space if you use the large fonts. I am guessing that they are applicable only on large, chained arrays where there is plenty of "real estate" to display such large text characters; this is likely a legacy of their signage origins. Figure 4-17 shows you what I mean when I used the 10x20.bdf font in the following command:

```
sudo ./text-example -c 2 -f ./
fonts/10x20.bdf
```

The text-example program clears the display when either the ENTER key is pressed without any text entered or the text lines fill up the display and it "wraps around." You will need to press the CTRL-C (^C) key combination to exit the program.

Using a -C option along with the desired RGB values also may change the color and intensity of the display. I entered the following command to display some text with a pale pink color:

```
sudo ./text-example -c 2 -C 32, 0, 16 -f
./fonts/6x13.bdf
```

Figure 4-18 shows the result of this command.

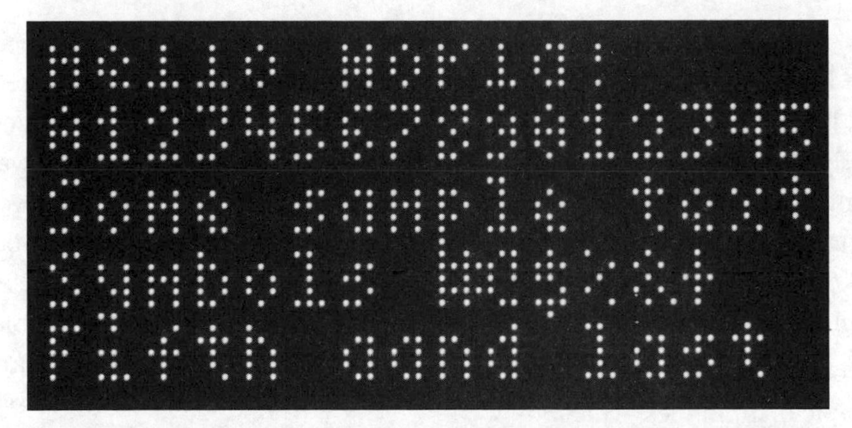

Figure 4-16 Text-example program result.

TABLE 4-3 Font Files

4x6.bdf	5x7.bdf	5x8.bdf	6x9.bdf	6x12.bdf
6x13.bdf	6x13B.bdf	7x13B.bdf	7x13O.bdf	7x14.bdf
7x14B.bdf	8x13.bdf	8x13B.bdf	8x13O.bdf	9x15.bdf
9x15B.bdf	9x18.bdf	10x20.bdf	helvR12.bdf	clR6x12.bdf

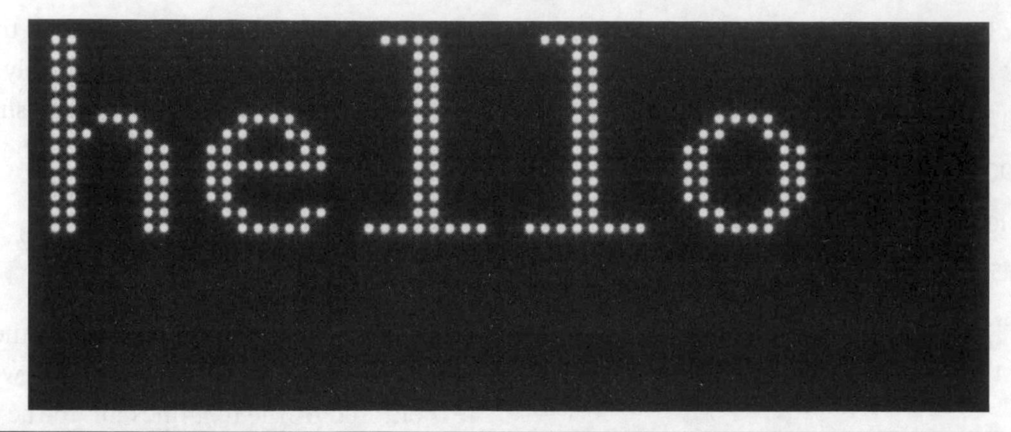

Figure 4-17 10 × 20 font example.

Figure 4-18 Changed color text display.

If you compare Figure 4-18 with earlier text displays, it should be apparent that the figure's pixels are not nearly as bright as the other examples, although it is hard to standardize the intensity of photographs that appear in this book. You can take my word that this display is much dimmer than previous ones.

Earlier in this chapter I mentioned that I would cover the topic of ppm files, which allow you to display your own creative artwork. Let's see what is involved with these graphic files.

ppm Files

ppm is short for "portable pixel map." It is an image-format scheme that is part of a larger set of similar imaging formats known as *Netpbm*. Netpbm is the most recent version of an original graphics format known as *portable bitmap format* (PBM), which was created by Jef Poskanzer in the 1980s. Jef developed PBM to allow monochrome images to be transferred by e-mail using plain ASCII text. PBM and all its descendants are human readable but are also very memory inefficient compared with pure binary image formats such as png or jpeg. I conducted the following test to determine the comparative sizes of a ppm file and a jpeg file that had the same content.

ppm to jpeg File Size Test

I first loaded the runtext.ppm file into the GNU Image Manipulation Program (GIMP) that

Figure 4-19 Runtext.ppm as imaged by GIMP.

I loaded into my MacBook Pro. Figure 4-19 shows the converted image of all the ASCII text contained in the ppm file.

I then used GIMP's export function to save runtext.ppm as a jpeg file, which I named runtext.jpeg. Figure 4-20 is an image captured from the MacBook's Preview application.

You can easily see that the images in Figures 4-19 and 4-20 are identical.

I next listed the sizes of both files that were stored on the MacBook desktop. Figure 4-21 is a snippet from the desktop file list.

The ppm file is about 48 kB in size, while the jpeg file is about 15 kB. This result convincingly demonstrates that the jpeg file is over three times as efficient in storing images as the ppm format. However, the `led-matrix` program cannot work with jpegs but must use the ppm format. In conducting this test, I only wanted to make you aware of the significant memory constraints that are inherent in using the ppm file format. I will show how to display a png or jpeg file directly in a later section, where I discuss how to use Python with this matrix.

I will also show you how to create your own ppm files using GIMP, but first I need to discuss the ppm file format.

ppm File Format

The ppm file format is relatively standardized using a few components, as listed next. Remember that the file is completely made of ASCII characters, which makes it easy to read directly. The components are listed in the order they appear in the file.

1. **A magic number.** For the files I use, it will be P6. The reason that it is called a *magic number* is that it has no relevance other than probably being a legacy artifact from the old days of pbm (see earlier).
2. **Pixel width.**
3. **Pixel height.**

Figure 4-20 Runtext.jpeg as imaged by the MacBook Preview app.

runtext.jpeg	15 KB	JPEG image
runtext.ppm	48 KB	TextEdit.app Document

Figure 4-21 File size listing.

4. **Maximum color value.** This number is 255 or 0xFF for R, G, or B. Note that it can be up to 65,535 for other ppm variants.
5. Image date in the form of ASCII triplets representing individual RGB pixel values.

Note that each of these components needs to be separated by some form of whitespace. This whitespace character is simply the carriage return (CR or 0x0A) for the ppm files I use. Figure 4-22 is the beginning portion of the runtext.ppm file, which shows these components.

This screenshot was taken from a Hex editor display because that makes the most sense when trying to view the image data. Notice that there is a line near the list beginning that starts with a # symbol. This shows that a comment follows, which is inserted for human consumption but is ignored by the matrix display program. As you may readily see from the listing, the width of the image in pixel units is 500; the height is 32 pixels, and the maximum value is 255.

I will next demonstrate how easy it is to create your own ppm file and display it on the matrix.

Creating a ppm File

You will need to install the GIMP application to create your own ppm file. Enter the following command to download and install GIMP:

```
sudo apt-get install gimp
```

You will have a choice regarding how to create a ppm file once you have installed GIMP. You can create a new image using the tools available in GIMP, or you can convert a preexisting image file. I actually tried both approaches to see what was involved in each approach. My first attempt was to use an existing gif file and convert it to a ppm file, which I detail next.

Converting an Existing Image File to ppm

I chose to use a cloud image named cloud.gif. This image is available from this book's website, where all the software shown in this book is also made available for download. Figure 4-23 shows the gif image that I used to create the converted ppm file.

Follow these steps to create the equivalent ppm file from this gif image:

1. Start GIMP, and open the cloud.gif file: File → Open → cloud.gif.
2. Start the export process: File → Export.
3. Enter a name with a .ppm extension: Name → cloud.ppm.
4. Export the file: Click on the Export button.

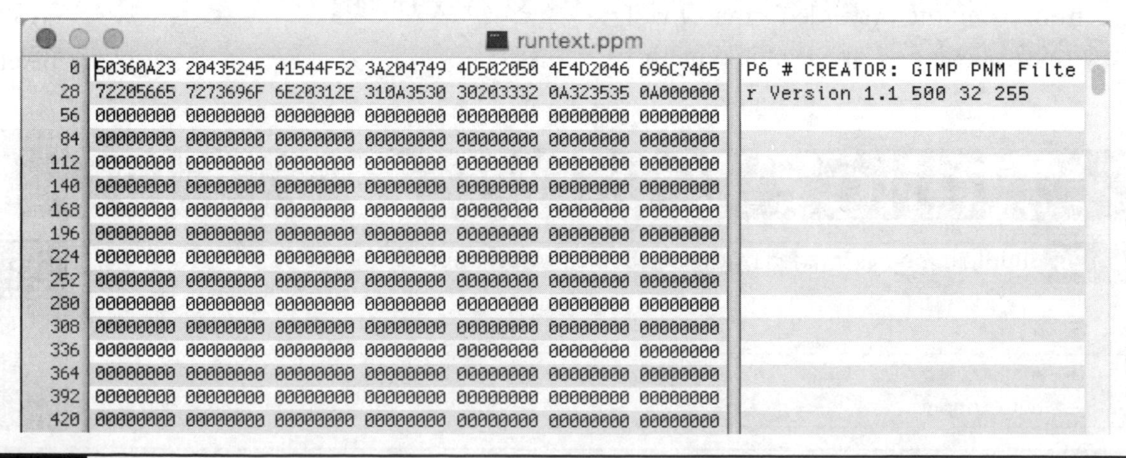

Figure 4-22 Beginning of runtext.ppm file.

Figure 4-23 Cloud gif image.

5. Select Raw for Data formatting: Select the Raw button on the export image as a pnb dialog.
6. Finish the export process: Click on the dialog's Export button.

You should next display the ppm file by entering the following command:

```
sudo ./led-matrix -d -c 2 -t 60 -D 1
cloud.ppm
```

Figure 4-24 shows the result of this command. There is a nice animation that takes place when this file is scrolled with the cloud repeatedly displayed.

The next section demonstrates how to create a very simple image and display it.

Creating an Original ppm File

The file I created for this demonstration was extremely simple and very boring, consisting of a single black filled-in circle displayed on a white background. I intentionally chose this image to create a ppm file because it would be easy to discern between the circle and the background in the data listing. Figure 4-25 shows the ppm image as displayed on the RGB LED matrix by entering the following command:

Figure 4-24 Cloud.ppm display.

```
sudo ./led-matrix -d -c 2 -t 60 -D 1
test1.ppm
```

The following text is an excerpt from a test1.ppm Hex listing that shows both the header data and the beginning of the top of the black circle. This transition is evident because the Hex data change from 0xFFs to 0x00, which reflects the change from white to black.

P6

CREATOR: GIMP PNM Filter Version 1.1

64 32

255

•

•

•

FF FF FF FF FF FF FF FF FF FF FF FF FF FF FF FF FF
FF FF FF FF FF FF FF FF FF FF 00 00 00 00 00 00 00
00 00 00 00 00 00 00 00 00 00 00 00 00 00 FF FF FF
FF FF FF FF FF FF FF FF FF FF FF FF FF FF FF FF FF
FF FF FF FF FF FF FF FF FF FF FF FF FF FF FF FF

•

•

I confirmed that this section of data precisely matches the seven black pixels located at the top of the black circle, as shown in Figure 4-25. I doubt that you will be able to ascertain this from the figure as reproduced in this book, but I do have the original digital image, which I was able to enlarge so as to be able to count the pixels. I also determined that the top row was 7 pixels long

Figure 4-25 Original ppm image.

by counting 21 zero bytes and then dividing that by 3, given that 3 bytes are assigned to each pixel.

The next section explores how to use the Python language with the matrix display.

Using Python with the RGB LED Matrix Display

I will start this discussion by stating that an RGB LED matrix display program written in Python will run slower and less efficiently than a comparable C/C++ program. Having said that, I will also point out that Python is an easier language than C/C++ in which to create your own programs or modify someone else's program. I will be using Henner's Python library along with his test program named matrixtest.py to demonstrate basic matrix operations such as clearing a display, filling one with a solid color, and displaying individual pixels. The Python library is named rgbmatrix.so, where the so extension is short for "shareable object." This library must be in the same directory as the calling program, matrixtest.py. What you probably don't realize is that the rgbmatrix.so library is in reality a collection of C functions with appropriate "hooks" to allow the C functions to operate seamlessly with a Python script. The Python program matrixtest.py is listed with comments so that you can see how it functions.

```
#!/usr/bin/python
# Simple RGBMatrix example, using only
Clear(), Fill() and SetPixel().
# These functions have an immediate
effect on the display
# No special refresh operation needed.
# Requires rgbmatrix.so to be present
in the same directory.

import time
from rgbmatrix import Adafruit_RGBmatrix

# Rows and chain length are both
required parameters:
matrix = Adafruit_RGBmatrix(32, 1)

# Flash screen red, green, blue (packed
color values)
# red screen first
matrix.Fill(0xFF0000)
# wait for 1 sec
time.sleep(1.0)
# green screen next
matrix.Fill(0x00FF00)
# wait for 1 sec
time.sleep(1.0)
# blue screen last
matrix.Fill(0x0000FF)
# wait for 1 sec
time.sleep(1.0)

# Show a RGB test pattern (separate R,
G, B color values)
for b in range(16):
    for g in range(8):
    for r in range(8):
        matrix.SetPixel(
            (b / 4) * 8 + g,
            (b & 3) * 8 + r,
            (r * 0b001001001) / 2,
            (g * 0b001001001) / 2,
             b * 0b00010001)
# wait for 10 secs
time.sleep(10.0)
# clear the screen
matrix.Clear()
```

Enter the following to run this Python program:

```
sudo python matrixtest.py
```

You should observe red, green, and blue screens flash for 1 second each, followed by a checkerboard pattern that stays on the matrix screen for 10 seconds. Figure 4-26 shows this last pattern.

There is one more Python program that I wish to discuss. It is named matrixtest2.py and uses the Python Imaging Library (PIL) to directly display images using common formats such as gif, png, and jpeg.

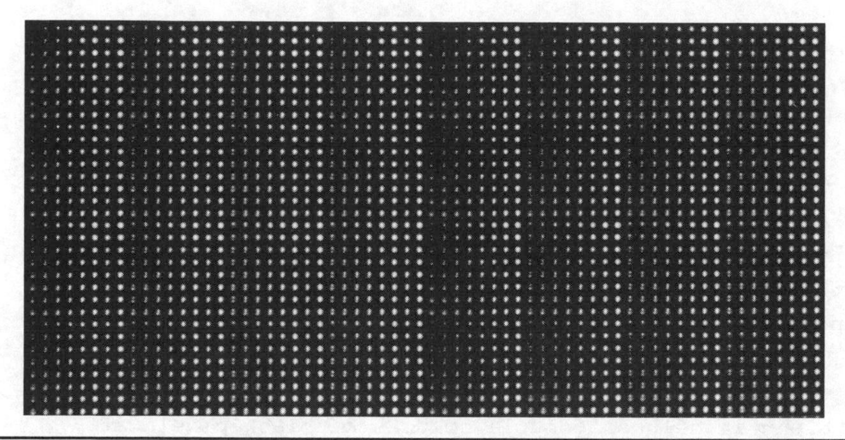

Figure 4-26 Checkerboard pattern display.

```
#!/usr/bin/python

# A more complex RGBMatrix example works with the Python Imaging Library,
# demonstrating a few graphics primitives and image loading.
# Note that PIL graphics do not have an immediate effect on the display --
# image is drawn into a separate buffer, which is then copied to the matrix
# using the SetImage() function (see examples below).
# Requires rgbmatrix.so present in the same directory.

# PIL Image module (create or load images) is explained here:
# http://effbot.org/imagingbook/image.htm
# PIL ImageDraw module (draw shapes to images) explained here:
# http://effbot.org/imagingbook/imagedraw.htm

import Image
import ImageDraw
import time
from rgbmatrix import Adafruit_RGBmatrix

# Rows and chain length are both required parameters:
matrix = Adafruit_RGBmatrix(32, 2) # chain now = 2; DJN

# Bitmap example w/graphics prims
image = Image.new("1", (32, 32)) # Can be larger than matrix if wanted!!
draw = ImageDraw.Draw(image) # Declare Draw instance before prims
# Draw some shapes into image (no immediate effect on matrix)...
draw.rectangle((0, 0, 31, 31), fill=0, outline=1)
draw.line((0, 0, 31, 31), fill=1)
draw.line((0, 31, 31, 0), fill=1)
# Then scroll image across matrix...
for n in range(-32, 33): # Start off top-left, move off bottom-right
    matrix.Clear()
    # IMPORTANT: *MUST* pass image ID, *NOT* image object!
    matrix.SetImage(image.im.id, n, n)
```

```
    time.sleep(0.05)

# 8-bit paletted GIF scrolling example
image = Image.open("cloud.gif")
image.load() # Must do this before SetImage() calls
matrix.Fill(0x6F85FF) # Fill screen to sky color
for n in range(32, -image.size[0], -1): # Scroll R to L
    matrix.SetImage(image.im.id, n, 0)
    time.sleep(0.025)

# 24-bit RGB scrolling example.
# The adafruit.png image has a couple columns of black pixels at
# the right edge, so erasing after the scrolled image isn't necessary.
matrix.Clear()
image = Image.open("adafruit.png")
image.load()
for n in range(32, -image.size[0], -1):
    matrix.SetImage(image.im.id, n, 1)
    time.sleep(0.025)

matrix.Clear()
```

Figure 4-27 shows a snapshot of the cloud gif scrolling past on the matrix screen.

This cloud is the same one as shown in Figure 4-24, but that one was in a ppm format. You can try all sorts of imaging experiments using this program as a beginner's programming template. I believe that you will find this approach to be a pleasant compromise between a more complex and efficient C/C++ program and a much less complex but slightly less efficient Python program.

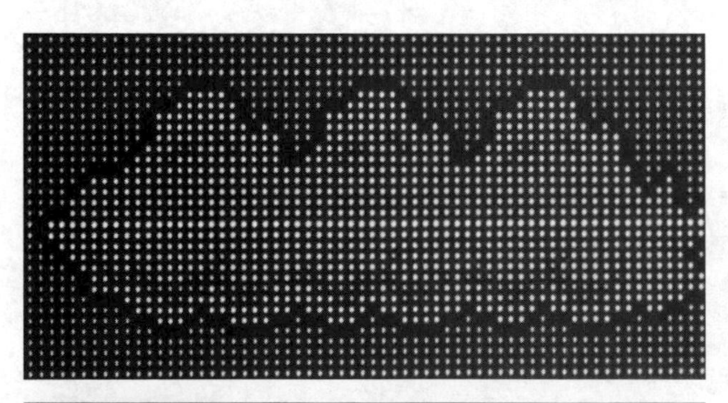

Figure 4-27 Cloud gif animation snapshot.

This last program concludes my discussion on this interesting topic of an RGB LED matrix controlled by a RasPi.

Summary

The chapter started by showing what a 32 × 64 RGB LED matrix display is and how it works. The HUB75 connection protocol was discussed next, along with the HAT RasPi interface board. I proceeded to demonstrate H. Zeller's C/C++ RGB LED matrix library, which includes many great programs that allow you to display both text and graphics on a matrix display. In addition, a number of demos were shown that may be quickly and easily run. I finished the chapter with a demonstration of two Python programs that you can easily modify to suit your own purposes to display text and/or graphics.

CHAPTER 5

Raspberry Pi Supercomputer Cluster

In this chapter, I will show you how to build your very own supercomputer with eight Model B+ RasPis. However, in all honesty, this project really isn't about building a "true" supercomputer but more about how to assemble a computer cluster and have it function efficiently to process computations in a distributed manner. If you are looking to solve problems really fast, I would suggest using an Intel Core i7 6-Core PC. It will be orders of magnitudes faster than this RasPi cluster. However, if you are looking to become familiar with how networked or distributed computers work, then this is the chapter for you.

I begin this chapter with a brief background discussion on what makes up a real supercomputer and the intended purposes of such computers. This should help to place the RasPi cluster in an appropriate perspective.

Brief Supercomputer Discussion and History

I will start this section by stating that supercomputers and parallel computing have forever been connected, and it is hard to discuss one without involving the other. *Parallel computing*, as the name implies, means that two or more computing processes are being executed simultaneously. Of course, now I need to define what is meant by a *process* because that term is used quite frequently in both parallel computing and supercomputing. In strict computer science terms, a process is *any instance of a program that is executing or running on a computer. In addition, all processes are composed of one or more threads.*

Chapter 5 Parts List

Sometimes processes are confused with processors, which are the hardware that runs the

Item	Model	Quantity	Source
RasPi Model B+	83-16317	8	mcmelectronics.com
Ethernet patch cables	Commodity	6 (1 ft) 2 (2 ft) 1 (8 ft)	amazon.com
Anker micro-USB to B connecting cables	Commodity	6 (1 ft) 2 (2 ft)	amazon.com
Plugable USB hub	USB3-HUB81x4	2	amazon.com
Netgear Ethernet 16-port 10/100 switch	SF116	1	amazon.com; staples.com
Machine screws and matching nuts	Commodity	See build drawings	Local home-improvement store
Polycarbonate and acrylic sheet material	Commodity	See build drawings	Local home-improvement store

programs. Just remember, processes are software, while processors are tangible hardware.

It will be important to keep this distinction in mind during later discussions because it will help to clarify what is happening with the RasPi cluster when running some parallel computations. I will also discuss threads in a later section.

Supercomputers first appeared in the computer industry in the 1960s, usually consisting of a few processors coupled together using a shared-memory architecture. The concept of several to many processors sharing a common memory is referred to as a *centralized structure*. The other supercomputer architecture is known as a *distributed architecture*, where there are separate processors with their own dedicated memories but interconnected using some type of communications system. Figure 5-1 shows two conceptual block diagrams illustrating both these architectures.

The RasPi cluster used in this chapter's project has a distributed architecture, which most favorably conforms to the RasPi design.

Problem Types Suitable for Parallel Computing

Generally speaking, any problem that can be compartmentalized is suitable for a parallel computer environment. *Compartmentalization*, in this context, means that a problem may be solved in "chunks" that are not interdependent. This allows multiple processors to function in an independent manner, working on portions of a problem that eventually will be combined in some fashion for an ultimate problem solution. Breaking a problem into actionable chunks has been the cornerstone of parallel computing for many years. This type of problem analysis is not an easy task and has been the subject of much research for many decades.

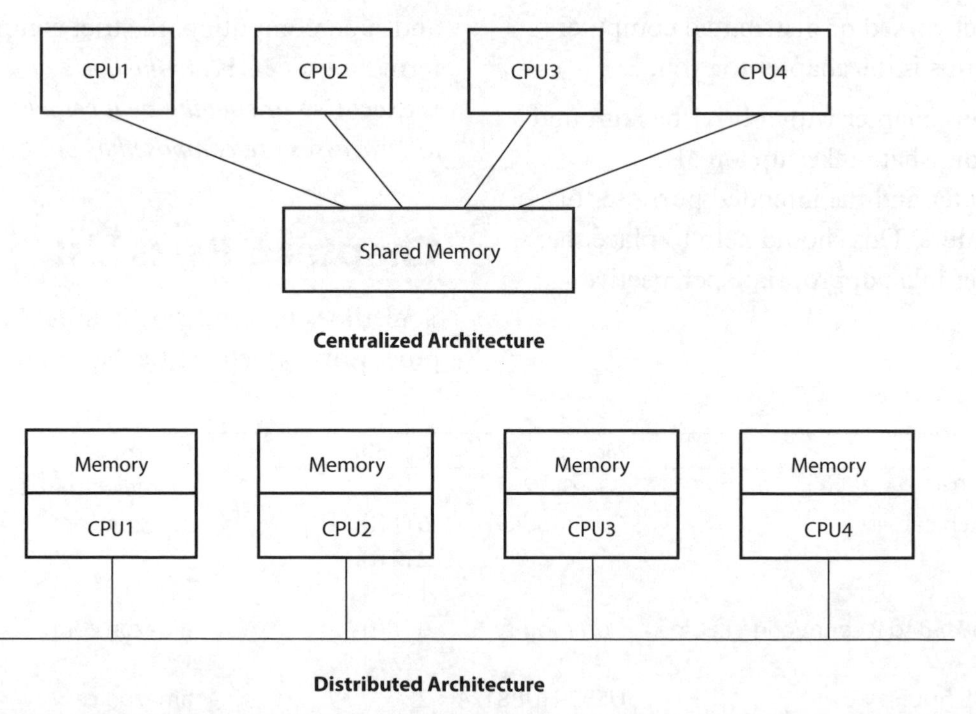

Figure 5-1 Centralized and distributed supercomputer architecture block diagrams.

Historically, the following problem areas have been conducive to solutions using parallel computing:

- Machine learning
- Modeling
- Medical research
- Quantum mechanics
- Weather forecasting
- Oil/gas exploration

One of the more impressive examples of distributed computing in medical research is the folding@home project. The following project description was excerpted from the folding@ home website:

> Proteins are biology's workhorses—its "nanomachines." Proteins help your body break down food into energy, regulate your moods, and fight disease. Before proteins can carry out these important functions, they assemble themselves, or "fold." While protein folding is critical and fundamental to virtually all of biology, much of the process remains a mystery.
>
> When proteins do not fold correctly (misfolding), there can be serious health consequences, including many well-known diseases, such as Alzheimer's, mad cow (BSE), CJD, ALS, AIDS, Huntington's, and Parkinson's disease, as well as many cancers.

This project is essentially a distributed supercomputer architecture in which the computing nodes are home PCs. Each one of the 138,000 PCs involved with the folding@home project works on a chunk of a huge protein database as distributed by a web-based central server. The client portion of the program is designed only to be functional during PC idle time and is guaranteed to have no impact on normal PC operations. The data are sent back to the central server after they have been processed on the PC. A new chunk of unprocessed data is subsequently transmitted back to the PC.

The SETI@home project is another popular web-based distributed computing project that has been devoted to the search for extraterrestrial intelligence. In this case, home PCs compute what are known as *fast Fourier transforms* (FFTs) on chunks of raw data sent to them by a central server. It turns out that FFTs are well suited to parallel computations because each data chunk is independent of other similar chunks. SETI@home has over 3 million PCs involved in its computing project, making it one of the world's largest distributed computer systems.

Supercomputer Attributes

The most common descriptor or attribute concerning a supercomputer is gigaFLOPS (GFLOPS), or billion (giga) floating point operations per second. To put this attribute into a meaningful context, the latest Intel Core i7-5775R Quad-Core CPU chip with a 3.8-GHz clock rate is rated at 120 to 150 GFLOPS. Put in another way, the RasPi 2 Model B quad core has an approximate GFLOPS rating of 0.093 while operating at its standard 1-GHz clock rate. I would also caution that GFLOPS is just a relative performance indicator and does not necessarily represent real-world performance.

Modern supercomputers are now routinely rated using petaFLOPS, where peta equals a million times giga. China now boasts operating the Tianhe-2 (translated as "Sky River 2" or, more loosely, as "Milky Way 2") supercomputer, which is rated at 33.86 petaFLOPS. Meanwhile, it has been widely speculated that the United States National Security Agency (NSA) has its own supercomputers that operate well beyond 33 petaFLOPS but have not made this rating public for reasons of national security. China has also recently announced its intent to develop and deploy a 100-petaFLOPS supercomputer. Such a machine would have a serious impact on the ability of 256-bit Advanced Encryption Standard (AES) encryption algorithms to keep data safely encrypted.

The world's most powerful distributed supercomputer is reported to be the Bitcoin network with an incredible 287-exaFLOPS attribute, where exa = 1000 times peta. Thus the Bitcoin network would be rated at an insane 287,000 petaFLOPS. Now, this network is totally decentralized and cannot be repurposed to solve any problem other than its own Bitcoin application, so the world's secret spy agencies have nothing to fear from Bitcoin operations.

There are several other methods to rate supercomputers that involve standard types of tests and operations that are used routinely in the computing industry. These other methods are not relevant to this discussion and will not be pursued further. Now I will discuss the RasPi cluster.

RasPi Cluster

I chose to build an eight-node RasPi cluster using the Model B+ for each node. You do not need to use eight nodes to replicate this project but can use as few as two RasPis to satisfactorily demonstrate the project software. Using four RasPis is probably a reasonable compromise between cost and utility for readers who wish to complete this project but minimize their total outlay for project components. Table 5-1 is the bill of materials for this project as I built it, but it can be greatly reduced, as just mentioned.

Figure 5-2 shows a block diagram illustrating how all these components are interconnected.

The connections are very straightforward, and no soldering is involved in this project. Everything is plug-and-play with regard to setting up the physical components, which should go together very quickly. I decided to use two USB hubs to power the RasPis, assigning each one to power four RasPis. The particular hubs I used are rated with a 2.5-A overall current capacity, which means that over 600 mA is available to power each RasPi; this is more than adequate for these models considering that I will not be using WiFi adapters with them. You can also use individual power supplies with the RasPis if that's what you have available, and such an arrangement should not affect cluster

TABLE 5-1 Project Bill of Material for an Eight-Node RasPi Cluster

Manufacturer	Model	Quantity	Remarks
RasPi Foundation	B+	8	May substitute Model B or two Model B's
Netgear	SF116	1	16-port 10/100 switch
Plugable	USB3-HUB81x4	2	May use separate wall power supplies
Anker	Micro-USB to B connector cables	6 2	1-ft length 2-ft length
Commodity	Ethernet patch cables	6 2 1	1-ft length 2-ft length 8-ft length
Commodity	Polycarbonate and acrylic sheet material	Various	See build instructions
Commodity	Machine screws, nuts, spacers	Various	See build instructions

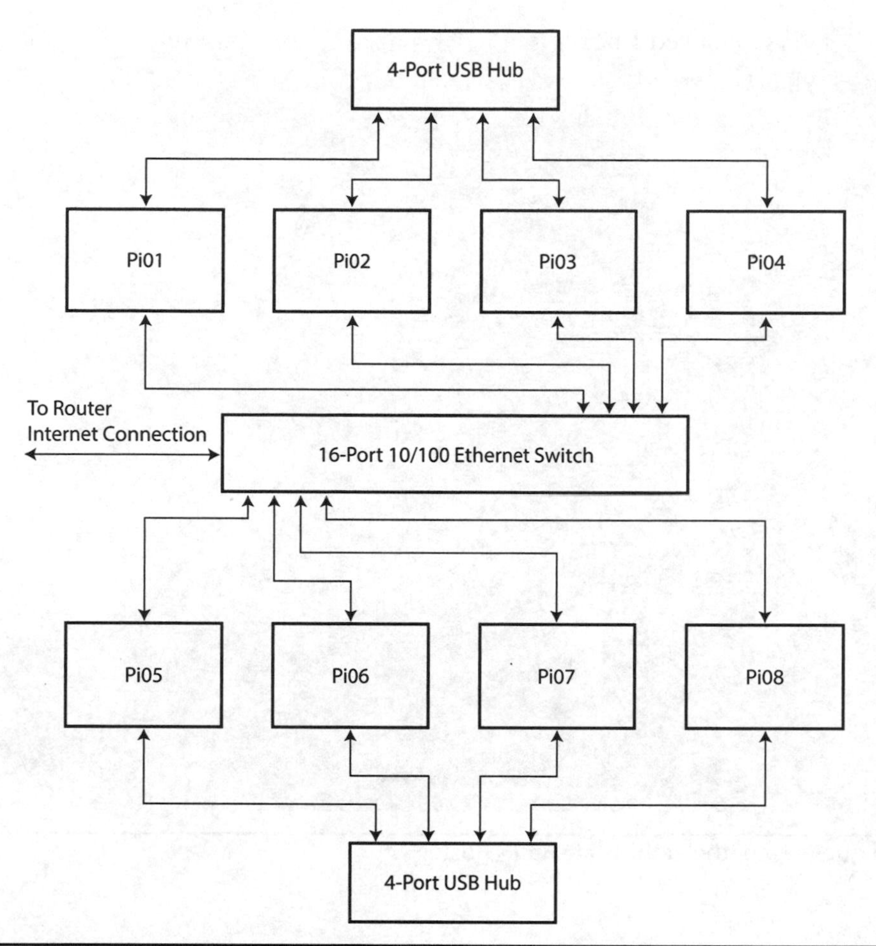

Figure 5-2 Eight-node RasPi cluster block diagram.

operations. Both 1- and 2-ft micro-USB to type B cables were used to interconnect the RasPis with the USB hubs to reduce cable clutter. Likewise, I used both 1- and 2-ft Ethernet patch cables to connect the RasPis with the switch. Note that it doesn't matter which switch ports you use to connect the RasPis because the network router will automatically assign IP addresses, assuming that you have DHCP enabled on your home router. You will also need a long Ethernet patch cable to connect between your home network and the switch, as shown in Figure 5-2.

Internet connectivity is useful but not essential for cluster operation after the initial setup. I will discuss the software setup in detail in a later section, but you should know that I initially used a stand-alone configuration for the first RasPi, which is considered the master. I discussed both stand-alone and headless configurations in Chapter 1, which you might want to review if you are a bit fuzzy on my terminology.

Building the somewhat elaborate stand will take the most time. The cluster stand build instructions are presented in the website build drawings for readers who desire to make one. Readers who just want to temporarily experiment with this project and then dismantle it can also easily set up the components on a tabletop without any stand. I chose to build a stand for presentation purposes because it is somewhat impressive, and I use it when I go to book sale events and Maker Faires. Figure 5-3 is an oblique front view of a fully assembled eight-node cluster.

Figure 5-3 Oblique view of the eight-node RasPi cluster.

If you examine this figure closely, you may see where the two inner polycarbonate sheets have been cut out at the top to accommodate a 5-inch square fan. I planned on using a fan to cool all the RasPis, but I found out that they operated without generating any heat, so the fan was unnecessary. However, feel free to add a fan for purely aesthetic reasons. Figure 5-4 clearly shows how all the Ethernet patch cables are connected between the eight RasPis and the switch.

None of these RasPis needs to be physically identified with its logical number, that is, Pi01, Pi02, etc., because the cluster software takes care of assigning identities, and all the nodes are identical RasPis. It is quite possible to hook up a LED to a GPIO pin on each RasPi and then remotely activate it to identify specific nodes, if desired. I do briefly discuss how to do this in the software section, which follows.

RasPi Cluster Software

As you can imagine, cluster software strongly depends on how the individual nodes communicate with each other. There are several existing schemes that implement cluster communications, and each has advantages and disadvantages. I will be using a standard protocol named the *Message Passing Interface* (MPI) for the RasPi cluster. It is robust and relatively simple to implement and operate. It also supports Python programming, which I feel is a huge bonus for my readers because that language is fairly simple to use but also quite powerful in the sense that significant problem domains can be modeled and solved using this cluster. I believe a basic background discussion on MPI will help you to understand how it works and how best to use it.

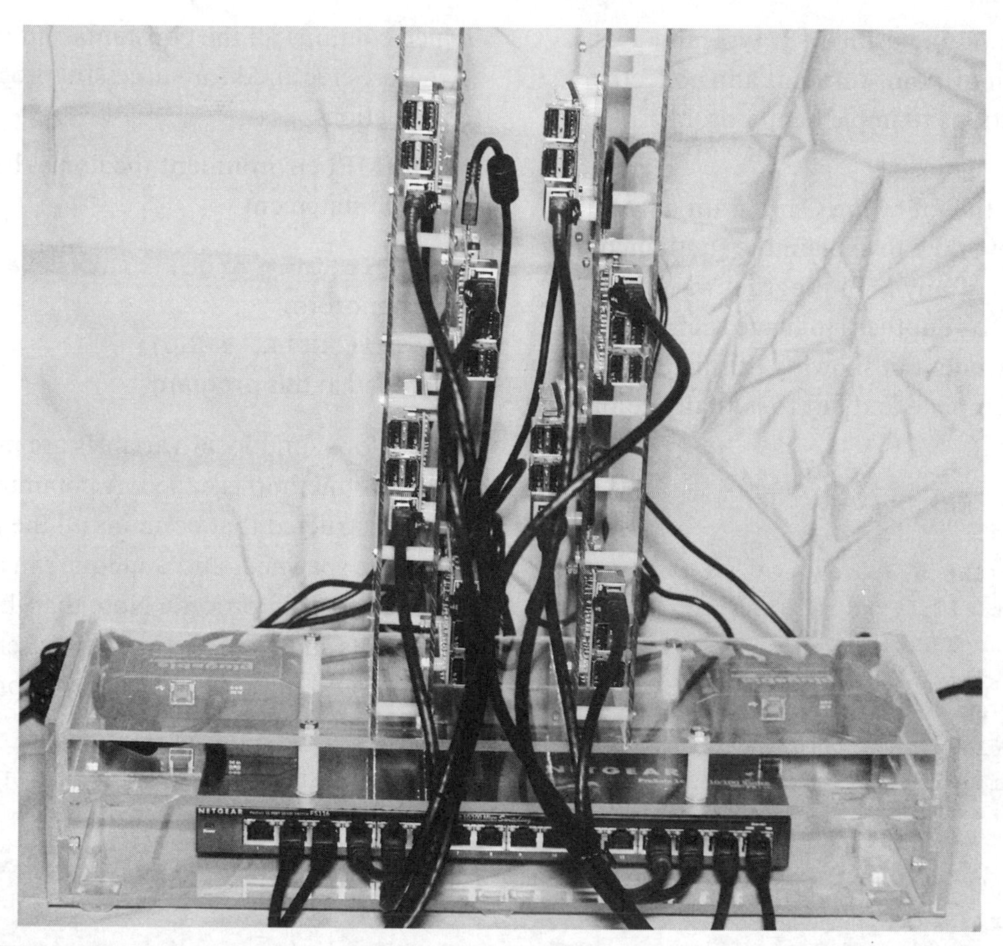

Figure 5-4 Front view of the eight-node RasPi cluster.

MPI Basics

I will start the discussion by stating that MPI is a definition of an interface, not an actual implementation. This means that there can be a variety of ways that developers can create the software that does the actual communication and message processing. The implementation I use in this project is named MPICH2 and is programmed using the C language. Three definitions are used in the interface, and you must be familiar with them before proceeding. One of these is the term `process`, which I earlier defined. The others are

- **Communicator.** This is a group of processes that have the ability to communicate with each other.
- **Rank.** This is a unique identifier assigned to each process that constitutes a communicator.

Communication has both send and receive operations. A given process uses another process's rank to send it a message. The sending process also may send an optional tag to help identify a specific message. The receiving process may acknowledge the message and then handle it appropriately. This type of operation is known as *point-to-point communication*.

Sometimes it may be necessary for one process to be able to communicate with all the other processes that make up the communicator. In this case, MPI provides a *collective communication operation*. This operation type is also known as a *broadcast message*.

Parallel computing using MPI typically uses a combination of point-to-point and collective communications to implement a problem solution.

I will use the following C program named mpi_hello_world.c to explain the fundamental components of any MPI program whether written in C or another language such as Python. My intent in showing this listing is to use it as a template and not to actually compile and execute it:

```
#include <mpi.h>;

int main(int argc, char** argv) {
  // Initialize the MPI environment
  MPI_Init(NULL, NULL);

  // Get the number of processes
  int world_size;
  MPI_Comm_size(MPI_COMM_WORLD, &world_
size);

  // Get the rank of the process
  int world_rank;
  MPI_Comm_rank(MPI_COMM_WORLD, &world_
rank);

  // Get the name of the processor
  char processor_name[MPI_MAX_PROCESSOR_
NAME];
  int name_len;
  MPI_Get_processor_name(processor_name,
&name_len);

  // Print off a hello world message
  printf("Hello world from processor %s,
rank %d"
    " out of %d processors\n",
    processor_name, world_rank, world_
size);

  // Finalize the MPI environment.
  MPI_Finalize();
}
```

The first step in creating a C language MPI program is to include the MPI header file mpi.h, which contains all the constants and method signatures required for successful program compilation.

The MPI environment then must be initialized with the statement

```
MPI_Init(int* argc, char*** argv)
```

Generic form

```
MPI_Init(NULL, NULL);
```

As used in this program

All global and local variables are initialized when this method is called. A communicator is also constructed that contains all the processes that were spawned, and a unique rank is assigned to each process. Note that the two arguments shown in the method parentheses are optional and may be replaced by appropriate null or nil values.

The next statement returns the communicator size:

```
MPI_Comm_size(MPI_COMM_WORLD, &world_
size);
```

The integer variable `world_size` has its value changed after this method call finishes. The constant `MPI_COM_WORLD` is defined in mpi.h and sets the appropriate integer value for this method. The value of `world_size` is a whole number that represents the number of processes contained in the communicator.

The next method determines the rank of a process involved with the communicator:

```
MPI_Comm_rank(MPI_COMM_WORLD, &world_
rank);
```

The integer variable `world_rank` has its value changed after this method call finishes. The value of `world_rank` is a whole number representing the unique value of a process in the communicator. Process ranks start at 0 and progress incrementally to encompass all the processes contained in the communicator.

The next MPI method used in this program returns the preassigned process name:

```
MPI_Get_processor_name(processor_name,
&name_len);
```

The process name is typically not required in most MPI programs but does come in handy for Hello World–type demonstrations. I will also show later that the RasPi hostname is also the process name for this cluster configuration.

The last MPI method closes out the MPI environment in an orderly manner:

```
MPI_Finalize();
```

It is always a good idea to terminate a program in a consistent way, ensuring that all data are transferred as required and any allocated memory is released.

This concludes the template introduction. I will next discuss how Python may be used with MPI.

Python and MPI

The package used to implement Python within the MPI environment is called *MPI for Python*, with a shorter descriptive name of *MPI4PY*. This package uses an object-oriented approach to establish message passing largely on C/C++ MPI implementation. Technically, MPI4PY translates the standard C/C++ MPI-2 bindings into Python-compatible code. This means that if you are familiar C/C++ code, you should have little to no problem handling the Python interface. However, I do not assume that readers are familiar with C/C++, so I will take it slow and try to be clear and precise with regard to the Python code.

Pickling

The odd name of this section refers to the process of converting a Python object from its normal state to a serialized state suitable for being sent as a MPI message. After all, MPI is concerned primarily with communications between nodes, and it wouldn't make much sense not to be able to send and receive Python objects or data primitives between processes. After a pickle is sent, it must be unpickled by the receiving process, otherwise known as *deserialization*.

The pickle object itself is constructed using pure Python. There is a variant called *cPickle* that uses C code for speed reasons because the pure Python version is much slower. It really doesn't matter which version is used in RasPi MPI4PY because the system clocks are fairly slow, and the transmission speed of Python objects should not materially affect overall operations.

Marshaling operations are also available in MPI4PY, which allows a developer to explicitly serialize or pickle selected Python objects. I will not be demonstrating any marshaling because it is a bit advanced for most readers and will not have any significant impact on any of the cluster programs demonstrated in this chapter. Just keep in mind that pickling and depickling are ongoing at any time Python programs are executing, and they do have some minor impact on execution speeds.

This concludes the background discussions on MPI and MPI4PY. It is now time to show how to set up the software.

Software Setup

I started the initial software setup by using one of the Model B+ Raspis in a stand-alone configuration and connected to the Internet. I found that I had problems in the setup using the headless configuration, especially in the file write permissions area. Just go back to Chapter 1 and review how to create a stand-alone configuration. I will also assume that you have downloaded and set up the latest Raspien image

TABLE 5-2 Raspi-config Guidance

Raspi-config Step Number	Title	Value	Remarks
1	Expand Filesystem	Yes	Maximize SD memory space
4	Internationalization Options/ Timezone	Select appropriate one for your location	
4	Internationalization Options/ Keyboard Layout	Select appropriate one for your own preference	Mine was US, generic, 105 keys
7	Overclock	Modest	800 MHz
8	Advanced Options/A2 Hostname	Pi01	Change from the default RasPi
8	Advanced Options/A3 Memory Split	16	Change from the default of 64
8	Advanced Options/A4 SSH	Enable	Needed for headless operation

on the RasPi. Again, just review all the steps on how to do this in Chapter 1. You will also need to follow this configuration guidance (as shown in Table 5-2) as you go through the raspi-config program.

Once last important item is to write down the IP address that appears just before the login screen prompt. You will need this for headless configuration setup.

The following MPI setup is presented in a step-by-step order. You should closely follow these steps as specified, and everything will go smoothly:

1. Always update and upgrade before proceeding. This ensures that the local repository database is in synch with the Internet databases and that all the latest packages are actually installed on the RasPi.
   a. `sudo apt-get update`
   b. `sudo apt-get upgrade`
2. Create a directory to hold the latest MPI package.
   a. `mkdir mpich2`
3. Change into the new directory.
   a. `cd ~/mpich2`
4. Download the latest MPI package. It was version 3.1 at the time this was written.
   a. `wget` (http://www.mpich.org/static/downloads/3.1/mpich-3.1.tar.gz)
5. Unzip the download.
   a. `tar xfz mpich-3.1.tar.gz`
6. Create a new directory in the home directory.
   a. `sudo mkdir /home/rpimpi/`
7. Create a new directory in the one that was just created.
   a. `sudo mkdir /home/rpimpi/mpi-install`
8. Create a new directory in the pi directory.
   a. `mkdir /home/pi/mpi-build`
9. Change into this last directory.
   a. `cd /home/pi/mpi-build`
10. Download and install the gfortran package.
    a. `sudo apt-get install gfortran`
11. Start the MPI configuration. Be patient; this step takes about 20 minutes. The following command is on one line:
    a. `sudo /home/pi/mpich2/mpi-3.1/configure -prefix=/home/rpimpi/mpi-install`

12. Compile the MPI package. Be very patient. This step takes over 2 hours to complete.
    a. `sudo make`
13. Finish the installation. This step takes about 8 minutes.
    a. `sudo make install`
14. Change directories to the next higher level.
    a. `cd ..`
15. Add the following line to the `.bashrc` configuration file using the nano editor. Add it right after the last line, which should be fi.
    a. `PATH=$PATH:/home/rpimpi/mpi-install/bin`
    b. The editor command line is `sudo nano .bashrc`.
16. Reboot the RasPi so that all the new configuration settings can take effect.
    a. `sudo reboot`
17. Enter this command to check whether the MPI installation and configuration were successful. You should see Pi01 returned as the hostname.
    a. `mpiexec -n 1 hostname`

The next part of the software configuration is the MPI4PY installation. You must carefully follow all the steps. I will point out any potential problems so that you may avoid those nasty issues.

1. You need to first download and install Python's developer package.
    a. `sudo apt-get install python-dev`
2. Now download the MPI4PY package.
    a. `wget` (https://mpi4py.googlecode.com/files/mpi4py-1.3.1.tar.gz)
3. Unzip the package.
    a. `tar -zxf mpi4py-1.3.1.tar.gz`
4. Change directories.
    a. `cd mpi4py-1.3.1`
5. Build the package. **Important:** Do not use `sudo` with this command or an error will result. This command takes about 17 minutes to complete.
    a. `python setup.py build`
6. Install the package. Note that you will need to use `sudo` with this command.
    a. `sudo python setup.py install`
7. Next, set up the path link by entering this command:
    a. `export PYTHONPATH=/home/pi/mpi4py-1.3.1`
8. Test the MPI4PY installation by entering the following:
    a. `mpiexec -n 5 python demo/helloworld.py`

Figure 5-5 shows the result of successful execution of this test command.

```
donnorris — pi@Pi01: ~/mpi4py-1.3.1 — ssh — 80×24
pi@Pi01 ~/mpi4py-1.3.1 $ mpiexec -n 5 python demo/helloworld.py
Hello, World! I am process 4 of 5 on Pi01.
Hello, World! I am process 1 of 5 on Pi01.
Hello, World! I am process 2 of 5 on Pi01.
Hello, World! I am process 3 of 5 on Pi01.
Hello, World! I am process 0 of 5 on Pi01.
pi@Pi01 ~/mpi4py-1.3.1 $
```

Figure 5-5 Successful MPI4PY test.

This test command ran a Python Hello World program on a single RasPi but with five separate processes spawned. These five processes were commanded to be started by the number 5 immediately entered after the -n option in the command line. You may have noticed in the figure that there was no particular sequence for these processes to complete. I have repeatedly run the command and have seen all sorts of random completion sequences. This demonstrates that MPI has no inherent synchronization mechanism built into the code.

If you change the number 5 to 1, the program will always display the following:

```
pi@Pi01 ~/mpi4py-1.3.1 $ mpiexec -n 1
python demo/helloworld.py
Hello, World! I am process 0 of 1 on
Pi01.
```

The 0 in the display refers to the world_ rank variable, while the 1 is the `world_size` variable or total number of processes in the communicator. Running the mpiexec program with one process is equivalent to running the same program in the normal manner on the RasPi, that is, one process as one thread with no message passing needed or even possible.

I show the helloworld.py program listing to indicate how Python handles the same key MPI components listed in the C template.

```
#!/usr/bin/env python
"""
Parallel Hello World
"""

from mpi4py import MPI
import sys

size = MPI.COMM_WORLD.Get_size()
rank = MPI.COMM_WORLD.Get_rank()
name = MPI.Get_processor_name()

sys.stdout.write(
  "Hello, World! I am process %d of %d
on %s.\n"
  % (rank, size, name))
```

As you can see, Python looks somewhat similar to the C version, but I think that it's a lot simpler and easier to understand. For instance, the `size` variable, which holds the total number of processes in the communicator, is assigned by the statement:

```
size = MPI.COMM_WORLD.Get_size()
```

The MPI object is constructed from the earlier import statement, which automatically links all the MPI4PY objects, one of which is `COMM_WORLD`, the communicator. Calling the `Get_size()` method with the `COMM_WORLD` object returns the total number of communicator processes.

The processor name is returned by the `Get_processor_name()` method called with the MPI object. You do not need either the size or the rank to obtain the name string. In this case, the processor name and the hostname are the same.

It will take you a bit of time to understand that the mpiexec program spawns or creates a set number of identical processes, as specified by the -n option in the command line. These processes must be identical because only one program is specified to be used in this situation, which is helloworld.py. This program is also situated in the demo directory, which is why the `demo/` path had to be prefixed to the filename. Later on in this chapter I will demonstrate a situation in which different program actions can be taken for a mpiexec command because separate and distinct nodes are involved.

I will next discuss how to clone the SD card such that you will not have to repeat all the preceding build, install, and configure steps. This next section will save you a lot of time.

Cloning the SD Card

It is a fairly simple process to copy or clone the SD card that you carefully prepared in the preceding section. All that's required is to run the Win32 Disk Imager program and insert the SD card into the PC. Next, enter a name for your image into the Image File text box. It can be any valid name as long as it has an .img extension. I chose raspi_cluster.img, as you can see in Figure 5-6.

Next, click on the Read button, assuming that the SD card is in the current slot. In the figure, the card slot is show as the D: drive. The Win32 Disk Imager program is pretty good at figuring out where the correct SD card is inserted. Then sit back and relax because the `read` operation can take a bit of time, mainly due to SD card memory size and rating. A 16-GB Class 10 card took about 15 minutes to completely read all the data. Much of it is empty, but the program has no way of knowing what is blank or empty data storage versus real data.

You're ready to create a clone after the image file is created. All you need to do is take out the SD card you just read and put in its place a fresh card with the same memory capacity as the one that you used to create the image file. The name of the image file still should be in the text box. If not, navigate to where the imager program stored the new image file. This will vary depending on where you last accessed an image file. In my case, the default directory was the Downloads folder. In any case, simply click on the Write button when you have the image file selected, and put a fresh SD card into the PC slot. The write process will take some time, as was the case with the read operation. Repeat the write process for as many remaining RasPis as you will use in the cluster. Don't be concerned that the hostname is the same for all these images. I will shortly show you how to change that as the cluster is finally set up.

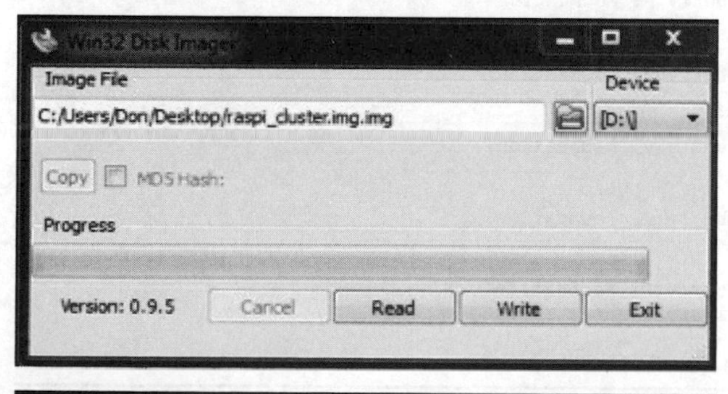

Figure 5-6 Cloning a SD card.

Cluster Software Setup

You will be ready to set up all the cluster nodes once all the SD cards have been created. You should disconnect the stand-alone configuration at this point and insert an SD into each of the cluster RasPis. Of course, ensure that no power is applied to any RasPi as you do this. Once done, power-on all the RasPis, and get ready to use a headless configuration to continue the setup. Again, review Chapter 1 to refresh your memory on how to connect to a RasPi in a headless configuration.

Your home network router will assign an IP address to every RasPi that it can reach via the wired Ethernet LAN. This is done through a process known as the *Dynamic Host Configuration Protocol (*DHCP), which should be automatic and not require any user intervention. You already know the IP address of the stand-alone RasPi you used for the initial setup. The simplest way to find out all the remaining RasPi IP addresses is to log onto the router. The router login varies with different brands, but a little research should reveal the correct address. My address for a Netgear router was 192.168.1.1. The user and default passwords should already be provided in the instruction manual or may be readily determined using an Internet search. Once you have logged on, you should click on the attached devices. You likely will see a Refresh button on the page, which you should click. You might have to refresh several times to see all the RasPis. The addresses that appear will vary depending on the router being used and its configuration. In my case, the lowest IP address I saw with a hostname of Pi01 was 192.168.1.26, which was also the stand-alone RasPi. The next was 192.168.1.28, and so forth. Your IP addresses likely will differ from mine, but that's fine. All you need to do is write the IP addresses down. You do not need to physically associate any RasPi with any IP address. The MPI software is completely indifferent to physical nodes versus logical IP addresses. The only situation where it would be important to physically associate a node with an address would be the case where you are reading a sensor from a node or controlling some type of actuator from the same or another node.

There is another way to determine all the remaining RasPis' IP addresses in case you cannot access your home router's attached devices table. SSH into your first RasPi using the IP address you wrote down when you first started the MPI software install. You will now need to install the `nmap` utility. `nmap` is short for "network mapper," and it is a very handy application for both network discovery and security analysis. I will only be using it to discover the remaining RasPi nodes. Enter the following command to install the `nmap` utility:

```
sudo apt-get install nmap
```

You will be ready to discover the remaining nodes after `nmap` is installed. Enter this next command to run a scan for these nodes:

```
sudo nmap -sP 192.168.1.1-254
```

You will likely have to change the starting IP address to suit your own network. Most likely the change will be the following:

```
sudo nmap -sP 192.168.0.1-254
```

Figure 5-7 shows the result of the `nmap` scan.

```
pi@Pi01 ~/mpi4py-1.3.1 $ sudo nmap -sP 192.168.1.1-254

Starting Nmap 6.00 ( http://nmap.org ) at 2015-06-23 20:05 EDT
Nmap scan report for 192.168.1.1
Host is up (0.00040s latency).
MAC Address: C4:3D:C7:9D:FD:D7 (Netgear)
Nmap scan report for 192.168.1.3
Host is up (0.0082s latency).
MAC Address: E4:F4:C6:F7:10:ED (Unknown)
Nmap scan report for 192.168.1.10
Host is up (0.00054s latency).
MAC Address: B8:27:EB:BC:74:67 (Raspberry Pi Foundation)
Nmap scan report for 192.168.1.26
Host is up.
Nmap scan report for 192.168.1.28
Host is up (0.00051s latency).
MAC Address: B8:27:EB:FD:C3:B3 (Raspberry Pi Foundation)
Nmap scan report for 192.168.1.31
Host is up (0.00038s latency).
MAC Address: B8:27:EB:96:7D:97 (Raspberry Pi Foundation)
Nmap scan report for 192.168.1.33
Host is up (0.00074s latency).
MAC Address: B8:27:EB:29:48:BF (Raspberry Pi Foundation)
Nmap scan report for 192.168.1.34
Host is up (0.00046s latency).
MAC Address: B8:27:EB:51:37:ED (Raspberry Pi Foundation)
Nmap scan report for 192.168.1.35
Host is up (0.00042s latency).
MAC Address: B8:27:EB:9F:18:45 (Raspberry Pi Foundation)
Nmap scan report for 192.168.1.36
Host is up (0.00047s latency).
MAC Address: B8:27:EB:07:CC:9D (Raspberry Pi Foundation)
Nmap scan report for 192.168.1.37
Host is up (0.00044s latency).
MAC Address: B8:27:EB:69:46:55 (Raspberry Pi Foundation)
Nmap done: 254 IP addresses (11 hosts up) scanned in 2.68 seconds
pi@Pi01 ~/mpi4py-1.3.1 $
```

Figure 5-7 nmap scan results.

You can clearly see all the RasPi nodes discovered in the scan. The Raspberry Pi Foundation tags positively identify them. I would recommend that you write all these IP addresses on a piece of paper because you will shortly need them for the next step in the cluster software setup.

Assigning Unique Hostnames

At this point in the setup, all the nodes have the same hostname, which is Pi01. This is a direct result of cloning the initial SD card. Believe it or not, you could leave them all with the same name because the IP addresses are the ones used by MPI to distinguish between the nodes. However, I would strongly recommend that you change the names as that helps you to clearly identify what each particular node is doing in the mpi environment. You may select any node to SSH into in order to change the hostname. The most logical approach is simply to use the next highest IP address and use that one as Pi02. In my case, this was 192.168.1.28 because 192.168.1.26 was the original, Pi01. You will then need to run the raspi-config utility after you log in. Figure 5-8 shows the raspi-config hostname dialog screen, which is found in the Advanced Options selection A2.

In this figure, I had already changed the name, which is the reason you can see the correct name in the box as well as at the top of the screenshot. You now need to repeat the hostname change for all the remaining RasPi nodes. I would also jot down the host name next to the IP address on the list that you already created. Note that some homes will occasionally reassign IP addresses to their attached devices, so it would be wise to occasionally check the recorded IP list against the real-world assignments. It will make no difference as far as the hostname assignment goes except that you would need to update your list. It does make a big difference with regard to the machinefile, which I discuss in the next section.

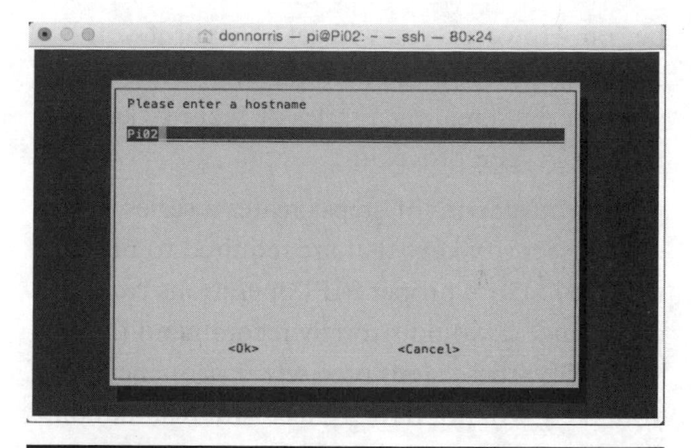

Figure 5-8 Advanced Options selection to change hostname.

Machinefile and Keygen

You will need to create what is known as a *machinefile* to have MPI access to any additional nodes other than the run it is currently running. The machinefile is simply a text file containing a list of all the node IP addresses, as shown in Figure 5-9 for my cluster.

The machinefile also must be present in the same directory where the mpiexec program will

```
donnorris — pi@Pi01: ~/mpi4py-1.3.1 — ssh — 80×24
pi@Pi01 ~/mpi4py-1.3.1 $ cat machinefile
192.168.1.26
192.168.1.28
192.168.1.31
192.168.1.33
192.168.1.34
192.168.1.35
192.168.1.36
192.168.1.37
pi@Pi01 ~/mpi4py-1.3.1 $
```

Figure 5-9 Machinefile.

be run. However, you will get an error if you attempt to run mpiexec with the machinefile without first creating a series of security keys using a keygen procedure.

This next series of steps creates a series of crypto security keys that are required to be installed before proper MPI operations can commence. I would strongly recommend that you follow these steps precisely in sequence or you likely will not have a successful installation. I will be using my assigned IP address and corresponding hostnames, so you likely have to adjust your commands to suit your own IP addresses.

1. SSH into Pi01 from your PC or Mac:

```
ssh pi@192.168.1.26
```

2. Enter the user name and password:

```
user: pi
password: xxxxxxxxx
```

3. Create a RSA key:

```
ssh-keygen
```

4. Press the ENTER key three times to accept the default selections.
5. Change to the home directory:

```
cd ~
```

6. Change to the hidden SSH directory:

```
cd .ssh
```

7. Copy the RSA data to a file named Pi01:

```
cp id_rsa.pub Pi01
```

8. SSH into Pi02 from Pi01:

```
ssh pi@192.168.1.28
```

9. Enter the user name and password:

```
user: pi
password: xxxxxxxxx
```

10. Create an RSA key:

```
ssh-keygen
```

11. Press the ENTER key three times to accept the default selections.
12. Change to the hidden SSH directory:

```
cd .ssh
```

13. Copy the RSA data to a file named Pi02:

```
cp id_rsa.pub Pi02
```

14. Securely copy the file Pi01 from the Pi01 node. Do not forget the single period at the end of the command line.

```
scp 192.168.1.26:/home/pi/.ssh/Pi01 .
```

15. Pipe the contents of Pi01 into a file named authorized_keys:

```
cat Pi01 >> authorized_keys
```

16. Exit from the current command:

```
exit
```

17. SSH into Pi03 from Pi02:

```
ssh pi@192.168.1.37
```

18. Enter the user name and password:

```
user: pi
password: xxxxxxxxx
```

19. Create an RSA key:

```
ssh-keygen
```

20. Press the ENTER key three times to accept the default selections.
21. Change to the hidden SSH directory:

```
cd .ssh
```

22. Copy the RSA data to a file named Pi03:

```
cp id_rsa.pub Pi03
```

23. Securely copy the file Pi01 from the Pi01 node. Do not forget the single period at the end of the command line.

```
scp 192.168.1.26:/home/pi/.ssh/Pi01 .
```

24. Pipe the contents of Pi01 into a file named authorized_keys:

```
cat Pi01 >> authorized_keys
```

25. Exit from the current command:

```
exit
```

26. SSH into Pi04 from Pi03:

```
ssh pi@192.168.1.31
```

27. Enter the user name and password:

```
user: pi
password: xxxxxxxxx
```

28. Create an RSA key:

```
ssh-keygen<
```

29. Press the ENTER key three times to accept the default selections.
30. Change to the hidden SSH directory:

```
cd .ssh
```

31. Copy the RSA data to a file named Pi04:

```
cp id_rsa.pub Pi04
```

32. Securely copy the file Pi01 from the Pi01 node. Do not forget the single period at the end of the command line.

```
scp 192.168.1.26:/home/pi/.ssh/Pi01 .
```

33. Pipe the contents of Pi01 into a file named authorized_keys:

```
cat Pi01 >> authorized_keys
```

34. Exit from the current command:

```
exit<
```

35. SSH into Pi05 from Pi04:

```
ssh pi@192.168.1.33
```

36. Enter the user name and password:

```
user: pi
password: xxxxxxxxx
```

37. Create an RSA key:

```
ssh-keygen
```

38. Press the ENTER key three times to accept the default selections.
39. Change to the hidden SSH directory:

```
cd .ssh
```

40. Copy the RSA data to a file named Pi05:

```
cp id_rsa.pub Pi05
```

41. Securely copy the file Pi01 from the Pi01 node. Do not forget the single period at the end of the command line.

```
scp 192.168.1.26:/home/pi/.ssh/Pi01 .
```

42. Pipe the contents of Pi01 into a file named authorized_keys:

```
cat Pi01 >> authorized_keys
```

43. Exit from the current command:

```
exit
```

44. SSH into Pi06 from Pi05:

```
ssh pi@192.168.1.34
```

45. Enter the user name and password:

```
user: pi
password: xxxxxxxxx
```

46. Create an RSA key:

```
ssh-keygen
```

47. Press the ENTER key three times to accept the default selections.
48. Change to the hidden SSH directory:

```
cd .s
```

49. Copy the RSA data to a file named Pi06:

```
cp id_rsa.pub Pi06
```

50. Securely copy the file Pi01 from the Pi01 node. Do not forget the single period at the end of the command line.

```
scp 192.168.1.26:/home/pi/.ssh/Pi01 .
```

51. Pipe the contents of Pi01 into a file named authorized_keys:

```
cat Pi01 >> authorized_keys
```

52. Exit from the current command.

```
exit
```

53. SSH into Pi07 from Pi06:

```
ssh pi@192.168.1.35
```

54. Enter the user name and password:

```
user: pi
password: xxxxxxxxx
```

55. Create an RSA key:

```
ssh-keygen
```

56. Press the ENTER key three times to accept the default selections.

57. Change to the hidden SSH directory:

```
cd .ssh
```

58. Copy the RSA data to a file named Pi07:

```
cp id_rsa.pub Pi07
```

59. Securely copy the file Pi01 from the Pi01 node. Do not forget the single period at the end of the command line.

```
scp 192.168.1.26:/home/pi/.ssh/Pi01 .
```

60. Pipe the contents of Pi01 into a file named authorized_keys:

```
cat Pi01 >> authorized_keys
```

61. Exit from the current command:

```
exit
```

62. SSH into Pi08 from Pi07:

```
ssh pi@192.168.1.36
```

63. Enter the user name and password:

```
user: pi
password: xxxxxxxxx
```

64. Create an RSA key:

```
ssh-keygen
```

65. Press the ENTER key three times to accept the default selections.

66. Change to the hidden SSH directory:

```
cd .ssh
```

67. Copy the RSA data to a file named Pi08:

```
cp id_rsa.pub Pi08
```

68. Securely copy the file Pi01 from the Pi01 node. Do not forget the single period at the end of the command line.

```
scp 192.168.1.26:/home/pi/.ssh/Pi01 .
```

69. Pipe the contents of Pi01 into a file named authorized_keys:

```
cat Pi01 >> authorized_keys
```

70. Exit from the current command:

```
exit
```

You now need to link all the keys by following these steps:

1. SSH into Pi01:

```
ssh pi@192.168.1.26
```

2. Enter the following secure copy command. Don't forget the period at the end of the command line.

```
scp 192.168.1.28:/home/pi/.ssh/Pi02 .
```

3. Pipe the file into the authorized_keys file:

```
cat Pi02 >> authorized_keys
```

4. SSH into Pi02:

```
ssh pi@192.168.1.28
```

5. Enter the following secure copy command. Don't forget the period at the end of the command line.

```
scp 192.168.1.37:/home/pi/.ssh/Pi03 .
```

6. Pipe the file into the authorized_keys file:

```
cat Pi03 >> authorized_keys
```

7. SSH into Pi03:

```
ssh pi@192.168.1.37
```

8. Enter the following secure copy command. Don't forget the period at the end of the command line.

```
scp 192.168.1.31:/home/pi/.ssh/Pi04 .
```

9. Pipe the file into the authorized_keys file:

```
cat Pi04 >> authorized_keys
```

10. SSH into Pi04:

```
ssh pi@192.168.1.31
```

11. Enter the following secure copy command. Don't forget the period at the end of the command line.

```
scp 192.168.1.33:/home/pi/.ssh/Pi05 .
```

12. Pipe the file into the authorized_keys file:

```
cat Pi05 >> authorized_keys
```

13. SSH into Pi05:

```
ssh pi@192.168.1.33
```

14. Enter the following secure copy command. Don't forget the period at the end of the command line.

```
scp 192.168.1.34:/home/pi/.ssh/Pi06 .
```

15. Pipe the file into the authorized_keys file:

```
cat Pi06 >> authorized_keys
```

16. SSH into Pi06:

```
ssh pi@192.168.1.34<
```

17. Enter the following secure copy command. Don't forget the period at the end of the command line.

```
scp 192.168.1.35:/home/pi/.ssh/Pi07 .
```

18. Pipe the file into the authorized_keys file.

```
cat Pi07 >> authorized_keys
```

19. SSH into Pi07:

```
ssh pi@192.168.1.36
```

20. Enter the following secure copy command. Don't forget the period at the end of the command line.

```
scp 192.168.1.35:/home/pi/.ssh/Pi08 .
```

21. Pipe the file into the authorized_keys file:

```
cat Pi08 >> authorized_keys
```

At this point, you should take a rest from all the preceding intensive setup commands.

The bad news: if your router decides to reassign the IP addresses that you used to authorize the RasPis, you will need to go through all the preceding steps and regenerate all new crypto (RSA) codes, create new machinefile and authorized_keys files, and finally relink everything. I realize that this is a huge pain, but it can only be avoided by assigning static IP addresses to the RasPis or creating an entirely new subnet, which will create new issues and problems, especially with your existing network. I choose to stay with the DHCP approach and deal with the occasional reassignments. The choice is entirely up to you.

I would recommend you do the following test to check whether all the preceding configurations were entered correctly. Enter the following while in Pi01's mpi4py-1.3.1 directory:

```
mpiexec -f machinefile -n 8 python demo/
helloworld.py
```

You should see a display as shown in Figure 5-10.

Congratulations, you are now in possession of a true multiprocessing parallel computer system. However, the real fun is just beginning.

```
donnorris — pi@Pi01: ~/mpi4py-1.3.1 — ssh — 80×24
pi@Pi01 ~/mpi4py-1.3.1 $ mpiexec -f machinefile -n 8 python demo/helloworld.py
Hello, World! I am process 0 of 8 on Pi01.
Hello, World! I am process 2 of 8 on Pi04.
Hello, World! I am process 1 of 8 on Pi02.
Hello, World! I am process 7 of 8 on Pi03.
Hello, World! I am process 4 of 8 on Pi06.
Hello, World! I am process 5 of 8 on Pi07.
Hello, World! I am process 3 of 8 on Pi05.
Hello, World! I am process 6 of 8 on Pi08.
pi@Pi01 ~/mpi4py-1.3.1 $
```

Figure 5-10 Hello World results from eight separate nodes.

It is time to progress from the relatively simple Hello World–type programs to something that is more meaningful and useful. Let's calculate the value pi.

Pi Calculations

As most readers know, pi is the ratio between a circle's circumference and its diameter. It is also a *transcendental number*. Technically, a transcendental number is a real number that is not the solution of any single-variable polynomial equation with all integer coefficients. Practically, *transcendental* means literally "never ending," so it is always an approximation, sometimes to an amazing degree. Figure 5-11 is an output from the WolframAlpha application showing pi to 1950 decimal points.

There are a variety of ways to approximate pi's value, some quite rough and others quite precise. Some of these methods are listed as follows:

```
Decimal approximation:                                   Fewer digits  More digits
3.141592653589793238462643383279502884197169399375105820974944592307816
40628620899862803482534211706798214808651328230664709384460955058223
17253594081284811174502841027019385211055596446229489549303819644288
10975665933446128475648233786783165271201909145648566923460348610454
32664821339360726024914127372458700660631558817488152092096282925409
17153643678925903600113305305488204665213841469519415116094330572703
65759591953092186117381932611793105118548074462379962749567351885752
72489122793818301194912983367336244065664308602139494639522473719070
21798609437027705392171762931767523846748184676694051320005681271452
63560827785771342757789609173637178721468440901224953430146549585371
05079227968925892354201995611212902196086403441815981362977477130996
05187072113499999983729780499510597317328160963185950244594553469083
02642522308253344685035261931188171010003137838752886587533208381420
61717766914730359825349042875546873115956286388235378759375195778185
77805321712268066130019278766111959092164201989380952572010654858632
78865936153381827968230301952035301852968995773622599413891249721775
28347913151557485724245415069595082953311686172785588907509838175463
74649393192550604009277016711390098488240128583616035637076601047101
81942955596198946767837449448255379774726847104047534646208046684259
06949129331367702898915210475216205696602405803815019351125338243003
55876402474964732639141992726042699227967823547816360093417216412199
24586315030286182974555706749838505494588586926995690927210797509302
95532116534498720275596023648066549911988183479775356636980742654252
78625518184175746728909777727938000816470600161452491921732172147723
50141441973568548161361157352552133475741849468438523323907394143334
54776241686251898356948556209921922218427255025425688767179049460165
34668049886272327917860857843838279679766814541009538837863609506800
64225125205117392984896084128488626945604241965285022210661186306744
278622039194945047123713786960956364371917287...
```

Figure 5-11 Pi to 1950 decimal points.

- **Graphical.** Wrap a string taut around a jar lid, and then measure its length. Next, measure the jar lid's diameter and divide that into the string length. The result is a rough approximation for pi.
- **Infinite-series calculations.** There are a few different series in this category.
  - **Gregory Leibniz:** Pi = (4/1) − (4/3) + (4/5) − (4/7) + (4/9) − (4/11) + · · ·
  - **Nilakantha:** Pi = 3 + 4/(2 × 3 × 4) − 4/(4 × 5 × 6) + 4/(6 × 7 × 8) − 4/(8 × 9 × 10) + · · ·
  - **Simulations:** There are several ways to numerically simulate pi's value.
    - **Dartboard.** Consider a dartboard contained in a square whose sides are exactly the circle's diameter, as shown in Figure 5-12. Now imagine throwing darts at the board in a completely random manner. Most darts will land in the circle, but some will land in one of the four shaded areas outside the circle but still in the square box. If you then count up the number of darts that landed in the box but not in the circle and divide that by the total number of darts thrown, you should approximate pi's value. Of course, increasing the number of darts thrown increases the overall approximation accuracy.
    - **Buffon's needle problem.** In this simulation, there will be small "needles" thrown across a rectangle filled with parallel lines. The lines are separated by distance *t*, as shown in Figure 5-13, excerpted from Wikipedia. The needle's length, as shown in the figure, is *l*, which must be set equal to *t* for this simulation to function properly. Then you randomly toss these needles onto the playing field and count the number of needles that cross any line. Divide this number into the total number thrown, and you again approximate pi. Increasing the number of needles thrown increases the accuracy of the pi estimation, as was the case for the dartboard simulation.
- **Trigonometric.** Several trigonometric algorithms can be used to calculate pi. A common one is the arcsine(x) function, with the variable x constrained to be a value between +1 and −1.

  Pi = 2 × {arcsin[sqrt(1 − x^2) + abs(arcsin)]}

I will next show you how to calculate pi using the RasPi cluster now that you have had a brief introduction to some of the popular approaches to this endeavor.

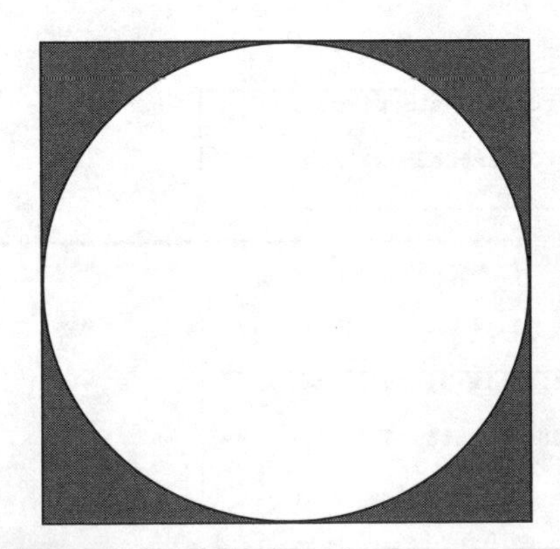

Figure 5-12 Dartboard.

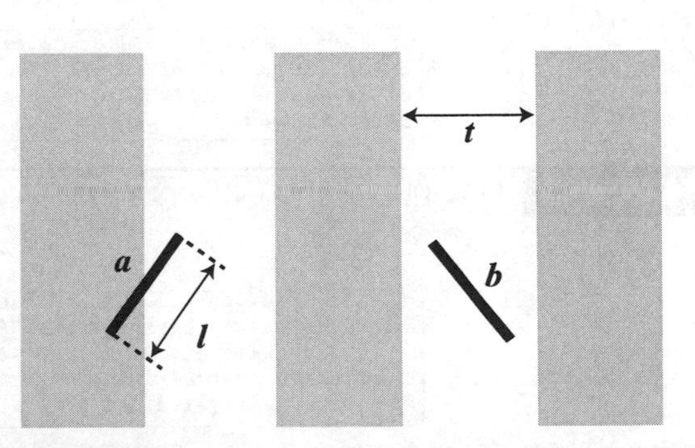

Figure 5-13 Buffon's needle playing field.

Calculating Pi Using the RasPi Cluster

A directory named compute-pi was created for you when you installed the MPI4PY package. This directory is located in the demo directory, which, in turn, is located in the mpi4py-1.3.1 directory. The compute-pi directory has three Python programs that calculate pi in three different ways:

- **cpi-cco.py.** Computes pi using collective communications operations.
- **cpi-dpm.py.** Computes pi using dynamic process management.
- **cpi-rma.py.** Computes pi using remote memory access.

I will only run the cpi-cco.py program because it is adequate to demonstrate cluster operations. You should try the other pi computation programs if they interest you.

cpi-cco.py

I will start executing this program using only one RasPi, namely, Pi01. SSH into Pi01 and cd down to the mpi4py-1.3.1 directory. Enter the following command to create one process for the pi calculation:

```
mpiexec -n 1 python demo/compute-pi/
cpi-cco.py
```

Enter **10000** when prompted for the number of intervals. Figure 5-14 shows the result of this command.

Enter a **0** for the number of intervals to quit the program. Next, enter the following command to create eight processes on Pi01 to collectively calculate pi:

```
mpiexec -n 8 python demo/compute-pi/
cpi-cco.py
```

Figure 5-15 shows the result of this command.

This last command took a little longer to run, and the results were a bit more accurate compared with the single-process calculation. Next, enter the following command to use all the RasPis in the computation:

```
mpiexec -f machinefile -n 8 python demo/
compute-pi/cpi-cco.py
```

Figure 5-16 shows the result of this command.

The results are identical to the eight processes as run on a single RasPi. If you think about it for a moment, this result should not surprise you because exactly the same number of computations are being done in both cases except that in the first case one RasPi is doing all the work, whereas for the second case, the computations are distributed among eight RasPis. Although it is hard to confirm with 10,000 intervals, I did note that the single-RasPi case did take longer than all eight RasPis working together. I next increased the intervals

```
pi@Pi01 ~/mpi4py-1.3.1 $  mpiexec -n 1 python demo/compute-pi/cpi-cco.py
Enter the number of intervals: (0 quits) 10000
pi is approximately 3.1415926544231341, error is 0.0000000008333410
Enter the number of intervals: (0 quits)
```

Figure 5-14 One-process Pi01 cpi-cco.py pi calculation.

```
pi@Pi01 ~/mpi4py-1.3.1 $  mpiexec -n 8 python demo/compute-pi/cpi-cco.py
Enter the number of intervals: (0 quits) 10000
pi is approximately 3.1415926544231247, error is 0.0000000008333316
Enter the number of intervals: (0 quits) 0
pi@Pi01 ~/mpi4py-1.3.1 $
```

Figure 5-15 Eight-process Pi01 cpi-cco.py pi calculation.

```
pi@Pi01 ~/mpi4py-1.3.1 $ mpiexec -f machinefile -n 8 python demo/compute-pi/cpi-cco.py
Enter the number of intervals: (0 quits) 10000
pi is approximately 3.1415926544231247, error is 0.0000000008333316
Enter the number of intervals: (0 quits) 0
pi@Pi01 ~/mpi4py-1.3.1 $ █
```

Figure 5-16 Eight-processes distributed to Pi01 through Pi08 cpi-cco.py pi calculation.

```
pi@Pi01 ~/mpi4py-1.3.1 $ mpiexec -n 8 python demo/compute-pi/cpi-cco.py
Enter the number of intervals: (0 quits) 1000000
pi is approximately 3.1415926535898899, error is 0.0000000000000968
Enter the number of intervals: (0 quits) 0
pi@Pi01 ~/mpi4py-1.3.1 $ mpiexec -f machinefile -n 8 python demo/compute-pi/cpi-cco.py
Enter the number of intervals: (0 quits) 1000000
pi is approximately 3.1415926535898899, error is 0.0000000000000968
Enter the number of intervals: (0 quits) 0
pi@Pi01 ~/mpi4py-1.3.1 $ █
```

Figure 5-17 Eight- versus single-RasPi Pi calculation.

from 10,000 to 1 million to try to measure the actual time for each situation. The results were precisely as I expected:

One RasPi = 112 seconds

Eight RasPis = 14 seconds

112/14 = 8, or an 8 times speedup

Figure 5-17 shows the actual pi calculations for each case. The results are identical, as may be seen in the figure.

I do wish to mention that this program uses parallel computations employing collective communications operations (CCOs), and it really makes no difference from a logical viewpoint whether all the processes are executed on one processor or on multiple processors. It does make a huge difference from a speed-of-execution viewpoint, as I just demonstrated.

Error Analysis I was curious to determine whether there was a relationship between the number of intervals used in the pi computations and the reported error. I therefore conducted a series of tests in which the number of intervals was increased by 10 times starting at 10 and going to 10 million. Table 5-3 shows the result of this experiment.

I wanted to plot these because it is a nice way to spot relationships, but I realized that the obvious disparity in the size of the table numbers would not lend itself to being plotted on a linear scale. Looking at the numbers, I realized that I could take the log to the base 10 for each of them and plot those numbers in a reasonable manner. Figure 5-18 shows the log-log chart for the intervals versus errors.

TABLE 5-3 Reported Error vs. Number of Intervals

Intervals	Error	Log of Error	Log of Intervals
10	8.33×10^{-6}	−5.079198618	1
100	8.33×10^{-8}	−7.079354999	2
1,000	8.33×10^{-10}	−9.079354999	3
10,000	8.33×10^{-12}	−11.079355	4
1,000,000	9.68×10^{-14}	−13.01412464	6
10,000,000	1.38×10^{-14}	−13.86012091	7

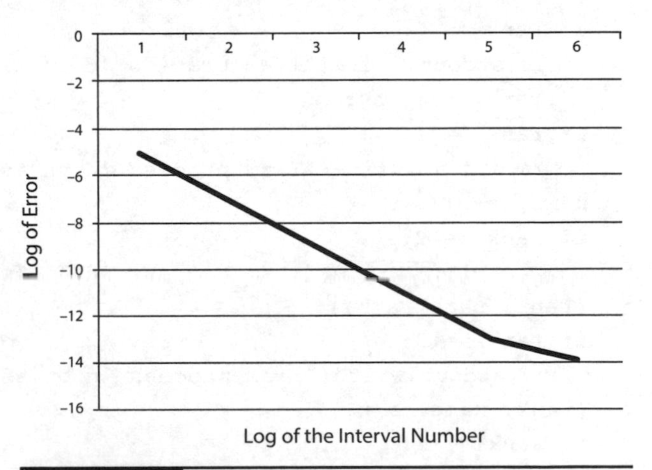

Figure 5-18 Interval versus error chart.

Looking at the chart reveals that there is almost a perfect linearity between the log of the number of intervals used versus the log of the reported error. What this all means is that if you use 10 times the number of intervals for a given trial, you will reduce the error 100 times. The chart also reveals that the relationship tends to flatten a bit after 1 million intervals, which means that the error improvement will decrease somewhat. In any case, this RasPi cluster computation for pi is very accurate, provided that you use at least a half-million intervals.

To this point, I have only run identical programs on the cluster RasPis. While this meets the main purpose for cluster operations, it is possible to run a program in which each node can have different functions depending on its rank. I will demonstrate this functionality in the next section.

Unique Functions for Cluster Operations

I will demonstrate this type of functionality by creating a program named `uniqueFunctions.py`, where nodes can do distinct things depending on their rank within the communicator. However, don't get too excited because the functions will only be simple arithmetic operations to demonstrate the concept. The program is as follows:

```
#!/usr/bin/env python
"""
uniqueFunctions.py
"""

from mpi4py import MPI
import sys

a = 12.0
b = 6.0

rank = MPI.COMM_WORLD.Get_rank()
name = MPI.Get_processor_name()

if rank == 0:
  sys.stdout.write("I am process %d on %s\nI only show raw values\na = %d b = %d\n\n"
% (rank, name,a,b ))
if rank == 1:
  sys.stdout.write("I am process %d on %s\nI sum raw values\na + b = %d\n\n" % (rank,
name, a+b))
if rank == 2:
  sys.stdout.write("I am process %d on %s\nI multipy raw values\na * b = %d\n\n" %
(rank, name, a*b))
if rank == 3:
  sys.stdout.write("I am process %d on %s\nI divide raw values\na / b = %f\n\n" %
(rank, name, a/b))
if rank == 4:
  sys.stdout.write("I am process %d on %s\nI square raw values\na*a + %d b*b =
%d\n\n" % (rank, name, a*a,b*b))
```

```
if rank == 5:
  sys.stdout.write("I am process %d on %s\nI cube raw values\na*a*a = %d b*b*b = 
%d\n\n" % (rank, name, a*a*a, b*b*b))
if rank == 6:
  sys.stdout.write("I am process %d on %s\nI subtract raw values\na - b = %d\n\n" % 
(rank, name, a-b))
if rank == 7:
  sys.stdout.write("I am process %d on %s\nI invert raw values\n1/a = %f 1/b = 
%f\n\n" % (rank, name, 1/a, 1/b)
```

I first manually loaded the code into Pi01 using the nano editor. The code was then run on Pi01 by entering the following command:

```
mpiexec -n 8 python uniqueFunctions.py
```

Figure 5-19 shows the result of running this program.

You can clearly see that each process handles its function according to its assigned rank. The `if` statements conditionally test the process rank and will run the indented code below the `if` statement when there is a match to the preassigned rank. For instance, the process of rank 5 will display the cubes for the `a` and `b` variables.

The next step is to load the program into the MPI4PY directory for all the remaining RasPis. The easiest way to do this is to use the `scp` command. For example, to load the program into Pi02, use the following command:

```
scp uniqueFunctions.py 192.168.1.28:/
home/pi/mpi4py-1.3.1/uniqueFunctions.py
```

To speed up this copy operation, you can press the up arrow to recall the last command and just change the appropriate digit or digits in the IP address for the next RasPi node.

```
donnorris — pi@Pi01: ~/mpi4py-1.3.1 — ssh — 127×38
pi@Pi01 ~/mpi4py-1.3.1 $ mpiexec -n 8  python uniqueFunctions.py
I am process 0 on Pi01
I only show raw values
a = 12 b = 6

I am process 3 on Pi01
I divide raw values
a / b = 2.000000

I am process 4 on Pi01
I square raw values
a*a + 144 b*b = 36

I am process 6 on Pi01
I subtract raw values
a - b = 6

I am process 1 on Pi01
I sum raw values
a + b = 18

I am process 5 on Pi01
I cube raw values
a*a*a = 1728 b*b*b = 216

I am process 7 on Pi01
I invert raw values
1/a = 0.083333 1/b = 0.166667

I am process 2 on Pi01
I multipy raw values
a * b = 72

pi@Pi01 ~/mpi4py-1.3.1 $
```

Figure 5-19 uniqueFunctions.py program results.

Enter this command to run the program after all the remaining RasPi nodes have been loaded with the program:

```
mpiexec -f machinefile -n 8 python
uniqueFunctions.py
```

Figure 5-20 shows the result of running the uniqueFunctions.py program on all eight RasPi nodes in the cluster.

You can easily see that each node has a unique rank, which also equates to a specific arithmetic function with regard to the `a` and `b` variables.

At this point, I will change topics to demonstrate some basic MPI operations, including point-to-point and collective communications.

Basic MPI Operations

It would be helpful to understand some of the basic MPI operations if you would like to write your own programs. I will cover some point-to-point and collective operations using Python. I also acknowledge Lisandro Dalcin's fine tutorial, "MPI for Python," from which the following programs were derived.

Point-to-Point Communications

Point-to-point communication, as the name implies, is the direct sending and receiving of messages between two processes. As I have demonstrated previously, both processes may be running on a single node or on multiple nodes. I will demonstrate both the single- and multiple-node cases in this section. I have listed a simple Python program next that is named *p2p.py*:

```
from mpi4py import MPI
comm = MPI.COMM_WORLD
rank = comm.Get_rank()
if rank == 0:
  data = {'a':7, 'b':3.14}
  comm.send(data, dest = 1, tag = 11)
elif rank == 1:
  data = comm.recv(source = 0, tag = 11)
  print data['a']
  print data['b']
```

```
donnorris — pi@Pi01: ~/mpi4py-1.3.1 — ssh — 127×38
pi@Pi01 ~/mpi4py-1.3.1 $ mpiexec -f machinefile -n 8 python uniqueFunctions.py
I am process 0 on Pi01
I only show raw values
a = 12 b = 6

I am process 2 on Pi04
I multipy raw values
a * b = 72

I am process 3 on Pi05
I divide raw values
a / b = 2.000000

I am process 4 on Pi06
I square raw values
a*a + 144 b*b = 36

I am process 6 on Pi08
I subtract raw values
a - b = 6

I am process 7 on Pi03
I invert raw values
1/a = 0.083333 1/b = 0.166667

I am process 5 on Pi07
I cube raw values
a*a*a = 1728 b*b*b = 216

I am process 1 on Pi02
I sum raw values
a + b = 18

pi@Pi01 ~/mpi4py-1.3.1 $
```

Figure 5-20 Cluster uniqueFunctions.py program results.

There is an MPI communicator object created named `comm`, by which the basic send and receive methods can be called. Within the process with a rank equal to zero, I created a dictionary list with two key-value pairs. The first key is `a`, with a value equal to 7, and the second key is `b`, with a value equal to 3.14. This dictionary is then sent in total to the process with a rank equal to 1. Note that there is an additional optional identifier named `tag` with a value of 11 that is also sent to the receiving process. I mentioned earlier that while tags are optional, they do help to sort out messages, especially if a process receives multiple messages from other processes.

The else-if statement (`elif`) will take effect for the process with rank equal to 1. The sent data are received within this conditional code block. Note that the source of the data is identified by its process rank along with the optional tag. The values for both dictionary list items are displayed next.

This program demonstrates basic send and receive operations permitted between two processes. I would like to point out that the data type being sent did not have to be identified beforehand. Python handled that chore very nicely using dynamic type casting. In other words, you can send a dictionary list, a numerical array, data primitives, string arrays, and so on, and Python will automatically handle the data appropriately. This is part of the pickling operation I discussed earlier.

While in the mpi4py-1.3.1 directory, enter the following to run p2p.py on a single RasPi node:

```
mpiexec -n 5 python p2p.py
```

Figure 5-21 shows the result of executing p2p.py on one RasPi.

You will need to copy p2p.py into all the remaining RasPi nodes to run the program on the cluster. Use the following command to securely copy p2p.py from Pi01 to Pi02:

```
scp p2p.py 192.168.1.28:/home/pi/
mpi4py-1.3.1/p2p.py
```

Just repeat this command, changing the IP address, to finish copying the file to all the remaining RasPi nodes, just as you did with the earlier example. Once all the files are in place, you can run p2p.py on the cluster by entering:

```
mpiexec -f machinefile -n 8 python p2p.py
```

Figure 5-22 shows the result of executing p2p.py on the cluster.

This completes my brief introduction to MPI point-to-point communications. There is much

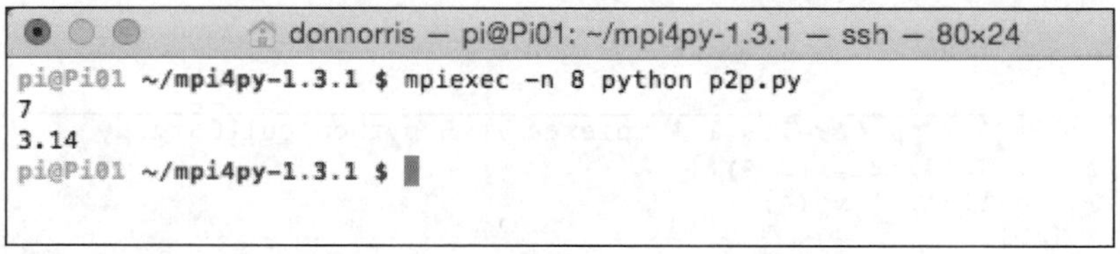

```
donnorris — pi@Pi01: ~/mpi4py-1.3.1 — ssh — 80×24
pi@Pi01 ~/mpi4py-1.3.1 $ mpiexec -n 8 python p2p.py
7
3.14
pi@Pi01 ~/mpi4py-1.3.1 $
```

Figure 5-21 p2p.py program display from a single node.

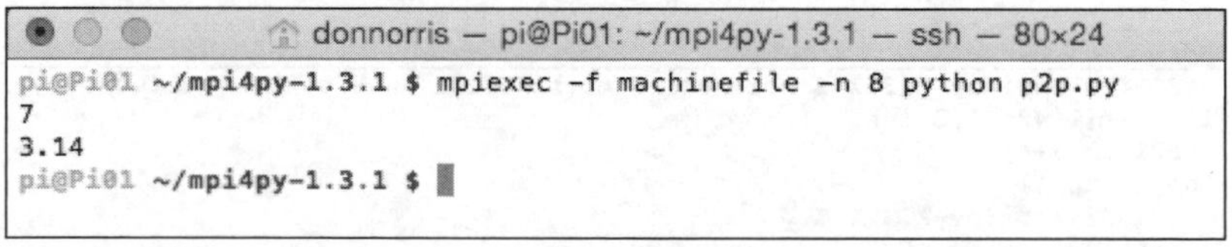

```
donnorris — pi@Pi01: ~/mpi4py-1.3.1 — ssh — 80×24
pi@Pi01 ~/mpi4py-1.3.1 $ mpiexec -f machinefile -n 8 python p2p.py
7
3.14
pi@Pi01 ~/mpi4py-1.3.1 $
```

Figure 5-22 p2p.py program display from the cluster.

more to this topic, and I would recommend Dalcin's tutorial as a good starting point to delve deeper into this topic.

The next basic MPI operation concerns collective communications, where one process sends messages to more than one process.

Collective Communications

It is often necessary for a process to send data to multiple processes in a single operation. This is known in MPI terms as *collective communications* and sometimes as *broadcast communications* when all communicator processes are involved. As I mentioned in the point-to-point discussion, all the involved processes may be running on a single node or multiple nodes. I will demonstrate collective communications for both the single- and multiple-node cases in this section. I have listed a simple Python program next that is named collComm.py. Enter the following to run collComm.py on a single RasPi node:

```
mpiexec -n 8 python p2p.py
```

Figure 5-23 shows the result of executing p2p.py on one RasPi.

You will need to copy collComm.py into all of the remaining RasPi nodes in order to run the program on the cluster. Use the following command to securely copy collComm.py from Pi01 to Pi02:

```
scp collComm.py 192.168.1.28:/home/pi/
mpi4py-1.3.1/collComm.py
```

Just repeat this command, changing the IP address, to finish copying the file to all the remaining RasPi nodes, just as you did with the point-to-point example. Once all the files are in place, you can run collComm.py on the cluster by entering:

```
mpiexec -f machinefile -n 8 python
collComm.py
```

Figure 5-24 shows the result of executing collComm.py on the cluster.

Surprise! I bet you were expecting to see a similar display to what was shown in Figure 5-23 instead of the jumbled mess shown in this figure. At this point, I think most readers will have figured out what happened. The Python `print` command is *nonblocking*, meaning that instances from the `print` command emanating from different processes can "jump" into the output character stream going to the display, resulting in the confused display. I must also confess that I did see this to a much smaller degree in the single-node case, but I simply repeated issuing the command many times over until I got the nice display shown in Figure 5-23. Thoughtful readers also might ask why there was

```
pi@Pi01 ~/mpi4py-1.3.1 $ mpiexec -n 5 python collComm.py
1 [1, 3.14, 21, (3+4j)]
2 ('hello', ' mpi')
3 [42]
pi@Pi01 ~/mpi4py-1.3.1 $
```

Figure 5-23 collComm.py program display from a single node.

```
pi@Pi01 ~/mpi4py-1.3.1 $ mpiexec -f machinefile -n 8 python collComm.py
13 2 [ [1(42, ']3.14h
, e21l, (3+4j)]
lo', ' mpi')
pi@Pi01 ~/mpi4py-1.3.1 $
```

Figure 5-24 collComm.py program display from the cluster.

no "confusion" in the uniqueFunctions.py program, which also had multiple output statements in the program. The answer is simple: I used the standard output `write` function in that program, which is *blocking*, meaning that each output statement must be completed before another is started. Blocking slows things down, but it does have a distinct advantage in this case where there are nonsynchronized output commands vying for a single display terminal.

This completes my brief introduction to MPI collective communications. There is a good deal more to this topic, and I would once again recommend Dalcin's tutorial as a good resource for a more in-depth study of collective communications. The next section describes an interesting way to monitor the network traffic ongoing in the cluster.

Monitoring Cluster Network Traffic

This section describes an application that will allow you to monitor on a real-time basis what is going on with all the packets "flying" around in the cluster. The application is named iptraf, and you only need to load on one RasPi. I would recommend using Pi01 because it is considered the master. Enter the following command to download and install iptraf:

```
sudo apt-get install iptraf
```

You will need administrative privileges to run it, so enter

```
sudo iptraf
```

I also opened a second terminal window in order to copy SSH into Pi01 to be able to run the cluster version of the uniqueFunctions.py program. This program generated some cluster activity, which was shown in the iptraf window. Figure 5-25 shows both the iptraf GUI window as well as the uniqueFunctions.py program display screen.

There are many options that may be set with this application to refine its monitoring. I found it quite interesting to just look at all the different TCP packet activity ongoing within and outside the cluster. It is also easy to spot problems such as loose or broken network connections. Many other network monitoring applications are available, but I have found iptrap to be fairly simple to use as well as being highly configurable.

This last section closes out this chapter on building and using a RasPi cluster.

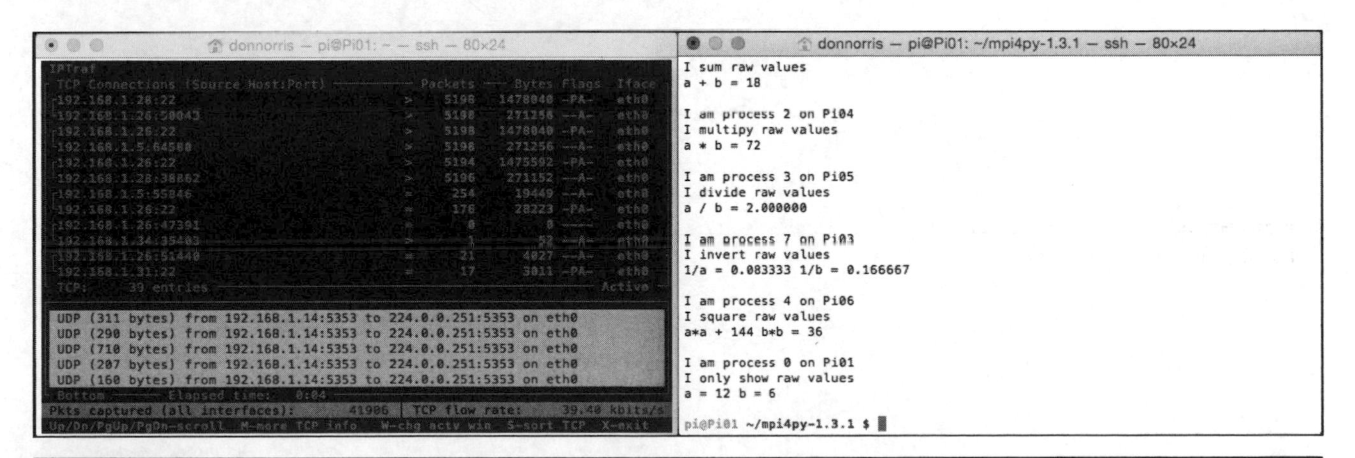

Figure 5-25 iptraf and uniqueFunctions.py screen displays.

Summary

This chapter began with a brief history of supercomputers and a review of terms commonly used in this technology. I next showed you how to physically build an eight-node RasPi cluster, including how to make an optional display stand. A general discussion of cluster software followed that focused on the Message Passing Interface (MPI) Standard. I followed this with detailed, step-by-step instructions on how to install both MPI and Python for MPI. I showed you how to clone a SD card, which greatly reduced the software installation time on the remaining cluster RasPis. All the RasPis were also set up with the necessary security files required for networked operations.

Hello World programs were run on a single RasPi, as well as on the whole cluster. I followed this demonstration with one that computed pi both on a single RasPi and on the whole cluster. I showed you how individual RasPis can run separate functions instead of exactly the same operations, which can provide utility for cluster computations. I also demonstrated some basic MPI operations for readers interested in distributed programming.

The chapter concluded with a brief demonstration on how to monitor network traffic using the iptrap application.

CHAPTER 6

RasPi-to-RasPi Communications Using MQTT

THE FOCUS OF THE CHAPTER, as the title suggests, is direct computer-to-computer communications or, as it is more commonly referred to, machine-to-machine (M2M) communications. There is no human intervention in this type of system because all the computers or machines, as they are generically termed, are set up to communicate with each other using an established protocol. I will be using two RasPis, one as a data source or publisher client and the other as a data sink or subscriber client. The data source RasPi system also will use a single TMP36 temperature sensor to continuously generate temperature data for this project. I will also be using Python for the publisher system and Java for the subscriber system. Using two languages helps to point out the heterogeneous nature of the Meassage Queuing Telemetry Transport (MQTT) Protocol, which allows for different systems to interact seamlessly. In this case, the hardware platforms will be the same, but the computer language software will differ.

Before I start the detailed discussion, I would like to acknowledge a fine blog article entitled, "Using Eclipse Paho's MQTT on BeagleBone Black and Raspberry Pi," which was written by a highly talented developer named D. J. Walker-Morgan. His article is great and is available at the Eclipse.org's talkingsmall blog at www.eclipse.org/paho/articles/talkingsmall/. I highly recommend you read it when have an opportunity to do so.

Paho and Eclipse.org

Paho is an open-source project sponsored by the Eclipse.org Foundation. This project is dedicated to providing scalable client implementations for both open and standard messaging protocols. The Paho Project is designed to provide an exciting infrastructure in support of new M2M and Internet-of-Things (IoT) applications. The home website is located at www.eclipse.org/paho/. At the heart of the Paho Project is a lightweight

Chapter 6 Parts List

Item	Model	Quantity	Source
Analog Devices Model TMP36 temperature sensor	136-TMP36	1	mcmelectronics.com
Microchip MCP3008 ADC	116-MCP3008	1	mcmelectronics.com
Raspberry Pi Model B+	83-16317	2	mcmelectronics.com

publish/subscribe message protocol named MQTT, which I describe in the next section.

MQTT

MQTT is the current name for this protocol, although it was originally named Message Queuing Telemetry Transport. I guess the project managers felt that the original name was a mouthful or it could be that there are no actual queues used in the protocol. In any case, it is now simply called MQTT.

MQTT is over 10 years old, having been created originally by the IBM Pervasive Software Group in conjunction with Arcom, which is now called Eurotech. IBM still supports MQTT, and the current version 3.1 specifications are available in pdf form from the IBM developerWorks website at www.ibm.com/develpoperworks/webservices/library/ws-mqtt/index.html.

MQTT is technically known as a *middleware application*, as can be seen in the block diagram for this project in Figure 6-1. It is important to realize that both publishers and subscribers are treated as client applications in this configuration type.

There is a block named *MQTT Broker* located between the publisher client and the subscriber client blocks. This broker may be thought of as a message dispatcher that ensures that the MQTT messages are properly sent from the client publishers to the correct client subscribers. In this way, subscriber clients do not have to constantly monitor all the network traffic looking for messages that are addressed to them. The broker takes over that function, and it also serves as an acknowledgment intermediary, which I explain in the section concerning quality of service.

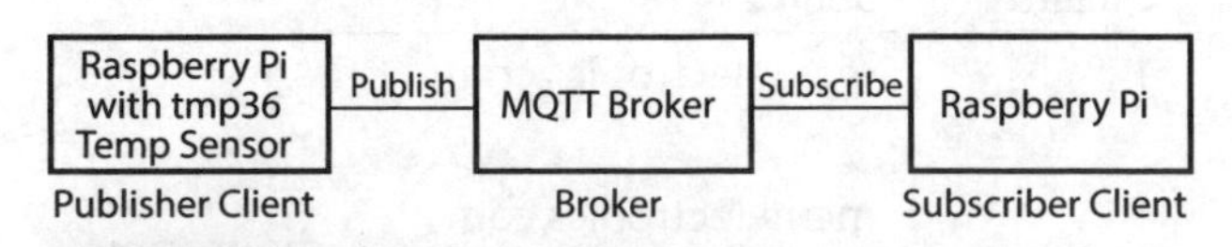

Figure 6-1 Project block diagram.

Table 6-1 lists some of the salient features of MQTT that make it so popular as a messaging protocol.

TABLE 6-1 MQTT Features

Feature	Description
API	Simple, only five methods required
Packets	Compact binary packets; capable of up to a 250-MB payload
Headers	No compressed headers needed
Verbose	Minimal text, much less than HTML

These features make MQTT very popular for M2M applications, including weather monitoring, stock ticker, smart power grid meters, and even Facebook messaging. It is also a very popular way for cellular services to implement message alerts.

Quality of Service (QoS)

QoS refers to the level assurance that MQTT provides regarding message delivery. There are three QoS levels:

- **Level 0.** This is also known as "fire and forget." At this level, the publisher sends off messages, and there is no attempt to acknowledge their reception by the broker on behalf of the publisher. It is obviously the quickest message-delivery method, but it is also the least reliable.
- **Level 1.** This is also known as "at least one." Here messages are sent and resent until the broker receives one acknowledgment from the subscriber. It does provide some assurance that the message did get through to its intended recipient. Level 1 is typically set as the default QoS for a MQTT messaging system.
- **Level 2.** This is also known as "exactly one." At this level, messages undergo a two-stage process in which there is a definitive

acknowledgment between the broker and subscriber ensuring that one and only one message copy was delivered. This QoS level is the slowest among the three levels owing to the additional processing overhead required to establish a high reliability level.

Wills

No, this section has nothing to do with legal probate but instead focuses on what happens when a client abnormally loses its connection with the broker. A *will* is a both a set of instructions and a prescribed message that is stored by the broker and will only be acted on if the connection between the broker and a client is unexpectedly broken. Basically, it is a dialog that states, "If you (the broker) cannot connect to me and I (the client) haven't cleanly disconnected, then carry out the preset instructions and also send out the stored message on my behalf." The will concept is implemented in Python by a `setWill` method and in Java by an object of the `MqttConnectionsOptions` class.

Using wills in MQTT improves both system robustness and reliability and ensures that messages either will be delivered or an error message will be created and distributed describing what went wrong.

Reconnecting

Connections will be broken, and MQTT has the inherent ability to reconnect using two system elements. The first is a logical flag known as the *clean flag*, which is set at a 1 (high value) for every fresh or new connection. The clean flag informs the client and broker that they must start the messaging process from the beginning because it represents a new connection. However, if the clean flag becomes false or 0 (low value), a second element comes into play. This is called the *client ID*, and we will see that it plays a key part in establishing an original connection when the test code is discussed. For now, let's assume that it had already been set to some `String` value when the connection broke. Now, assume that the connection is restored, as might happen when a client briefly loses power. MQTT will attempt to restore the connection to the same precise state because it recognizes that it still has same client ID String stored in its record structure, which existed for this particular connection when it first became disconnected. Note that various MQTT libraries, whether they be Python or Java, have different implementations for storing client IDs and messages so that the connections can be recovered without any message loss.

It is now time to demonstrate a temperature-monitoring application that uses MQTT to distribute single-valued data points between one publisher and one client.

RasPi MQTT Publisher-Client System

Before I demonstrate the temperature-monitoring program, I first need to describe the temperature-monitoring sensor, the analog-to-digital converter (ADC) chip, and the serial peripheral bus. All three are required for the RasPi to be able to generate the temperature data.

TMP36 Temperature Sensor

The basic temperature sensor I will use in this project is an Analog Devices Model TMP36, shown in Figure 6-2. It is housed in a standard TO-92 plastic form factor that is also common to most transistors. The TMP36 is far more complex than a simple transistor in that in contains circuits to both sense ambient temperature and convert that temperature to an analog voltage. The functional block diagram is shown in Figure 6-3.

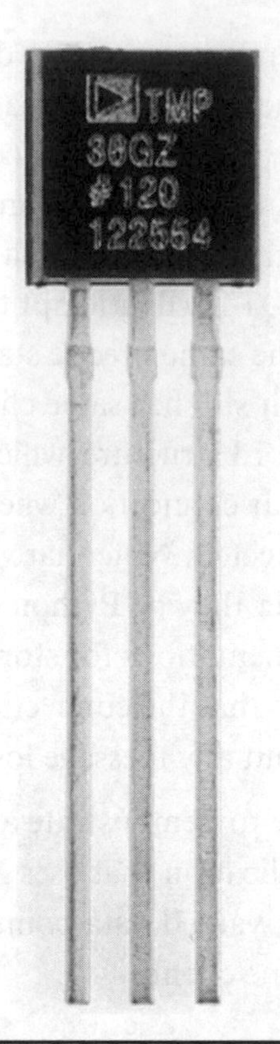

Figure 6-2 Analog Devices Model TMP36 temperature sensor.

The TMP36 only has three leads, which are shown in a bottom view in Figure 6-4.

Table 6-2 provides details concerning these three leads, including important limitations.

The voltage representing the temperature depends on the TMP36 supply voltage, which must be considered when converting the Vout voltage to the equivalent real-world temperature. I account for this in the software that converts the Vout voltage to an actual temperature.

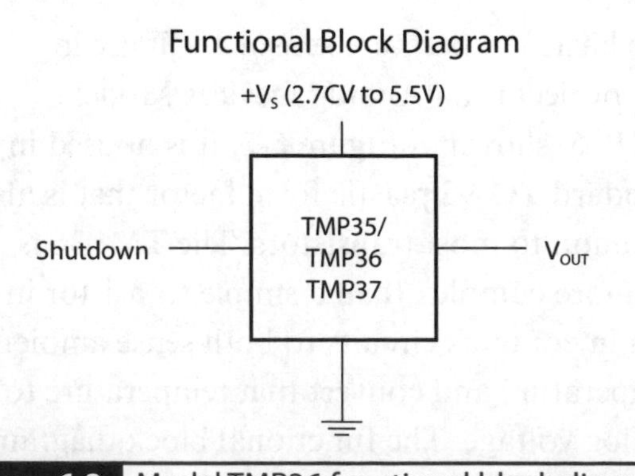

Figure 6-3 Model TMP36 functional block diagram.

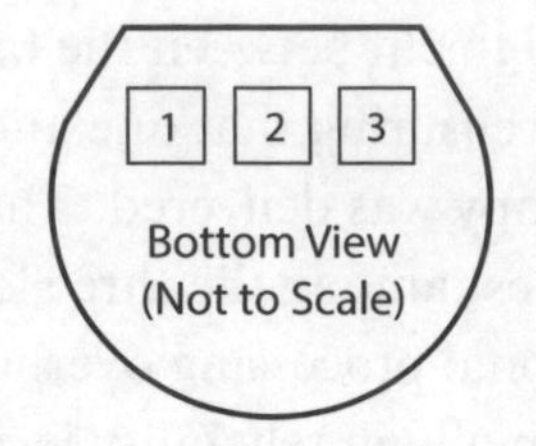

Figure 6-4 TMP36 bottom view showing external leads.

TABLE 6-2 TMP36 Pin Details

Pin Number	Description	Remarks
1	$+V_S$	Supply voltage; ranges from 2.7 to 5.5 V
2	Vout	Analog voltage representing temperature; maximum voltage depends on the supply voltage
3	GND	Common reference used by both the supply and Vout pins

Figure 6-5 is a graph of the Vout voltage versus temperature using a 3-V supply voltage.

The actual temperature-measurement range for the TMP36 is –40 to +125°C, with a typical accuracy of ±2°C and a 0.5°C linearity. All in all, these are not too shabby specifications considering that the cost of the TMP36 is

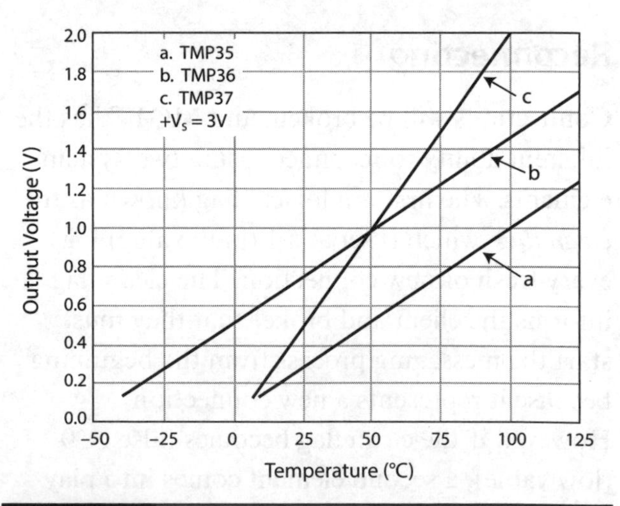

Figure 6-5 Graph of Vout voltage versus temperature for a +VS = 3 V.

typically less than US$2. The TMP36 range, accuracy, and linearity are well suited for a home temperature-monitoring system.

Analog-to-Digital Conversion

The RasPi does not contain any means by which analog signals may be processed, as most readers already know. This means that some type of analog-to-digital converter (ADC) must be used before the RasPi can deal with the temperature signals.

I used a Microchip Model MCP3008, which is described in the Microchip datasheet as a 10-bit SAR ADC with SPI data output. This means that the MCP3008 uses a Successive Approximation Register (SAR) technique to create a 10-bit digital result that in turn is outputted in a serial data stream using the Serial Peripheral Interface (SPI) Protocol, which is discussed after the sidebar. The very inexpensive MCP3008 ADC chip has impressive

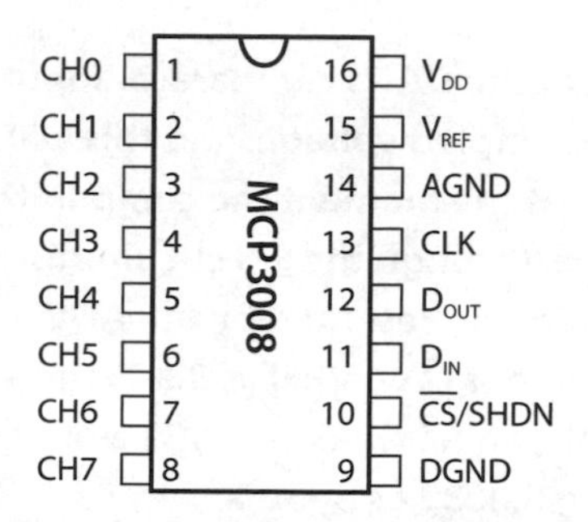

Figure 6-6 MCP3008 package form and pin-out.

specifications despite its very low cost. Figure 6-6 shows the package form and pin-out for this chip.

The MCP3008 chip as used in this project is in a dual-in-line package (DIP), which means that either a custom printed circuit board (PCB) or a solderless breadboard must be used to interface it with the RasPi. I discuss how to connect the RasPi to the MCP3008 after the sidebar. I encourage you to read the sidebar if you are interested in how the MCP3008 accomplishes the ADC process.

Inner Workings of the MCP3008 ADC

I will refer to the MCP3008 functional block diagram shown in Figure 6-7 throughout this discussion. The analog signal is first selected from one of eight channels that may be connected to the input channel multiplexer. Using one channel at a time is called *operating in a single-ended mode*. The MCP3008 channels can be paired to operate in a *differential mode* if desired. A single configuration bit named *SGL/DIFF* selects single-ended or differential operating modes. Single-ended is the mode used in this project.

The selected channel is than routed to a sample and hold circuit that is one input to a comparator. The other input to the comparator is from a digital-to-analog converter (DAC) that receives its input from a 10-bit SAR.

Basically, the SAR starts at 0 and rapidly increments to a maximum of 1023, which is the largest number that can be represented with 10 bits. Now each increment increases the voltage

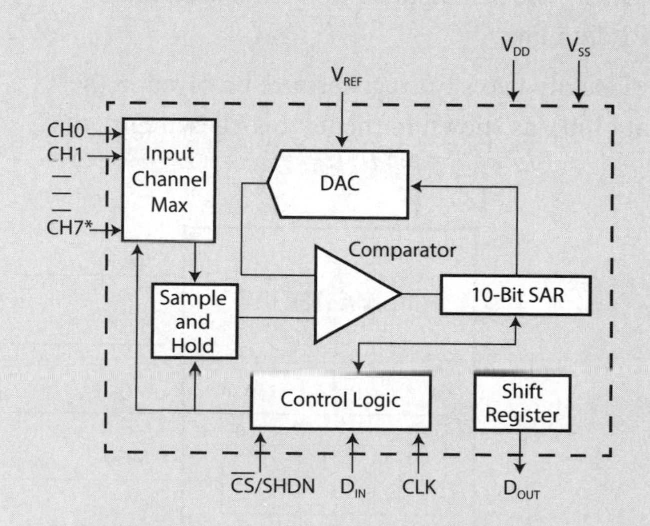

Figure 6-7 MCP3008 functional block diagram.

appearing at the DAC's comparator input. The comparator will trigger when the DAC voltage precisely equals the sampled voltage, and this will stop the SAR from incrementing. The digital number that exists on the SAR at the moment the comparator "trips" is the ADC value. This number is then outputted, one bit at a time, through the SPI circuit (discussed next). All this takes place between sample intervals. The actual voltage represented by the ADC value is a function of the reference voltage VREF connected to the MCP3008. In our case, VREF is 3.3 V; therefore, each bit represents 3.3/1024 or approximately 3.223 mV. For example, an ADC value of 500 would represent an actual voltage of 1.612 V, which was computed by multiplying 0.003223 by 500.

Serial Peripheral Interface

The SPI is one of several data communication channels that the RasPi supports. It is a synchronous serial data link that uses one master device and one or more slave devices. A minimum of four data lines are used with SPI, and Table 6-3 shows the names associated with the master (RasPi) and slave (MCP3008) devices.

TABLE 6-3 SPI Data-Line Descriptions

Master Device: RasPi	Slave Device: MCP3008	Remarks
SCLK	CLK	Clock
MOSI	Din	Master out, slave in
MISO	Dout	Master in, slave out
CS/SHDN	SS	Slave select

Figure 6-8 is a simplified block diagram showing the principal components used in an SPI data link.

Usually two shift registers are involved in the data link, as shown in the figure. These registers may be hardware or software depending on the devices involved. The RasPi implements its shift register in software, while the MCP3008 has a hardware shift register. In either case, the two shift registers form what is known as an *interchip circular buffer arrangement* that is the heart of the SPI.

Data communications are initiated by the master by first selecting the required slave. The RasPi selects the MCP3008 by bringing the SS line to a low state or 0 VDC. During each clock cycle, the master sends a bit to the slave, which reads it from the MOSI line. Concurrently, the slave sends a bit to the master, which reads it from the MISO line. This operation is known as *full-duplex communication*, that is, simultaneous reading and writing between master and slave.

The clock frequency used depends primarily on the slave's response speed. The MCP3008 can easily handle bit rates of up to 3.6 MHz if powered at 5 V. Because we are using 3.3 V, the

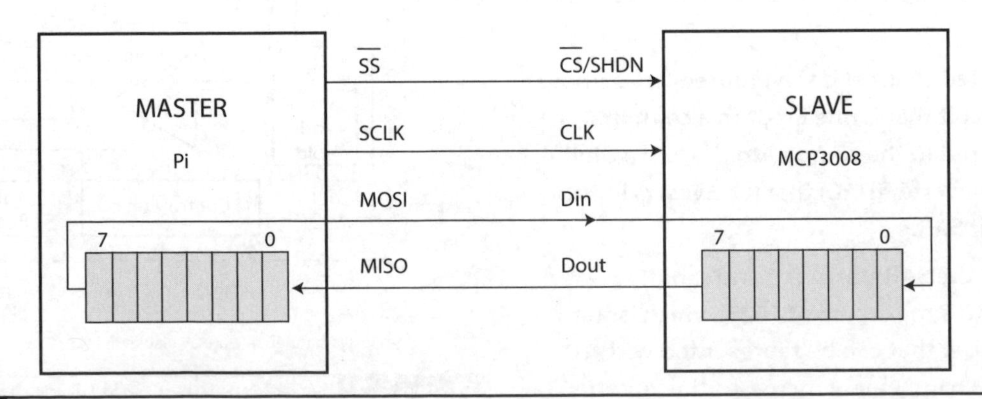

Figure 6-8 SPI simplified block diagram.

maximum rate is a bit less at approximately 2 MHz. This is still very fast and will process the RasPi input without losing any data.

The first clock pulse received by the MCP3008 with its CS held low and Din high constitutes the *start bit*. The SGL/DIFF bit follows next and then 3 bits that represent the selected channel(s). After these 5 bits have been received, the MCP3008 will sample the analog voltage during the next clock cycle.

The MCP3008 then outputs what is known as a *low null bit* that is disregarded by the RasPi. The following 10 bits, each sent on a clock cycle, are the ADC value with the most significant bit (MSB) sent first down to the least significant bit (LSB) sent last. The RasPi will then put the MCP3008 CS pin high, ending the ADC process.

Connecting and Testing the MCP3008 with the RasPi

The MCP3008 is connected to the RasPi using the Pi Cobbler prototype tool along with a solderless breadboard. The Pi Cobbler is available from a variety of sources, but it must be assembled, which will require some soldering. Instructions are available on the Adafruit website that show you, step-by-step, how to assembly the Pi Cobbler. Soldering is a fun activity, provided that you have the right equipment and skill. I recently acquired a comparatively inexpensive digitally controlled soldering workstation, which is shown in Figure 6-9. It may be set to precise temperatures that enable very nice solder joints to be made with ease and repeatability.

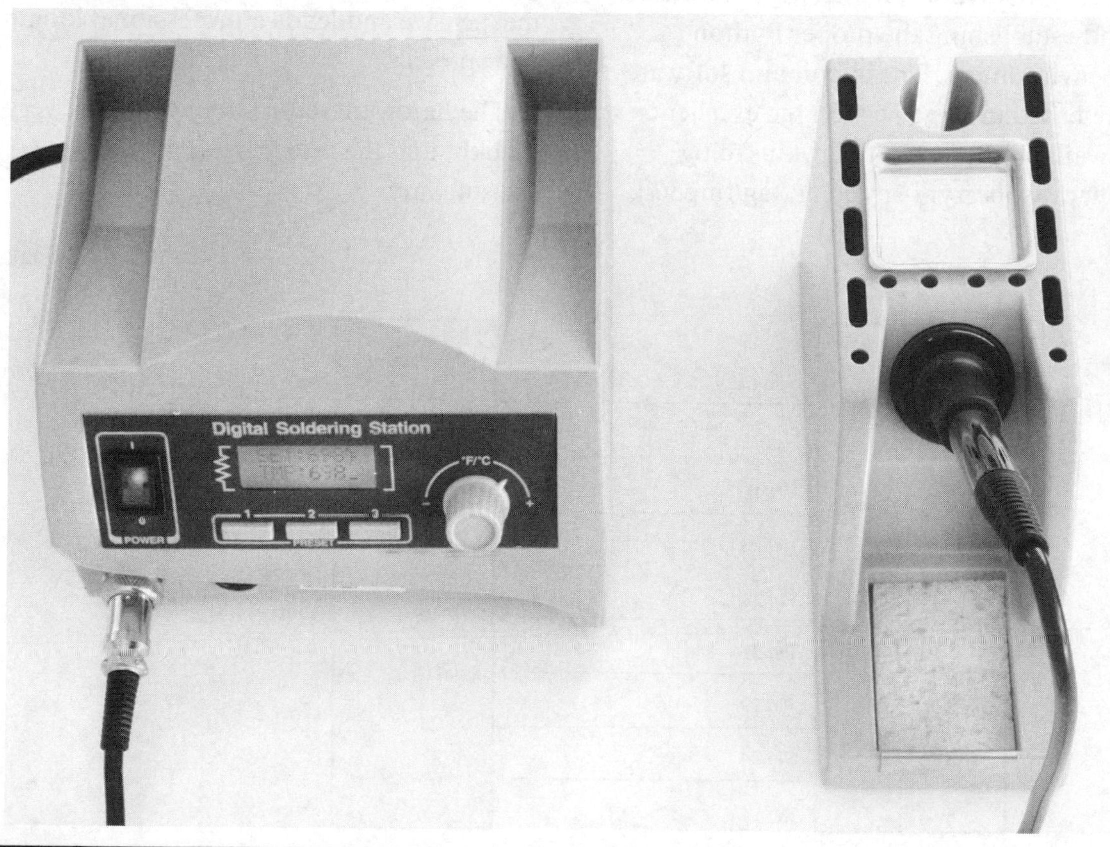

Figure 6-9 Digital soldering station.

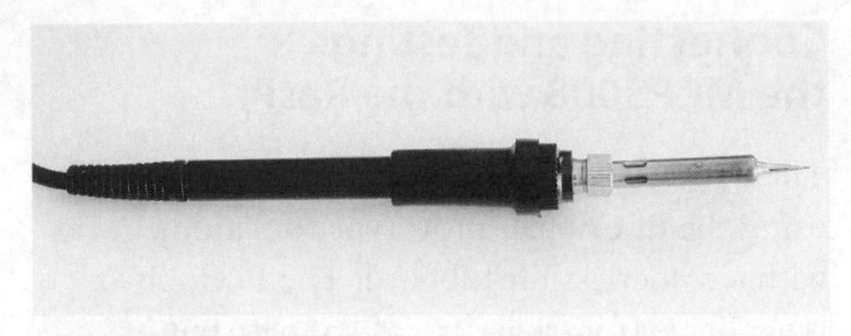

Figure 6-10 Sharp-pointed soldering iron.

Of course, the station is only as effective as the soldering iron that connects to it. Figure 6-10 shows the very sharply pointed soldering iron that came with the soldering station. The sharp point allows for some very closely spaced solder joints to be made while avoiding those troublesome solder bridges. I also used a 60/40 rosin-core solder, which I found to be very effective.

Initial Test

Initial testing involves both creating a hardware circuit and establishing the proper Python software environment. The circuit and software setups are based in large part on the excellent tutorial available from Matt Hawkins' blog (http://www.raspberrypi-spy.co.uk/tag/tmp36/), in which he discusses both the MCP3008 and the TMP36 sensor as well as the Python software.

Hardware Setup

I will first discuss the hardware circuit because it is relatively straightforward. Figure 6-11 shows the test schematic for the Pi Cobbler, MCP3008, and TMP36. I connected the TMP36 Vout lead to the MCP3008 Channel 0 input, which is pin 1.

The actual physical setup is shown in Figure 6-12. On the left side of the breadboard you can see the TMP36 sensor connected with three jumper wires to the breadboard. Incidentally, I find using commercial jumper wires very useful and more reliable than using homemade jumpers constructed from hookup wire. There is almost nothing more frustrating than to find that a poor wiring connection due to a broken jumper wire was responsible for a nonfunctioning circuit. Besides, a set of jumper wires is quite inexpensive and lends a professional look to your project.

The hardware setup should proceed very quickly, and the next portion of the test concerns the software.

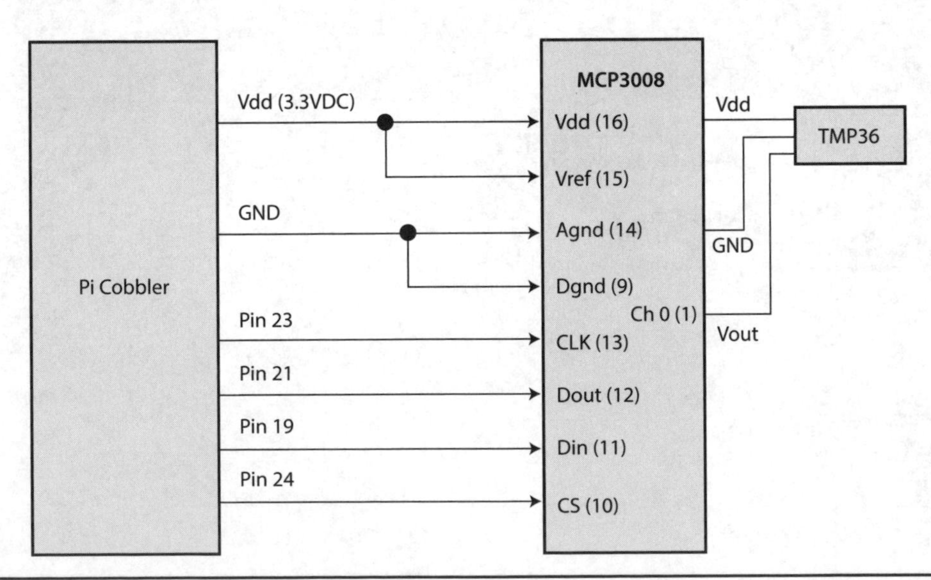

Figure 6-11 Test schematic.

Figure 6-12 Physical test setup.

Software Setup

The SPI hardware circuits that are part of the RasPi must be enabled before executing any code that relies on those circuits. Initially, you should check to determine whether the native SPI device is available. Enter the following, and check to see whether there is a `spi_bcm2708` in the list that is displayed:

```
lsmod
```

If there is, skip the next procedure; otherwise, edit the raspi-blacklist.conf as follows:

1. Enter
   a. `sudo nano /etc/modprobe.d/raspi-blacklist.conf`.
   b. Add the number symbol (#) in front of the line blacklist `spi-bcm2708`.
   c. Press CTRL-O to save and CTRL-X to exit the nano editor.
2. Enter
   a. `sudo nano /boot/config.txt`.
   b. Add this line to the bottom of the file: `dtparam=spi=on`.
   c. Press CTRL-O to save and CTRL-X to exit the nano editor.
3. Reboot the RasPi by entering **sudo reboot**.

Try the `lsmod` command again, and you should see the `spi-bcm2708` device listed.

You now need to load the Python libraries that will allow programs to be run by the SPI circuits you just enabled. First, install the Python development libraries by entering

```
sudo apt-get install python-dev
```

After this install finishes, you need to create a special directory in which to locate and run the SPI Python programs. From the Home directory, which should be at `/home/pi`, enter the following:

```
mkdir py-spidev
```

Next, change into the newly created directory:

```
cd py-spidev
```

Now download a Python script that will automatically create the necessary SPI development environment:

```
sudo wget https://raw.github.com/doceme/
py-spidev/master/setup.py
```

Next, download an additional file that is required before the setup can begin:

```
sudo wget https://raw.github.com/doceme/
py-spidev/master/spidev_module.c
```

Now run the script and create the SPI development environment by entering:

```
sudo python setup.py install
```

The following test program displays a continuous stream of temperature values generated by the TMP36 sensor. The program is named *TMPSensor.py* and is available for download on this book's companion website. The code follows the MCP3008 ADC configuration guidelines and SPI protocols, as discussed earlier.

```
#!/usr/bin/python

import spidev
import time
import os

#open the SPI bus
spi = spidev.SpiDev()
spi.open(0,0)

#define a function to read the MCP3008
ADC value
#channel must be an integer between 0
and 7
def ReadChannel(chan):
  if((chan < 0) or (chan > 7)):
    return -1
  adc =spi.xfer2([1, (8 + chan) << 4, 0])
  data = ((adc[1]&3) << 8) + adc[2]
  return data

#define a function that converts data to
voltage levels
#round to a specified number of decimal
places
def ConvertVolts(data, places):
  volts = (data * 3.3) / 1023
  volts = round(volts, places)
  return volts

#define a function to calc temperature
from TMP36 data
#round to a specified number of decimal
places
def ConvertTemp(data, places):

  #ADC Value Temp (°C) Volts
  # 0          -50         0.00
  # 78         -25         0.25
  # 155          0         0.50
  # 233         25         0.75
  # 310         50         1.00
  # 465        100         1.50
  # 775        200         2.50
  # 1023       280         3.30

  temp = ((data * 330)/1023) – 50
  temp = round(temp, places)
  return temp

#define temp channels
temp_chan0 = 0

#define time between readings
delay = 1

#print column headers
print "temp_level temp_volts temp"

#main loop
while True:
```

```
    #read the temp channel
    temp_level = ReadChannel(temp_chan0)
    temp_volts = ConvertVolts(temp_level, 2)
    temp = ConvertTemp(temp_level, 2)

    #display results
    print "--------------------------------"
    print temp_level, " ", temp_volts,
" ", temp

    #delay before taking next measurement
    time.sleep(delay)
```

Run this program by entering:

```
sudo python TMPSensor.py
```

Figure 6-13 is a screen shot of a portion of the program output with the TMP36 sensor measuring ambient room temperature.

Adding MQTT Features to the Application

You first need to load the appropriate MQTT client implementation library before adding the messaging features into the application. Please follow these steps to load the Python library that will be used in this project. Also note that I will be using the Debian commands in all the following instructions.

1. The Raspian Linux distribution must first be updated to ensure that all dependencies will be located in the appropriate repositories. Enter

```
apt-get update
```

2. Download the source code from GitHub using this command:

```
git clone git://git.eclipse.org/
gitroot/paho/org.eclipse.paho.mqtt.
python.git
```

NOTE **If you have difficulty in doing a direct git clone, you may also go to http://git.eclipse.org/c/paho/org.eclipse.paho.mqtt.python.git/ and download one of the following compressed files:**

org.eclipse.paho.mqtt.python-1.0.zip
org.eclipse.paho.mqtt.python-1.0.tar.gz
org.eclipse.paho.mqtt.python-1.0.tar.bz2

3. Use the extraction application that matches the compressed file extension you downloaded, that is, WinZip or 7Zip for the zip file. The same source directory should be created after extraction as was used for the clone operation.

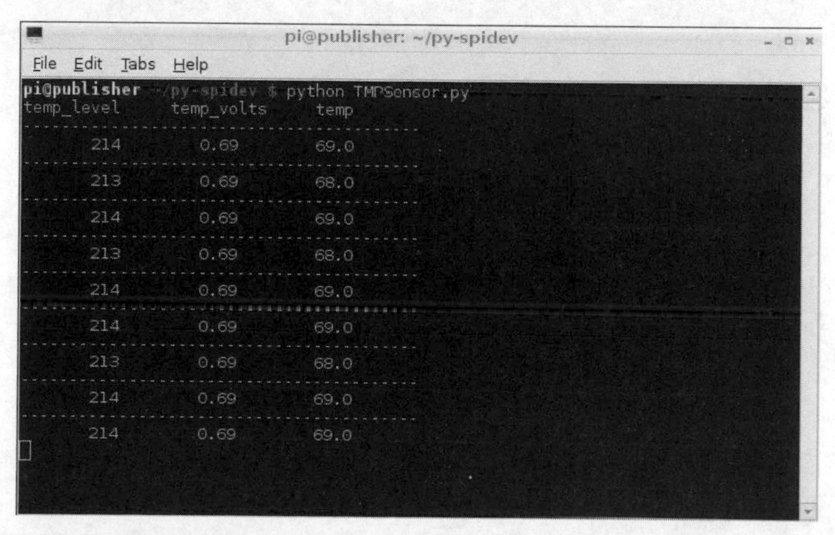

Figure 6-13 Initial test results.

4. Change into the source directory:

```
cd org.eclipse.paho.mqtt.python/
```

5. Compile the source code using a build script already available in the directory:

```
sudo make
```

6. Install all the compiled files:

```
sudo make install
```

The Python MQTT client should now be ready to be added to the TMPSensor.py program. However, I will first cover some basic concepts that you should have clear in your mind before going on to the complete application.

The publisher client must establish a logical connection to the broker before any messages can be passed. This is done with the following statements: `import paho.mqtt.client as mqtt` sets up a client reference named *mqtt*; `mqttc = mqtt.Client()` instantiates an MQTT client object named *mqttc*; and `mqttc.connect("m2m.eclipse.org", 1883, 60)` goes out to the Internet and connects with an MQTT broker at the website m2m.eclipse.org on port 1883. The 60 refers to a 60-second ping, which is a *keep-alive*, meaning it is sent when no other activity is happening on the connection. `mqttc.loop_start()` starts a separate execution thread that handles incoming messages from the broker.

The following two statements contain references to what are known as *topics* and *subtopics*:

```
mqttc.publish("raspiexample123/tmp36/
v","%.2f" % temp_volts);

mqttc.publish("raspiexample123/tmp36/
f",""%.2f" % temp);
```

In these statements, `"raspiexample123"` refers to a root topic created on the broker, which also contains the subtopics `"tmp36"`, `"v"`, and `"f"`. Real-time millivolt data are stored in the `"v"` subtopic, while real-time Fahrenheit temperature data are stored in the `"f"` subtopic. I will shortly demonstrate how to retrieve these real-time data from the broker. But first you should enter the following modified TMPSensor.py program, which I named *mqttTMPSensor.py* to reflect the new messaging capabilities:

```
#!/usr/bin/python
import spidev
import time
import os
import paho.mqtt.client as mqtt
#instantiate a mqqt object and setup the connection to the broker
mqttc = mqtt.Client()
mqttc.connect("m2m.eclipse.org", 1883, 60)
mqttc.loop_start()
#open the SPI bus
spi = spidev.SpiDev()
spi.open(0,0)

#define a function to read the MCP3008 ADC value
#channel must be an integer between 0 and 7
def ReadChannel(chan):
  if((chan < 0) or (chan > 7)):
    return -1
```

```
   adc =spi.xfer2([1, (8 + chan) << 4, 0])
   data = ((adc[1]&3) << 8) + adc[2]
   return data

 #define a function that converts data to voltage levels
 #round to a specified number of decimal places
 def ConvertVolts(data, places):
   volts = (data * 3.3) / 1023
   volts = round(volts, places)
   return volts

 #define a function to calc temperature from TMP36 data
 #round to a specified number of decimal places
 def ConvertTemp(data, places):

   #ADC Value Temp (°C) Volts
   # 0         -50         0.00
   # 78        -25         0.25
   # 155        0          0.50
   # 233        25         0.75
   # 310        50         1.00
   # 465        100        1.50
   # 775        200        2.50
   # 1023       280        3.30

   temp = ((data * 330)/1023) – 50
   temp = round(temp, places)
   return temp

 #define temp channels
 temp_chan0 = 0

 #define time between readings
 delay = 5

 #print column headers
 print "temp_level temp_volts temp"

 #main loop
 while True:
   #read the temp channel
   temp_level = ReadChannel(temp_chan0)
   temp_volts = ConvertVolts(temp_level, 2)
   temp = ConvertTemp(temp_level, 2)

   #publish to the broker

   mqttc.publish("raspiexample123/tmp36/v","%.2f"%
 temp_volts);
```

```
mqttc.publish("raspiexample123/tmp36/f",""%.2f"%temp);

#display results
print "-----------------------------------"
print temp_level, " ", temp_volts, " ",
temp, " mqtt"

#delay before taking next measurement
time.sleep(delay)
```

The program is run by entering:

```
python mqttTMPSensor.py
```

You should see exactly the same terminal display that was shown when the TMPSensor.py program was run in the earlier test, except that I added the word `mqtt` to the end of each display line to help me distinguish between the two program outputs. Figure 6-14 is a terminal screen shot for this program.

The data in the mqttTMPSensor.py program are also being sent to the broker located at m2m.eclipse.org and listening on port 1883. I believe that some discussion at this point regarding the broker website would be helpful for your overall understanding of the role that the MQTT broker plays in this messaging scheme.

MQTT Brokers

The web server located at m2m.eclipse.org is a public sandbox hosted by the Eclipse Foundation as part of its open-source IoT project. This web server's software itself is based on the Mosquito Project created and maintained by Roger Light, a highly talented UK developer. The sandbox server allows free public access to an actual MQTT broker where developers may test their software. There are no restrictions at this site, and just as in an African waterhole, all are welcome to use it, but beware of any predators that might be lurking nearby. This metaphor means that your data, which are being sent to the broker, can be accessed by anyone who is also concurrently on the site. This usually

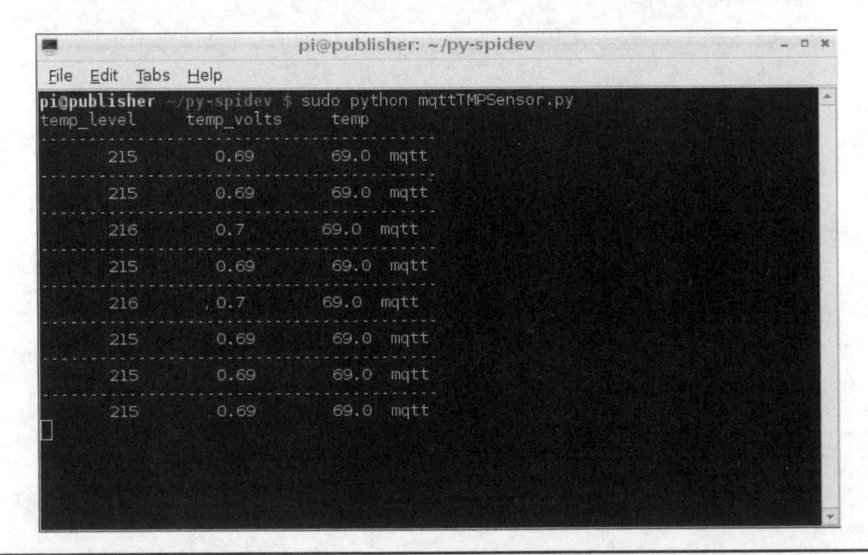

Figure 6-14 Terminal screen shot showing the mqttTMPSensor.py program output.

is not a problem because most developers are typically well behaved.

There are a number of other freely available MQTT brokers beside m2m.eclipse.org. Table 6-4 lists all the brokers that were reported as available at the time of this writing. All offer standard MQTT broker support, while some provide additional services such as SSL, a dashboard, or an HTTP bridge, as noted in the "Remarks" column.

The HTTP bridge is one of the features in the m2m.eclipse.org broker that will allow us to check whether the mqttTMPSensor application data are actually being received by the broker. To use the HTTP bridge, first, ensure that the mqttTMPSensor publisher client is running, and then, using a browser either on the RasPi or another computer, go to this website: http://eclipse.mqttbridge.com/raspiexample123/tmp/f. Once in the website, you should see only a single number such as 70, which represents a temperature reading taken from the TMP36 sensor. Figure 6-15 is a screen shot of the HTTP bridge website while I was running the mqttTMPSensor application.

You may have noticed that the order in which the HTTP bridge URL was constructed specifies

Figure 6-15 HTTP bridge screen shot for the `f` subtopic.

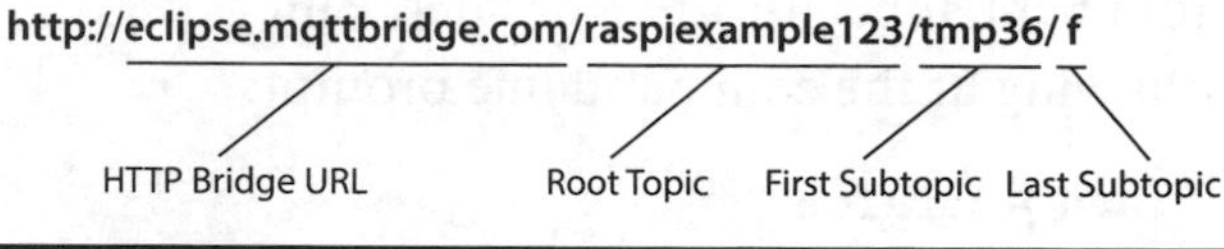

Figure 6-16 HTTP bridge URL with topics.

the root topic and all the branch subtopics leading to the desired one to be displayed, as detailed in Figure 6-16.

Going to the website http://eclipse.mqttbridge.com/raspiexample123/tmp36/f will allow you to retrieve the temperature data because the final subtopic is `f`, which matches the published subtopic descriptor. You should examine the publish statement in the mqttTMPSensor.py code listing to confirm this fact.

It is time to examine the RasPi client now that the publisher client is fully operational.

TABLE 6-4 Public MQTT Brokers

Address	Ports	Additional Services	Remarks
m2m.eclipse.org	1883, 80	HTTP bridge	Xively stats, topics, Mosquito info web page
test.mosquitto.org	1883, 8883 (SSL), 8884 (SSL), 80	HTTP bridge	Xively stats, topics, Mosquito info web page, SSL support
dev.rabbitmq.com	1883	Dashboard	
broker.mqttdashboard.com	1883, 8000	Dashboard	Stats, HiveMQ info web page, SSL not yet avail
q.m2m.io	1883		Requires registration before use
www.cloudmqtt.com	18443, 28443 (SSL)		Mosquito info web page, SSL support; requires registration; paid site but there is a free plan

RasPi Subscriber Client

The MQTT subscriber client will be implemented in Java rather than Python, which emphasizes the MQTT platform agnostic approach. You should first ensure that Oracle's Java JDK is already installed on the RasPi. It should be if you are using a Wheezy distribution from September 13, 2013 or later. Enter the following at the command line prompt:

```
java –version
```

Figure 6-17 is the result of this command line query as to the Java version installed on the RasPi. Your version may very well be different because upgraded Java versions are likely to be included in the Wheezy distribution with time.

Using a callback method is key to how the MQTT subscriber client functions. A *callback method* is one that is triggered by an *event*, which is the arrival of a message at the broker for this situation. Callback methods are specified in the MQTTCallback interface, which is implemented by this subscriber client class named PahoMqttSubscribe. The following statement shows the class declaration along with the interface implementation:

```
public class PahoMqttSubscribe
implements MqttCallback
```

The client class also requires a supporting library, which is in the form of a Java archive file named mqtt-client-0.4.0.jar. This jar file will need to be downloaded from the Eclipse.org website. Instructions on how to download it will be discussed in another section. This statement is the import for the MQTT client library:

```
import org.eclipse.paho.client.mqttv3.*;
```

An empty client constructor and a reference to the client are created by these statements:

```
public PahoMqttSubscribe() { }
MqttClient client;
```

The following main method contains only one method call in this minimal demo project. This method call also incorporates an instantiation of the PahoMqttSubscribe class.

```
public static void main(String[] args) {
 new PahoMqttSubscribe().doDemo();
}
```

The `doDemo()` method call made in the main method is where the application forever loop is located. The first action that happens when this method is first entered is an instantiation of the client object, which is directed to the desired broker website. A connect command follows the instantiation:

```
client = new MqttClient("tcp://m2m
.eclipse.org:1883,
MqttClient.generateClientId());
client.connect();
```

Notice that one of the arguments in the instantiation statement sets up a unique client ID. All clients connecting to a broker require a unique ID, which is typically constructed from metadata elements that the broker can discern from the initial TCP connection.

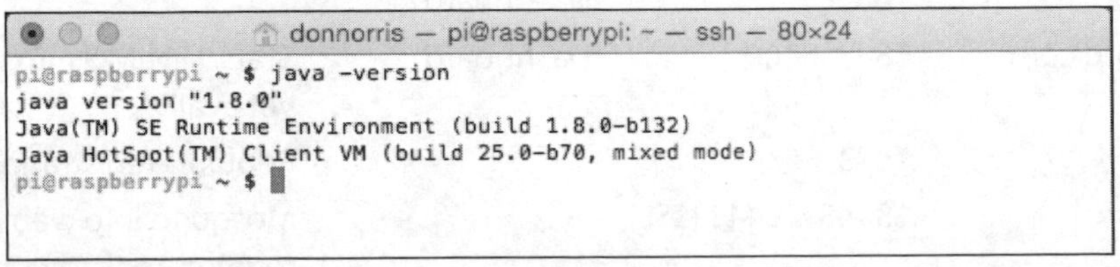

```
donnorris — pi@raspberrypi: ~ — ssh — 80×24
pi@raspberrypi ~ $ java -version
java version "1.8.0"
Java(TM) SE Runtime Environment (build 1.8.0-b132)
Java HotSpot(TM) Client VM (build 25.0-b70, mixed mode)
pi@raspberrypi ~ $
```

Figure 6-17 Java version query.

The next step in the doDemo method is to establish the callback method, which I mentioned previously. This method will be called when a message to which the client is subscribed is received by the broker:

```
client.setCallback(this);
```

The client must next inform the broker which topic it desires to subscribe to:

```
client.subscribe("raspiexample123/
tmp36/f");
```

Additional actions are normally added after the subscribe statement. In this minimal demo, there is only a forever loop containing a 1-second sleep statement. The sleep statement is contained by try/catch statements, which are needed for this execution sequence. Obviously, real-time control application statements also would be placed here as desired.

```
while(true) {
  try {
  Thread.sleep(1000):
 }
  catch(InterruptedException e) { }
}
```

The remaining item that is missing in this class definition is the actual callback method. This method is named messageArrived, and it takes two arguments, a String for the topic and an MqttMessage type representing the subtopic value. The MqttMessage type value is also known as the *payload*. The only action that the callback method will perform in this demo is to print the topic and the payload.

```
public void messageArrived(String topic,
MqttMessage message) throws Exception {
  System.out.println(topic + " " + new
String(message.getPayload()));
}
```

A pro-forma action is also required for this class definition to be complete and able to be compiled. Because the PahoMqttSubscribe class implements the MqttCallback interface, it is required to provide an implementation for all the methods specified by the interface. One method, messageArrived has already been implemented. There are two other methods that must be implemented. These are shown next as empty or null implementations.

```
public void connectionLost(Throwable
cause) { }
public void deliveryComplete(IMqttDelive
ryToken token) { }
```

The Java MQTT API does contain applications that will provide real implementations for the preceding callback methods. They are not needed in this demo, but you should know that they are available.

All this code is shown next as a complete class definition named PahoMqttSubscribe.java. You should use the nano editor to enter it or download it from this book's companion website.

```
import org.eclipse.paho.client.mqttv3.*;

public class PahoMqttSubscribe implements MqttCallback {

  MqttClient client;

public PahoMqttSubscribe() {}

public void messageArrived(String topic, MqttMessage message) throws Exception
{
```

```
    System.out.println( topic + " " + new String(message.getPayload()));
  }

  public void connectionLost(Throwable cause) {}
  public void deliveryComplete(IMqttDeliveryToken token) {}

  public static void main(String[] args) {
    new PahoMqttSubscribe().doDemo();
  }

  public void doDemo() {
    try {
      client = new MqttClient("tcp://m2m.eclipse.org:1883", MqttClient.
generateClientId());
      client.connect();
      client.setCallback(this);

      client.subscribe("raspiexample123/tmp36/f");

      while(true) {
        try { Thread.sleep(1000); }
        catch(InterruptedException e) {}
      }
    }
    catch(MqttException e) { e.printStackTrace(); }
  } // end of doDemo()
  } // end of class def
```

Do not compile the code at this point because it will not work owing to the missing jar file. Please use the following steps to download the jar, compile the source file, and execute the class file:

1. Enter the following while in a RasPi terminal window:

```
curl -O https://repo.eclipse.org/
content/repositories/paho-releases/
org/eclipse/paho/
mqtt-client/0.4.0/mqtt-client-
0.4.0.jar
```

2. Ensure that the jar file is in the same directory as the class source file PahoMqttSubscribe.java. Enter the following to compile the source file:

```
javac -cp mqtt-client-0.4.0.jar
PahoMqttSubscribe.java
```

3. Note that the `-cp` option in the command signifies the class path to use to find the required library file, which is mqtt-client -0.4.0.jar for this case. In addition, the library file also must be located in the current file, which also holds the source file, because there is no directory prefixed to the library file. You will shortly see a case where multiple libraries are involved, which will require an expanded class path specification.

4. Ensure that the RasPi publisher is running the mqttTMPSensor.py program. Also check that you are using the m2m.eclipse.org

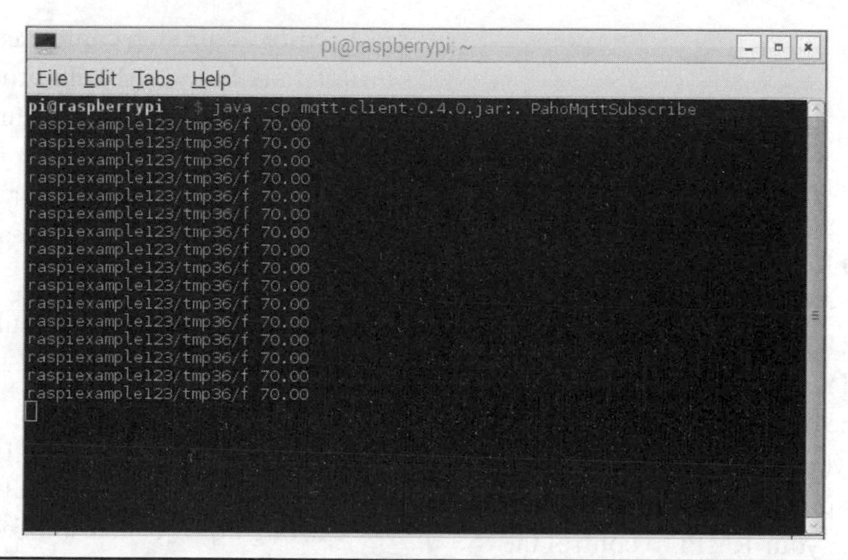

Figure 6-18 RasPi subscriber client terminal screen shot.

broker. Enter the following command to execute the class file:

```
java -cp mqtt-client-0.4.0.jar:.
PahoMqttSubscribe
```

NOTE Don't forget to enter the semicolon and period that follow the .jar extension. The program will not run unless you have those in the command.

Figure 6-18 is a screen shot taken from the RasPi terminal window showing the data streaming from the MQTT broker.

This last step completes the initial M2M demonstration project. To recap, I showed you how to first set up a RasPi Model B+ as a publisher client, which streamed temperature data to a MQTT broker. The RasPi was running a Python program for this function. I next showed you how to set up a RasPi 2 Model B as a subscriber client using a Java program. This subscriber RasPi was connected to the same MQTT broker as the publisher RasPi and thus was able to receive the data messages from that RasPi via the broker. This was made possible by a MQTT callback method named `messageArrived`. The next part of this M2M demonstration project is to expand the subscriber client Java class such that it can undertake some automatic actions based on the received data messages.

MQTT Two-Phase Thermostat

The two-phase thermostat in this section's title refers to a unit that can either start heating or start cooling depending on the measured temperature in the monitored space. In this section, I will show you how to establish two setpoints that will cause cooling, heating, or no action based on the received MQTT temperature data. The `PahoMqttSubscribe` Java class will be modified to incorporate this new controller application. I also renamed the class as `PahoMqttSubscribe1` to distinguish it from the original, noncontroller version. Another major change to the original class that's needed is to incorporate GPIO to control LEDs that will indicate the thermostat's state. In the next section, I will show how to install the Pi4J library that will enable GPIO control using Java.

Pi4J Library

The Raspberry Pi development community is quite fortunate to have a talented developer named Robert Savage who freely made available a fairly complete Java class library that implements GPIO functionality. This library includes both high-level application-type classes and many low-level driver classes. The library is named Pi4J and is available for download at www.pi4j.com. The download and installation of this library on your RasPi are crucial to the success of this project. Please follow this procedure to set up your RasPi to control the GPIO pins using Java:

1. The first step is to download the Pi4J library. I found that the simplest way to do this was to first download the SNAPSHOT release named pi4j-1.1-SNAPSHOT.deb onto my laptop from the website https://code.google.com/p/pi4j/downloads/list. You can then copy it into the RasPi's home directory using a thumb drive and the RasPi's File Manager application or use the `scp` utility and transfer it directly from the laptop.
2. Once it is in the Home Directory, enter the following command to install this SNAPSHOT into the appropriate locations on the RasPi:

```
sudo dpkg –i pi4j-1.1-SNAPSHOT.deb
```

NOTE **dpkg is a package manager application designed to unpack and install Debian-formatted packages, that is, software packages with a .deb file extension.**

3. After the installation is completed, a new directory (pi4j) will have been created with two new subdirectories (lib and examples) within it as follows:
   a. /opt/pi4j/lib
   b. /opt/pi4j/examples

The preceding step completed the Pi4J installation, but you should proceed with the next few steps to create all the needed class files and be ready to run the example program, which will, in turn, confirm that the library functions as expected and is usable for program development.

Change into the examples subdirectory by entering the following:

```
cd /opt/pi4j/examples
```

Once in the directory, enter the following, which automatically builds all the class files from the existing downloaded source files:

```
./build
```

There were 34 example source files in the download that I made. That number is subject to change as the developers who control the website add and subtract depending on comments received from the active Pi4J community.

The `./build` command causes a script to run that iterates through all the example source code files to produce corresponding class files. The actual compile command is shown next, and you must use it to compile your own source file:

```
sudo javac –classpath .:classes:/opt/pi4j/lib/'*' –d . <sourcefilename>
```

It is very important that you pay attention to all the symbols and whitespace in this command because leaving anything out or misaligning will cause the compile to fail, as I found out much to my frustration.

Enter the following to execute or run a class file:

```
sudo java –classpath .:/classes:/opt/pi4j/lib/'*' <classfilename>
```

I would strongly suggest that you try to compile and run one of the example programs named BlinkGpioExample.java that are near the top of the example directory file list. But first, I do need to explain how the GPIO pins are identified within this program because this will have a direct impact on how you connect the LEDs when using the `PahoMqttSubscribe1` program. The Pi4J library is based in large part on another RasPi framework known as WiringPi. WiringPi is a great programming tool, but unfortunately, it added its own way of identifying the GPIO pins, which, as many readers will know, is already a bit confusing because there are two other ways of identifying the same pins. The two ways are the approach taken by Broadcom, the chip manufacturer, and the Raspberry Pi Foundation's labeling. Incidentally, the manufacturer is usually identified using the initials *bcm*. To help you clarify this confusion, I have included Figure 6-19, which is a good reference for the different GPIO pin labels.

This figure relates to both the Model B's 26-pin P1 connector and the Model B+'s 40-pin J8 GPIO connector because the first 26 pins on the Model B identically match the Model B+'s pins. The shaded area on the 40-pin connector identifies this overlap.

The following code snippet was excerpted from the `BlinkGpioExample` program to show you a specific instance of this pin labeling:

```
final GpioPinDigitalOutput led1 =
gpio.provisionDigitalOutputPin(RaspiPin
.GPIO_01);
final GpioPinDigitalOutput led2 =
gpio.provisionDigitalOutputPin(RaspiPin
.GPIO_03);
```

Don't worry, I am not going through this code but will focus only on the two pin labels shown in the code, namely, `GPIO_01` and `GPIO_03`. If you refer to Figure 6-19, you will see the entries GPIO_GEN0 and GPIO_GEN3 are on the same lines as GPIO17 and GPIO22, respectively. GPIO_GEN0 is `GPIO_01` and GPIO_GEN3 is `GPIO_03`. They are also connected to physical

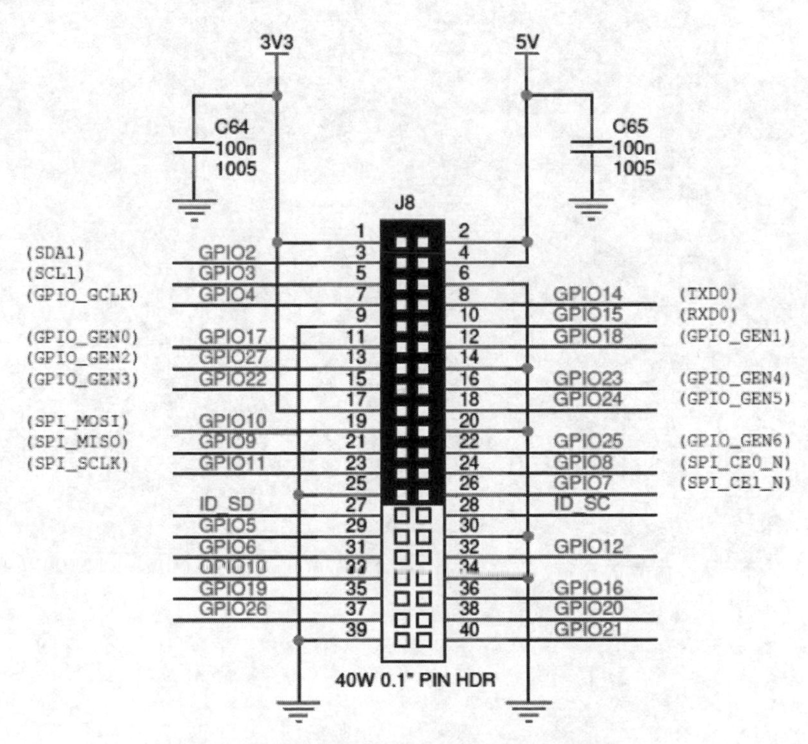

Figure 6-19 GPIO pin cross-reference.

pin numbers 11 and 15, respectively. I realize that this is all as clear as mud, but I can assure you that as you continue to work with the RasPi, the pin labeling will become second nature. Just be careful when connecting devices to the RasPi.

Figure 6-20 shows a physical setup in which I connected a LED with a series with a 220-Ω resistor to pin 15 or GPIO03. Note that I used two female jumper wires to make the connections between the GPIO03 and ground pins. I could have used a Pi Cobbler, but sometimes it is just as easy to "do it quick and dirty."

This LED connection allowed me to check that the `BlinkGpioExample` program functioned as expected. You will see some print statements appear on the terminal screen when the program runs, but it is always comforting to see an actual LED blink.

You will first need to compile the `BlinkGpioExample` program by entering the following:

```
javac -classpath .:classes:/opt/pi4j/
lib/'*' BlinkGpioExample.java
```

This will create a class file name BlinkGpioExample.class in the current directory.

Next, enter this to run the class file:

```
java -classpath .:classes:/opt/pi4j/
lib/'*' BlinkGpioExample
```

Note that you should not enter the .class extension in the class file name because that will cause an error.

Figure 6-20 LED with a series resistor connected to `GPIO_03`.

The next section shows the modifications I made to the `PahoMqttSubscribe` file to incorporate the control functions.

Modified Subscriber Client File with Added Control Functions

The modified Java class is now named `PahoMqttSubscribe1` and is listed next:

```
import org.eclipse.paho.client.mqttv3.*;
import com.pi4j.io.gpio.GpioController;
import com.pi4j.io.gpio.GpioFactory;
import com.pi4j.io.gpio.GpioPinDigitalOutput;
import com.pi4j.io.gpio.RaspiPin;

public class PahoMqttSubscribe1 implements MqttCallback
{

  MqttClient client;

  final static GpioController gpio = GpioFactory.getInstance();
  final static GpioPinDigitalOutput pinH =
  gpio.provisionDigitalOutputPin(RaspiPin.GPIO_01,"PinH");
  final static GpioPinDigitalOutput pinL =
  gpio.provisionDigitalOutputPin(RaspiPin.GPIO_03,"PinL");


public PahoMqttSubscribe1() { }

public void messageArrived(String topic, MqttMessage message) throws Exception
{
  String msg = new String(message.getPayload());
  System.out.println( topic + " " + msg );
  Double dValue = Double.parseDouble(msg);
  int iValue = dValue.intValue();
  if(iValue >= 80) {
    pinH.high();
   pinL.low();
    System.out.println("Above 80 F");
  }
  if(iValue <= 60) {
    pinH.low();
    pinL.high();
    System.out.println("Below 60 F");
  }
  if(iValue > 60 && iValue < 80) {
    pinH.low();
    pinL.low();
    System.out.println("Between 60 and 80 F");
  }
}
```

```
public void connectionLost(Throwable cause) {}
public void deliveryComplete(IMqttDeliveryToken token) {}

public static void main(String[] args) {
  new PahoMqttSubscribe1().doDemo();
}

public void doDemo() {

  try {
    client = new MqttClient("tcp://m2m.eclipse.org:1883",
MqttClient.generateClientId());
    client.connect();
    client.setCallback(this);

    client.subscribe("raspiexample123/tmp36/f");

    while(true) {
      try { Thread.sleep(1000); }
      catch(InterruptedException e) {}
    }
  }
  catch(MqttException e) { e.printStackTrace(); }
} // end of doDemo()
} // end of class def
```

The changes to the original Java class have been enclosed in boxes to help you identify the new additions.

I have really only added some new functionality to the `messageArrived` method in which the payload value is compared with two preset values to determine which GPIO pins are set to a high value. The logic is simple: if the payload value is higher than 80, turn on `GPIO_01`, which theoretically could be connected to a relay module controlling an air conditioner for cooling purposes. If the payload value is lower than 60, turn on `GPIO_03`, which likewise is theoretically connected to a relay module controlling a heater. Of course, if the temperature is between 60 and 80, do nothing because this is the desired comfort zone. The setpoints of 60 and 80 are purely arbitrary, but I did need some concrete values to test the system.

Enter the following command to compile the `PahoMqttSubscribe1` source file:

```
javac -cp mqtt-client-0.4.0.jar:/opt/
pi4j/lib/'*' PahoMqttSubscribe1.java
```

Enter this command to run this class file:

```
java -cp mqtt-client-0.4.0-jar:/opt/
pi4j/lib/'*':. PahoMqttSubscribe1
```

Again, play particular attention to the colon and period that are part of this command.

You are now ready to test this thermostat application.

Two-Phase Thermostat Test

I changed the earlier circuit connected to the publisher client RasPi to facilitate this two-phase thermostat test. The TMP36 sensor was replaced with a 10-kΩ potentiometer connected between 3.3 V and ground. The potentiometer tap was connected to the ADC channel 1, as can be seen in the schematic in Figure 6-21. I used the simple LED circuit I showed you earlier and alternately attached it to `GPIO_01` and `GPIO_03`, as shown in the figure.

Replacing the sensor in this way allowed me to quickly change the input voltage such that I could set it at 0.77 V to simulate 81°F or 0.65 V to simulate 59°F, where either temperature will trigger a state change. If you examine the code listing, you will see that I added some `println` statements in the control logic that allowed me to display the GPIO control states on the subscriber client RasPis terminal window. Figure 6-22 shows the system terminal display while in operation and the state changes as I set the potentiometer to a voltage simulating 81°F. Incidentally, I used a volt-ohmmeter (VOM) to measure the voltage going into channel 1 of the ADC. I also changed the measurement interval to 5 seconds to give myself enough time to change the simulated temperature without the intentionally changed temperatures scrolling off screen. The LED connected to `GPIO_01` also switched on to show that it was entering the cooling state.

Figure 6-23 shows another system terminal display where I set the potentiometer to a voltage simulating 58°F. The LED connected to `GPIO_03` also switched on to show that it was entering the heating state.

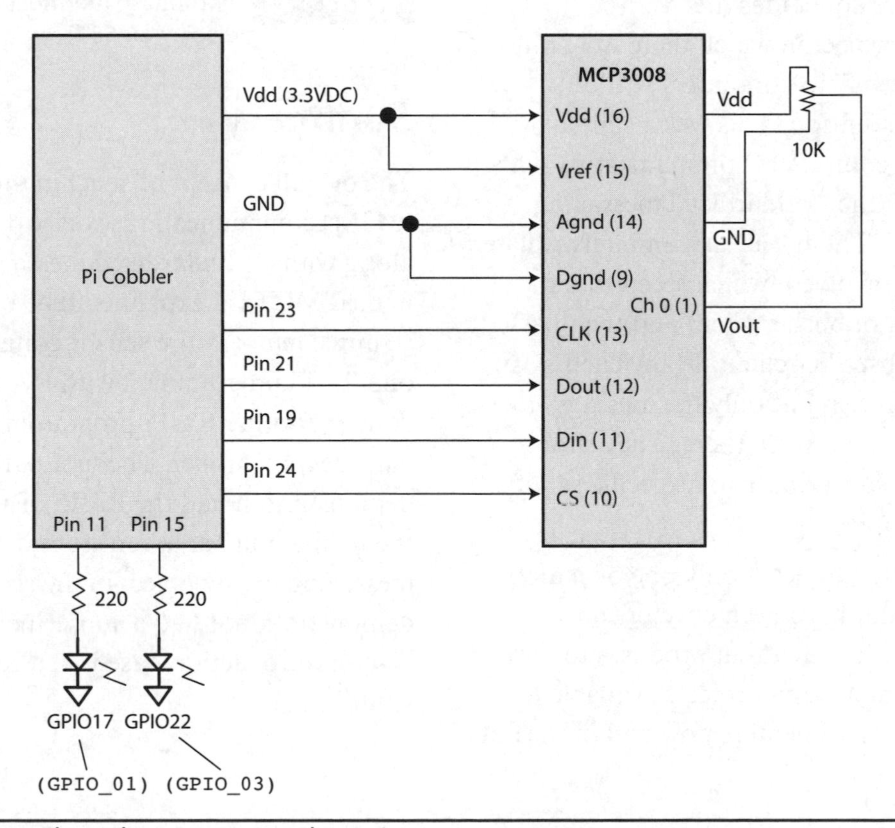

Figure 6-21 Two-phase thermostat test schematic.

```
raspiexample123/tmp36/f 81.00
Above 80 F
raspiexample123/tmp36/f 81.00
Above 80 F
raspiexample123/tmp36/f 81.00
Above 80 F
raspiexample123/tmp36/f 81.00
Above 80 F
raspiexample123/tmp36/f 81.00
Above 80 F
raspiexample123/tmp36/f 81.00
Above 80 F
raspiexample123/tmp36/f 81.00
Above 80 F
raspiexample123/tmp36/f 81.00
Above 80 F
raspiexample123/tmp36/f 81.00
Above 80 F
raspiexample123/tmp36/f 81.00
Above 80 F
raspiexample123/tmp36/f 81.00
Above 80 F
raspiexample123/tmp36/f 81.00
Above 80 F
```

Figure 6-22 Two-phase thermostat terminal display for 81°F.

This last figure concludes this M2M demonstration project in which there were only computers "talking" to computers without any human intervention. This was a simple example of two computers communicating with each other using the standardized messaging MQTT protocol. There was also an intermediate message broker involved, which received data messages from a publisher client and then passed them on to a subscriber client. Many clients can subscribe to a broker, but only the messages they are interested in are sent to them. They show their interest by subscribing to a specific set of topics and subtopics.

This messaging project is only one of many M2M projects that have been developed to date. It is an exciting area that promises to have many new and interesting projects available for developers and experimenters now and in the not too distant future.

```
raspiexample123/tmp36/f 58.00
Below 60 F
raspiexample123/tmp36/f 58.00
Below 60 F
raspiexample123/tmp36/f 58.00
Below 60 F
raspiexample123/tmp36/f 58.00
Below 60 F
raspiexample123/tmp36/f 58.00
Below 60 F
raspiexample123/tmp36/f 58.00
Below 60 F
raspiexample123/tmp36/f 58.00
Below 60 F
raspiexample123/tmp36/f 58.00
Below 60 F
raspiexample123/tmp36/f 58.00
Below 60 F
raspiexample123/tmp36/f 58.00
Below 60 F
raspiexample123/tmp36/f 58.00
Below 60 F
raspiexample123/tmp36/f 58.00
Below 60 F
raspiexample123/tmp36/f 58.00
Below 60 F
raspiexample123/tmp36/f 58.00
Below 60 F
```

Figure 6-23 Two-phase thermostat terminal display for 58°F.

Summary

The overall concept of machine-to-machine (M2M) communications was introduced initially along with a standardized message protocol named MQTT. I explained that I would use a simple temperature sensor connected to one RasPi programmed with Python to send data to another RasPi programmed with Java via a MQTT broker. The first part of the demonstration had the RasPi displaying only the temperature data sent to it without any messaging taking place. In the second part of the demonstration, I had a subscriber RasPi execute some control action based on a received data value.

CHAPTER 7

Software-Defined Radio

SOFTWARE-DEFINED RADIO (SDR) is a relatively new technology but one that many people use daily but are not aware of. SDR is a key component of most modern cell phones. If SDR did not exist, then the same would likely be true of cell phones and a lot more technology on which modern societies depend. This chapter's purpose is to make you aware of the basic SDR functions through some fun experimental projects using a RasPi and a few inexpensive SDR parts. However, before I launch into the projects, I believe a brief background discussion on how SDR functions should enrich your experience.

Basic Radio Concepts

Most readers are already likely familiar with these basic concepts, but a quick review never hurts. All radio communications rely on a carrier wave, which is typically a constant-frequency, constant-amplitude electromagnetic radiofrequency (RF) wave (shown graphically in Figure 7-1).

The waveform shown in the figure is also called a *continuous wave* (CW), and it "carries"

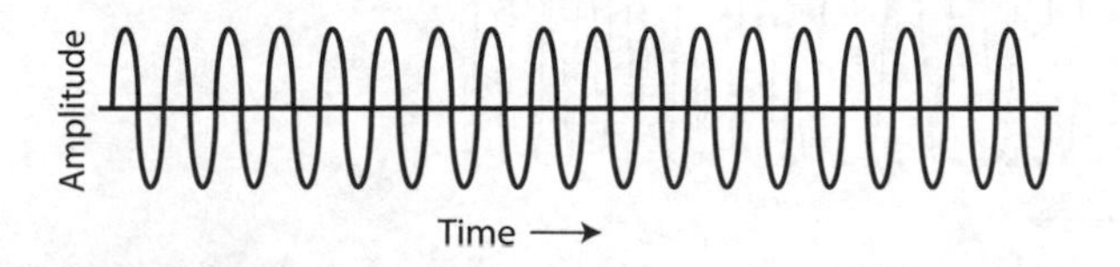

Figure 7-1 Radio carrier wave.

information via a process known as *modulation*. Technically, the CW carrier can transmit information alone simply by being switched on and off in a predefined manner representing the information or data. This operational mode is commonly known as *Morse code* or the more familiar "dit-dahs" associated with amateur radio operations.

Chapter 7 Parts List

Item	Model	Quantity	Source
RTL-SDR dongle R820T2 RTL2832U	Generic	1	amazon.com
Raspberry Pi 2 Model B	83-16530	1	mcmelectronics .com
PiTFT		1	See Chapter 2
5-V cell phone external battery	Generic	1	amazon.com

Three very common modulation techniques are used in RF communications:

- **Amplitude modulation (AM).** This modulation method changes the amplitude of the RF carrier in direct correspondence to the data being transmitted.
- **Frequency modulation (FM).** This modulation method changes the

instantaneous frequency of the RF carrier in direct correspondence to the data being transmitted.

- **Phase modulation (PM).** This modulation method changes the instantaneous phase of the RF carrier in direct correspondence to the data being transmitted.

There are many other modulation schemes in current use, but almost all are some combination of the basic modulation schemes just detailed. FM will be used for this chapter's projects because FM is the scheme used by the data transmitters being monitored. Figure 7-2 shows a carrier wave that was frequency modulated by two digital pulses.

The waveform modulation is apparent in the figure by the close spacing of the peaks and troughs of the CW carrier, while the digital pulses are in their high state. This change in frequency is called *deviation* and is directly proportional to the amplitude of the data signal. For digital signals, the deviation typically will be 0 Hz for the low portion and approximately 5000 to 10,000 Hz for the high portion. This type of FM is two-state modulation, with either a 0- or 10-kHz deviation appearing in the waveform. I will not be discussing how to implement digital FM transmission because this chapter's project is focused only on receiving and decoding radio waves using SDR techniques.

Heterodyning and Demodulation

The process of receiving and retrieving data sent by digital radio involves two techniques known as *heterodyning* and *demodulation*. Heterodyning involves the use of a radio tuner whose purpose is to convert the very high-frequency (VHF) radio wave to a much lower-frequency wave known as an *intermediate frequency* (IF). This IF is then sent to the demodulator. Demodulation is the precise counterpart to modulation, which encoded the original transmitted RF wave. As such, you would require an FM demodulator to decode the received FM radio wave.

In an analog FM receiver, an initial RF mixer is used to initially translate the VHF radio waves to a much lower IF, usually about 10.7 MHz for consumer FM radio. You may have heard of super-het radios, which are simply radios containing multiple heterodyne stages used to increase receiver selectivity. Figure 7-3 shows a block diagram of a typical analog FM receiver RF front end.

The LO shown in the figure is short for "local oscillator," whose purpose is to generate a RF signal precisely equal to the carrier frequency. The mixer shown in the figure generates sum and difference frequencies when the received RF signal is "mixed" with the LO frequency. Only the difference frequency is used because it contains the useful data signal. The RF amplifier shown in the figure amplifies the relatively weak RF signal coming from the antenna before it

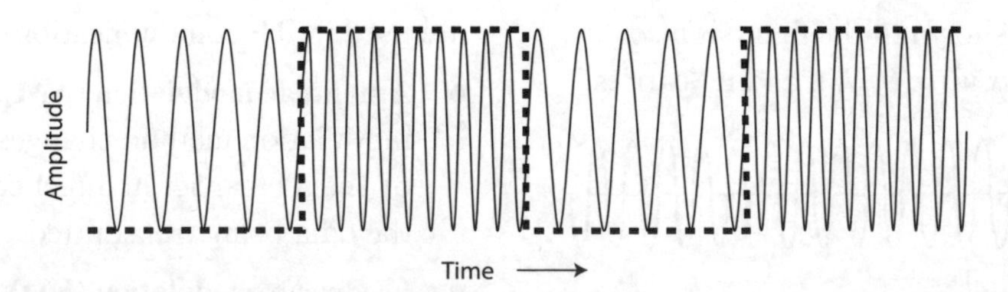

Figure 7-2 Digital pulse FM waveform.

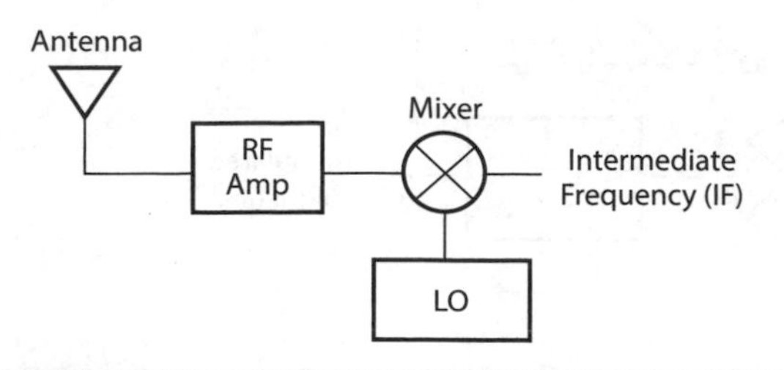

Figure 7-3 Analog FM receiver RF front-end block diagram.

is put into the mixer. This greatly improves the signal-to-noise ratio, which is effectively the radio's dynamic range. A wide dynamic range enables the radio to handle both strong and weak signals efficiently. You will shortly see that SDR dongles also have RF amplifiers, but they are termed *low-noise amplifiers* (LNAs).

The IF from the tuner is next sent to the demodulator. This signal contains a replica of the original data in the form of two analog signals: the I and Q waves, as shown in Figure 7-4, which is a simplified block diagram of an analog demodulator.

The I wave is an in-phase replica of the original data, while the Q wave is a quadrature or 90° out-of-phase replica of the same signal. Both I and Q waveforms are needed by the SDR to complete the demodulation process and re-create the original data.

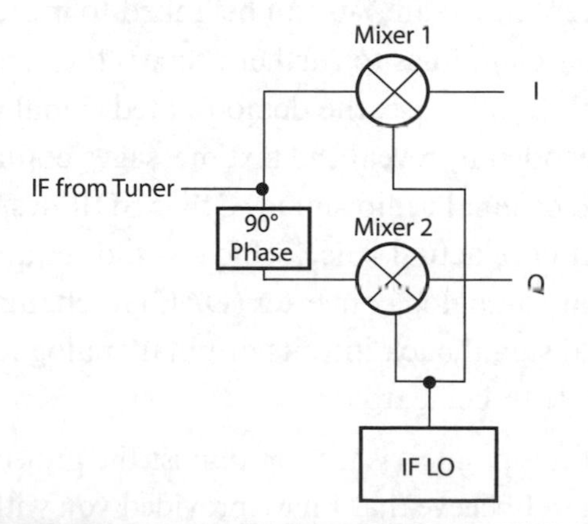

Figure 7-4 Block diagram of an analog demodulator generating I and Q waves.

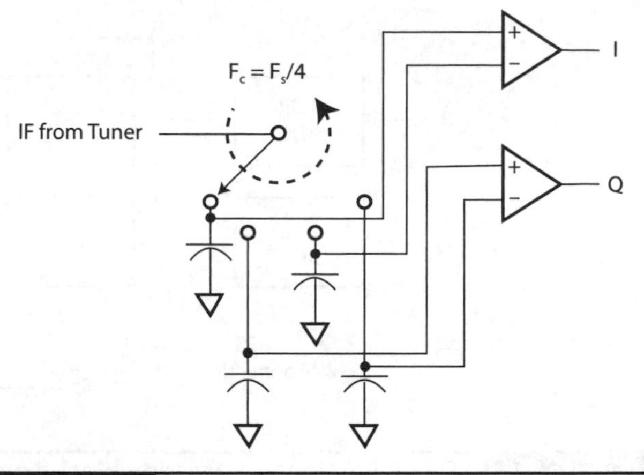

Figure 7-5 Block diagram of a Tayloe detector.

While the classic demodulation circuit is still widely used in modern radios, SDR typically employs another circuit that is highly effective and is a good match to meet SDR requirements. This circuit is called the *Tayloe detector* and is shown in the block diagram in Figure 7-5.

I will not provide a detailed description of the operation of a Tayloe detector other than to say that the incoming IF signal is sampled and "held" for a very brief time such that the I and Q signals can be extracted from the received waveform. The switching or commutation speed denoted as F_c in the figure is set as one-quarter of the sampling frequency F_s. F_s, in turn, is set at the IF, which is typically at or below 10 MHz for SDR dongles. The switching and sampling actions are accomplished in a single digital demodulator chip, which I discuss in a later section.

The next section details how the SDR uses the I and Q signals to re-create the original data signal.

Signal Reconstruction Using I and Q Waveforms

The next step in the SDR processing sequence is to convert the I and Q waveforms into their digital representations. This is accomplished

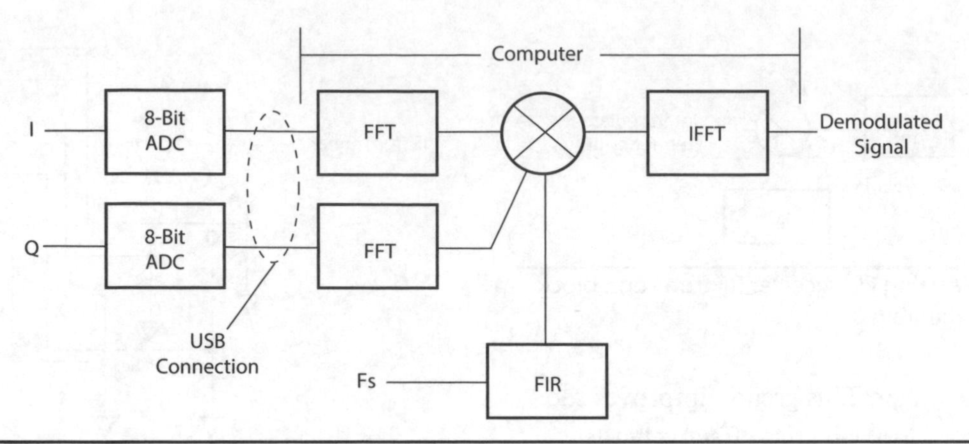

Figure 7-6 SDR digital signal-processing block diagram.

by using two 8-bit analog-to-digital converters (ADCs), as shown in Figure 7-6.

The two newly digitized waveforms are then sent serially over a USB connection to a computer, which is running the actual SDR program. The digitization and conversion of the I and Q data streams are also done in the same chip described in the preceding section.

The blocks shown in the figure's computer-designated area represent logical processing modules and not any actual hardware. This virtualization of hardware is the essential feature of SDR, which makes it so versatile. The abbreviations in the figure are detailed as follows:

- **ADC**: analog-to-digital converter
- **FFT**: fast Fourier transform
- **IFFT**: inverse fast Fourier transform
- **FIR**: finite-impulse filter
- F_s: sampling frequency

I will not discuss how this block diagram functions because it is not critical to implementing the project. I will say that the logical blocks will change according to the type of signal being processed. This diagram is suitable for digital, narrow-band FM demodulation, while a completely different diagram might be used for single-sideband voice demodulation. Such a difference only requires changes in the program code but no changes as far as the I and Q signals are concerned.

In addition, the associated data displays are completely software dependent, which again makes it so easy to accommodate the various types of radio signals to be handled by the SDR system. This chapter's project will use a simple console display to show the received text messages. Other displays can and usually are far more sophisticated. I refer readers to investigate Flex Radio's PowerSDR software to see an excellent example of a professional SDR control and display program. It is freely available and has a demo mode by which you can run it without attached hardware.

The demodulated signal line shown at the far right in Figure 7-6 can be linked to more software modules to further process the received signal. In our case, the demodulated signal will be decoded to reveal the text messages contained in the original radio signal. Other SDR systems might have actual voice, which would require a digital-to-analog converter (DAC) to change the digital signal back into its original analog form suitable to be heard.

At this point, it is time to discuss the project because I believe that I have provided you with a sufficient background in SDR technology to understand what happens when you actually run it.

SDR Dongle

The key piece of hardware required for this SDR project is actually a digital TV tuner device, affectionately termed an *SDR dongle*. Figure 7-7 shows the SDR dongle used in this project.

How the dongle TV tuner became an inexpensive SDR is a bit of a serendipitous story. Back in 2010, a fellow named Eric Fry was experimenting with this dongle type using a packet sniffer, hoping to get a digital TV broadcast data stream (DAB+) working with a Linux application. He confirmed that the I and Q signals generated by the dongle were suitable for not only DAB+ but also for the more general VHF signals not necessarily related to TV broadcasts. Eric and Antii Palosaari confirmed that the Realtek 2832U chip contained in the TV tuner dongle was perfectly capable of creating a wide range of I and Q signals suitable for use in SDR applications. Things progressed rapidly from that point, thanks to many smart and clever open-source developers. Today, quite a few SDR programs are available, including one made specifically for the RasPi, which I will use in this project.

Returning to Figure 7-7, you will notice that there is a small antenna connector located on the side. This is a U.FL micro coaxial connector, which is designed to be used with the small, collapsible stick antenna that is normally supplied with the dongle when you buy the package. Figure 7-8 shows this antenna with its cable connector.

Figure 7-7 SDR dongle.

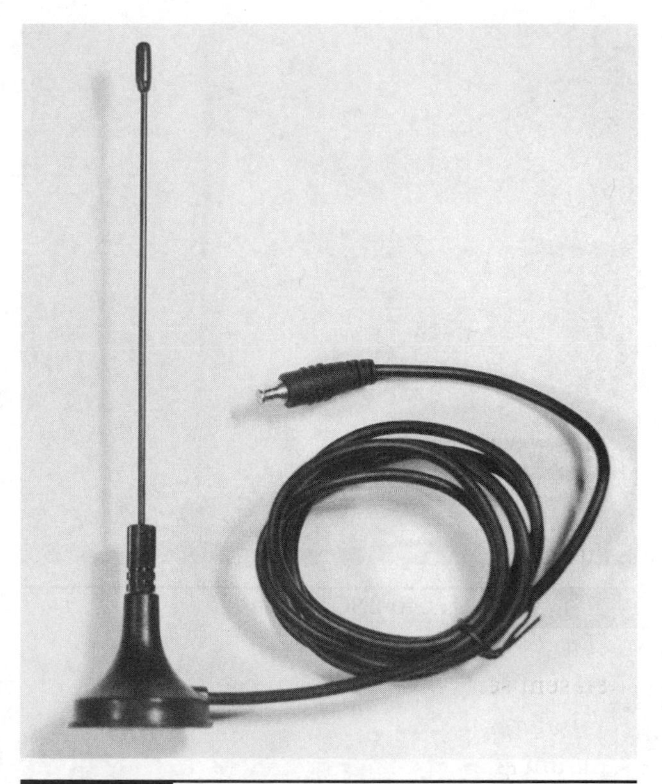

Figure 7-8 Collapsible stick antenna.

Now the small stick antenna might be fine when used in a strong signal area, but I found it to be a very poor performer where I live, in a somewhat rural area. I therefore purchased a U.FL-to-BNC adapter, which allowed me to connect an excellent outdoor VHF antenna to the dongle. Figure 7-9 shows this adapter, which I highly recommend along with an outdoor antenna because it will help to turn your SDR dongle into a viable receiver.

All SDR dongles have two main components: the tuner chip and the demodulation chip. All SDR dongles use the Realtek RTL2832U demodulation chip, which leaves the tuner chip as the principal difference between dongles. Some are better than others depending on your specific requirements. Table 7-1 lists all the available tuner dongles at the time of this writing along with their covered frequency bands.

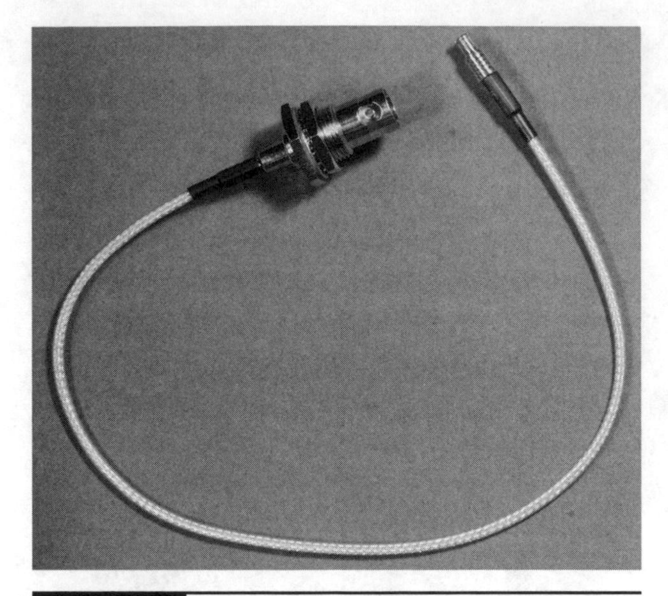

Figure 7-9 U.FL-to-BNC adapter.

TABLE 7-1 SDR Tuners

Tuner	Frequency Range
Elonics E4000	52–2200 MHz (gap 1100–1250 MHz)
Rafael Micro R820T	24–1766 MHz
Rafael Micro 828D	24–1766 MHz
Fitipower FC0013	22–1100 MHz
Fitipower FC0012	22–948.6 MHz
FCI FC2580	146–308 MHz 438–924 MHz

Table 7-2 is list of all the available dongles along with their tuner.

NOTE **Data for Tables 7-1 and 7-2 were obtained from the Osmocom.org website.**

The vendor ID (VID) and product ID (PID) are also listed in Table 7-2 to help verify

TABLE 7-2 Available SDR Dongles

Device Name	Tuner	VID	PID
Generic RTL2832U	All	0x0bda	0x2832
Ezcap USB 2.0 DVB-T/DAB/FM dongle	E4000	0x0bda	0x2838
Terratec Cinergy T Stick Black (rev 1)	FC0012	0x0ccd	0x00a9
Terratec NOXON DAB/DAB+ USB dongle (rev 1)	FC0013	0x0ccd	0x00b3
Terratec Cinergy T Stick RC (rev 3)	E4000	0x0ccd	0x00d3
Terratec NOXON DAB/DAB+ USB dongle (rev 2)	E4000	0x0ccd	0x00e0
Compro Videomate U620F	E4000	0x185b	0x0620
Compro Videomate U650F	E4000	0x185b	0x0650
GTek T803	FC0012	0x1f4d	0xb803
Lifeview LV5T Deluxe	FC0012	0x1f4d	0xc803
Twintech UT-40	FC0013	0x1b80	0xd3a4
Dexatek DK DVB-T Dongle (Logilink VG0002A)	FC2580	0x1d19	0x1101
Dexatek DK DVB-T Dongle (MSI DigiVox Mini II v3.0)	Not determined	0x1d19	0x1102
Dexatek Technology DK 5217 DVB-T Dongle	FC2580	0x1d19	0x1103
Genius TVGo DVB-T03 USB Dongle (v B)	Not determined	0x0458	0x707f
Gigabyte GT-U7300	FC0012	0x1b80	0xd393
Dikom USB-DVBT HD	Not determined	0x1b80	0xd394
Peak 102569AGPK	FC0012	0x1b80	0xd395
Sveon STV20 DVB-T USB and FM	FC0012	0x1b80	0xd39d

the dongle in use. Both the VID and PID are enumerated when the USB connection is made and may be read from the display console when the appropriate command is issued. I plugged the generic SDR dongle I purchased into my MacBook Pro and issued this command in a terminal window:

```
system_profiler SPUSBDataType
```

Figure 7-10 is portion of the resulting display concerning the dongle.

The VID was 0x0bda, confirming it was a Realtek chip, and the PID was 0x2832, confirming that it was also a genuine RTL2832U. I had purchased a generic SDR dongle package on eBay, and I had previously determined that it had a Rafael Micro R820T tuner. I will also demonstrate how to determine the VID and PID using the RasPi when I discuss software installation.

A good antenna, an SDR dongle, and a RasPi are all the hardware needed to build your own SDR receiver. Of course, installing the proper software also will go a long way toward ensuring that you have a great SDR experience. The next section shows you how to install a great software package named rtl-sdr, which will allow you to monitor some status data sent from commercial aircraft.

```
RTL2838UHIDIR:

  Product ID: 0x2838
  Vendor ID: 0x0bda  (Realtek Semiconductor Corp.)
  Version: 1.00
  Serial Number: 00000001
  Speed: Up to 480 Mb/sec
  Manufacturer: Realtek
  Location ID: 0x1d110000 / 6
  Current Available (mA): 500
  Current Required (mA): 500
```

Figure 7-10 USB enumeration for a generic SDR dongle.

rtl-sdr and GNU Radio Software Installation

To start, I will show how to install the rtl-sdr software package. It has essentially all that is needed to get the SDR dongle up and running. After completing the rtl-sdr installation, I will show how to install the GNU Radio software package, which contains some enhanced packages that provide additional functionality to the SDR experience. You should install rtl-sdr first because it has some important configuration steps that are needed to ensure that the SDR dongle works properly with the RasPi.

rtl-sdr Software Package Installation

Please carefully follow all the following steps in order to install the `rtl-sdr` software library. Note that you may have already completed some of these steps from an earlier project, but that's okay. The RasPi will simply report that the particular software package is already installed and no action was accomplished. You also will need the RasPi connected to the Internet to download the required software libraries.

1. `sudo apt-get update`
2. `sudo apt-get install git-core`
3. `sudo apt-get install git`
4. `sudo apt-get install cmake`
5. `sudo apt-get install libusb-1.0-0-dev`
6. `git clone git://git.osmocom.org/rtl-sdr.git`
7. `cd rtl-sdr`
8. `mkdir build`
9. `cd build`
10. `cmake ../-DINSTALL_UDEV_RULES=ON`
11. `make`
12. `sudo make install`

13. `sudo ldconfig`
14. `cd ~`
15. `sudo cp ./rtl-sdr/rtl-sdr.rules /etc/udev/rules.d/`

You will next need to create a new configuration file to deactivate the TV DVB driver that was installed by default when the `rtl-sdr` package was loaded. Technically, this is known as a *Dump1090 failure*, but the fix is relatively easy. Follow these steps:

1. `cd /etc/modprobe.d`
2. `sudo nano no-rtl.cong` # (Create a new file using a nano named no-rtl.conf.)
3. Add these lines to the new file:

```
blacklist dvb_usb-rtl28xxu
blacklist rtl2832
blacklist rtl2830
```

4. Save and close nano.

Next, plug the SDR dongle into the RasPi, and enter the following command to check the USB enumeration:

```
lsusb
```

This command displays the following regarding the SDR dongle:

```
Bus 001 Device 004: ID 0bda:2838 Realtek
Semiconductor Corp. RTL2838 DVB-T
```

This result confirms that the RasPi successfully enumerated the dongle and that the VID and PID matched what was expected. You now need to create a new rule that will allow non-root users to access the dongle. You do this by creating a new rules file using the nano editor:

1. `sudo nano /etc/udev/rules.d/20.rtlsdr.rules`
2. Add this line into the file:

```
SUBSYSTEM=="usb",
ATTRS{idVendor}=="0bda",
ATTRS{idProduct}=="2832",
GROUP="adm", MODE="0666",
SYMLINK+="rtl-sdr"
```

3. Save file and exit nano editor.
4. `sudo reboot`

After the reboot completes, enter this next command to determine whether the `rtl-sdr` installation was successful:

```
rtl_test
```

You should see what is shown in Figure 7-11 if you had a good installation.

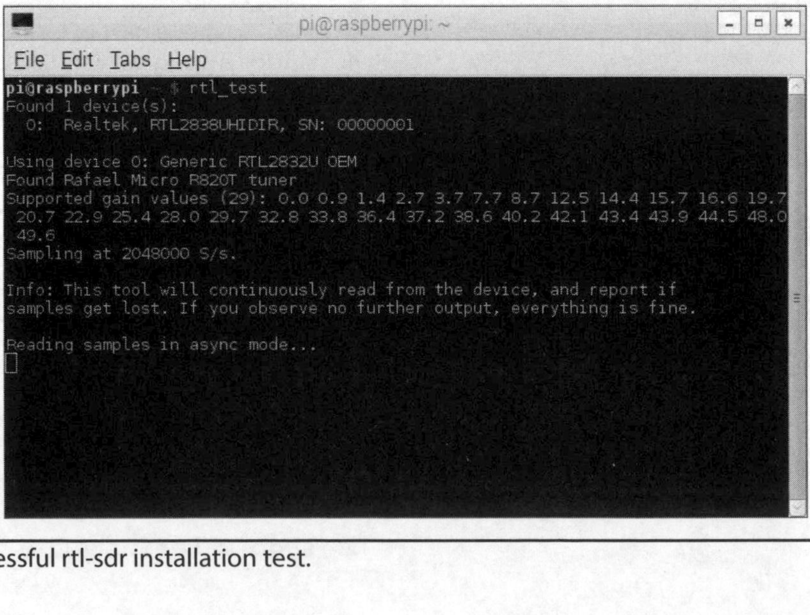

```
pi@raspberrypi ~ $ rtl_test
Found 1 device(s):
  0:  Realtek, RTL2838UHIDIR, SN: 00000001

Using device 0: Generic RTL2832U OEM
Found Rafael Micro R820T tuner
Supported gain values (29): 0.0 0.9 1.4 2.7 3.7 7.7 8.7 12.5 14.4 15.7 16.6 19.7
 20.7 22.9 25.4 28.0 29.7 32.8 33.8 36.4 37.2 38.6 40.2 42.1 43.4 43.9 44.5 48.0
 49.6
Sampling at 2048000 S/s.

Info: This tool will continuously read from the device, and report if
samples get lost. If you observe no further output, everything is fine.

Reading samples in async mode...
```

Figure 7-11 Successful rtl-sdr installation test.

Note that no antenna is needed for this test because it only checks whether samples are being sent from the dongle to the RasPi. The next test will require an antenna because you will try to tune to a local FM station. You probably can use the small stick antenna for this test because most likely you will have available a very strong signal from a local FM station. The following is the command I entered to conduct this test. Obviously, the FM station frequency you enter will differ unless you happen to live in my neighborhood.

```
rtl_fm -f 97.5e6 -M wbfm -s 200000 -r
48000 - | aplay -r 48k -f S16_LE
```

NOTE **The FM frequency I used was 97.5. Change this to suit your requirement, but keep the e6 suffix because it represents a million. All the other command elements should be unchanged.**

After I entered this command, I immediately heard the local FM station audio coming from my monitor's speakers because the RasPi was connected using an HDMI cable to the monitor. If you do not have this arrangement, then you should use the RasPi's headphone jack to listen to the audio output. Enter the following command to ensure that the audio is being directed to the headphone jack:

```
amixer cset numid=3 1
```

The last test I conducted was really more of a proof-of-performance check. This test involves trying to determine the maximum sample rate that could be set before any data packets are dropped or lost. This next command sets the sample rate at 3.2×10^6 samples per second, which is above the maximum sample rate that most SDR dongles can effectively handle:

```
rti-sdr -s 3.2e6
```

Figure 7-12 shows that packets were being consistently lost, which would result in poor to nil data reception.

I next reran the test using lower and lower sample rates until packets were no longer being dropped. I determined that a sample rate of 2.76×10^6 was the maximum that could be used with no data being dropped or lost. This sample rate is probably very typical for commodity SDR dongles and more than adequate to meet all `rtl-sdr` software package requirements.

At this point, you are ready to install the GNU Radio package, which will enable the SDR dongle to receive a great variety of signals.

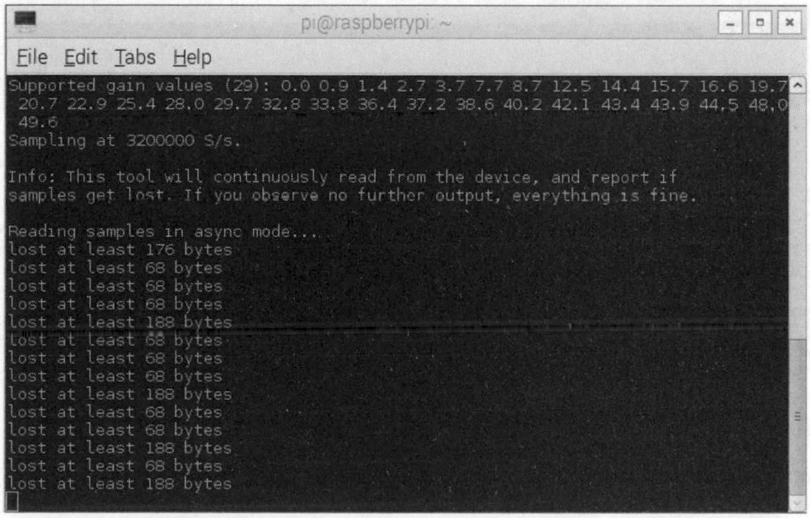

```
pi@raspberrypi: ~
File Edit Tabs Help
Supported gain values (29): 0.0 0.9 1.4 2.7 3.7 7.7 8.7 12.5 14.4 15.7 16.6 19.7
 20.7 22.9 25.4 28.0 29.7 32.8 33.8 36.4 37.2 38.6 40.2 42.1 43.4 43.9 44.5 48.0
 49.6
Sampling at 3200000 S/s.

Info: This tool will continuously read from the device, and report if
samples get lost. If you observe no further output, everything is fine.

Reading samples in async mode...
lost at least 176 bytes
lost at least 68 bytes
lost at least 68 bytes
lost at least 68 bytes
lost at least 188 bytes
lost at least 68 bytes
lost at least 68 bytes
lost at least 68 bytes
lost at least 188 bytes
lost at least 68 bytes
lost at least 68 bytes
lost at least 188 bytes
lost at least 68 bytes
lost at least 188 bytes
```

Figure 7-12 Maximum sample data rate test output.

GNU Radio Software Package Installation

The GNU Radio software package is not contained in the current Debian Wheezy release but instead is a part of a testing release named *jessie*. Fortunately, it is easy to access the jessie packages from Wheezy simply by adding the jessie repository to the Wheezy sources list. You do this by using the nano editor and following these steps:

1. `sudo nano /etc/apt/sources.list`
2. Add this line into the file:

```
deb http://archive.raspbian.org/
raspbian/ jessie main
```

3. Save the file and exit the nano editor.
4. `sudo apt-get update`

You will now need to download and install the GNU Radio runtime and development software packages. Please be patient because it does take a while to complete this 351-MB installation. Enter the following command to start the install:

```
sudo apt-get install gnuradio
gnuradio-dev
```

After the install finishes, you still have one more package to install, which contains some higher-level GNU Radio applications. Enter this command:

```
sudo apt-get install gr-osmosdr
```

Please notice that the great folks at Osmocom created these applications to enhance the SDR experience. In fact, enter the following command to see one of these applications run:

```
osmocom_fft
```

Figure 7-13 shows a screenshot from the RasPi while this application was running.

This application runs a fast Fourier transform (FFT) on the sampled data streaming from the SDR dongle. FFTs transform time-sampled data into their equivalent frequency-domain

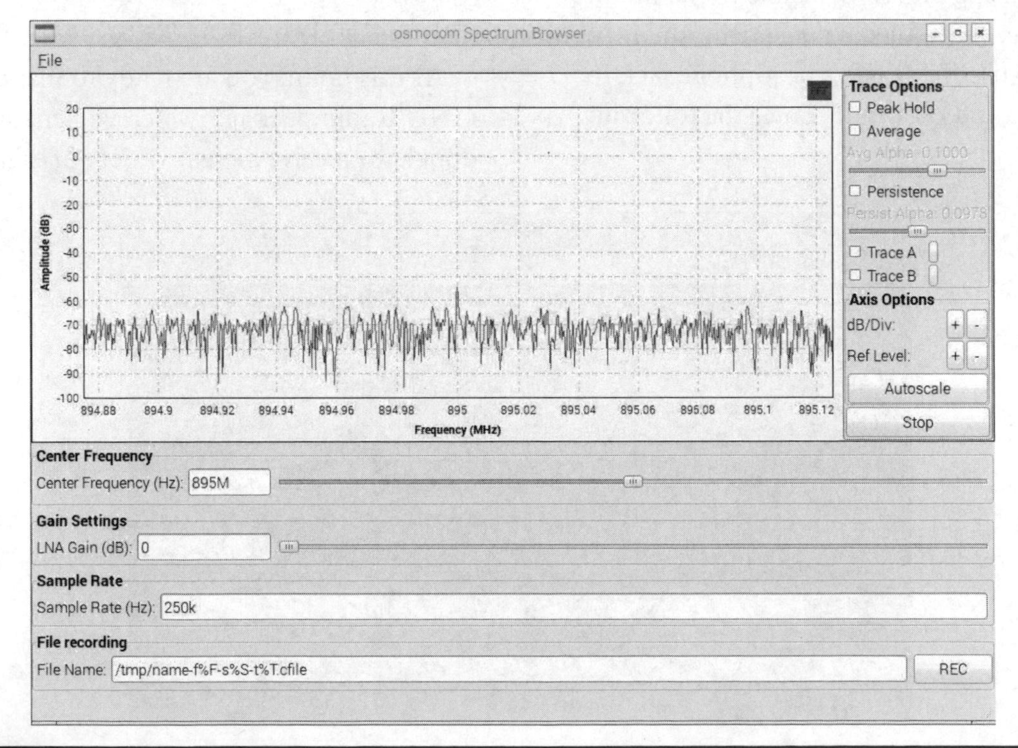

Figure 7-13 osmocom_fft screenshot.

data points. The figure shows a region of these frequency data points or spectrum centered at 895 MHz with a split bandwidth of ±125 kHz or 250 kHz of total bandwidth. This is a relatively narrow bandwidth when you consider that the carrier frequency is 895 MHz. Using such a narrow bandwidth allows for some very close scrutiny of the RF wave, but more important, it provides excellent selectivity, which helps to "pull" true data signals from the noise.

I have included the following sidebar for readers who wish to learn a bit more about FFTs. Feel free to skip it and just realize that time and frequency data are two sides of the same coin. You can't have one without the other.

A classic sine waveform signal is shown in Figure 7-14, where the vertical axis is amplitude measured in voltage and the horizontal axis is time in milliseconds (ms). The *period* of this signal, or time between positive peaks, is shown in the figure as 1 ms. This period measurement means that the signal repeats itself 1000 times per second. The inverse of the period is called the sine wave's *frequency*, which in this case would be equal to 1000 Hz. Hertz (Hz) used to be known as cycles per second (cps), but I think that it is far better to use the Hz suffix. This frequency is plotted in Figure 7-15 using a horizontal scale of hertz, while the vertical axis is in decibels (dB) referenced to 1 V, which is typical for frequency spectrum plots.

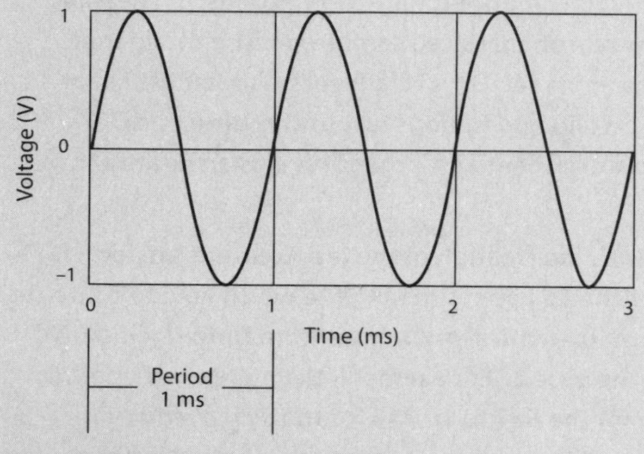

Figure 7-14 Sine wave time plot.

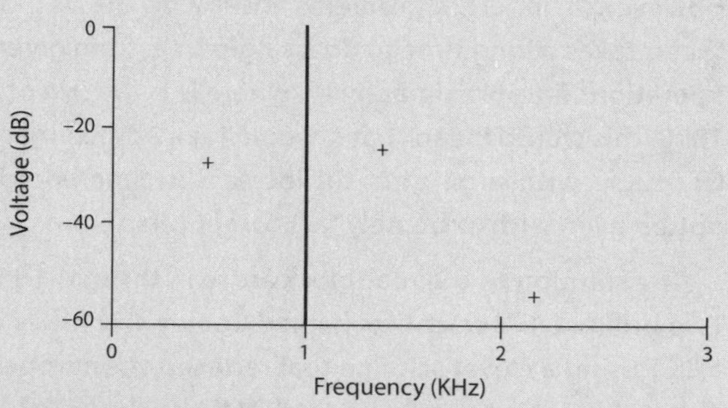

Figure 7-15 Sine wave frequency plot.

The famous mathematician Jean-Baptiste Joseph Fourier discovered many years ago that any arbitrary continuous mathematical waveform could be represented by an infinite series of sine waves. Of course, if the initial waveform was only at a single frequency, then obviously only one sine wave would be necessary, as shown in Figure 7-15. However, if you were to represent a perfect square waveform using sine waves, then you indeed would need an infinite number of sine waves to accomplish that feat. Figure 7-16 is a replica of a spectrum analyzer plot for a 1-kHz square wave showing the initial sine waves needed in this series up to 20 kHz.

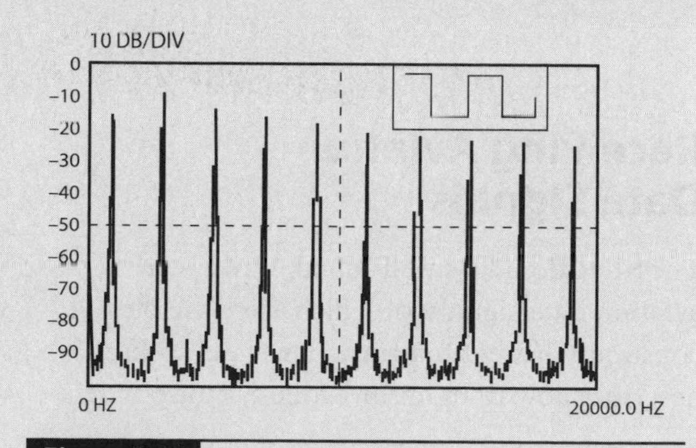

Figure 7-16 Square wave spectrum.

The 1-kHz square wave spectrum starts with a 1-kHz sine wave and then an additional 3-kHz sine wave, 5-kHz sine wave, and so on. Note that all the sine wave frequencies beyond the first one are at odd harmonics of the initial sine wave, which is also known as the *fundamental harmonic*. In addition, the amplitudes of all the odd harmonics decrease as their frequency increases. Spectrum displays are very useful and can provide much information about a signal that is not obvious from viewing a time-based display of that signal.

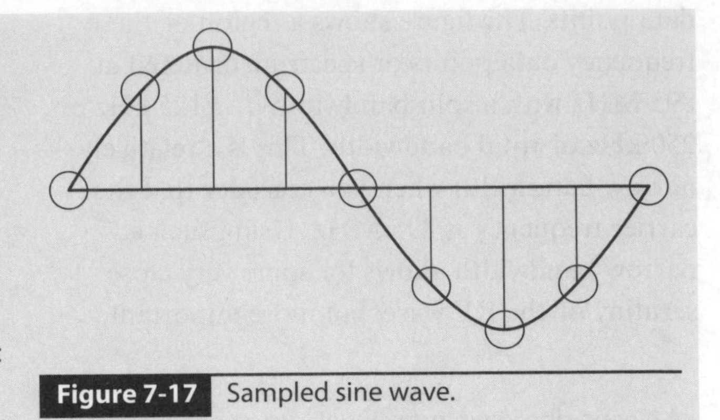

Figure 7-17 Sampled sine wave.

The data stream coming from the SDR dongle is in the form of digital numbers representing sampled values of the analog waveform. Figure 7-17 shows how these samples relate to the original waveform.

The problem now is how to convert these digitized samples into a matching digitized Fourier transform. The answer lies in the use of an algorithm known as the *discrete Fourier transform* (DFT). Through a series of trigonometric multiplications and summations, a sampled time series can be converted into its DFT. However, an inherent problems arises with the DFT in that it is computationally very expensive, meaning that it takes a long time to do a single transform given any reasonably sized sample set. The number of operations is approximately N^2, where N is the size of the sample set. For a relatively small sample set of 1000, this would mean that it would take approximately 1 million operations just to compute one DFT. Of course, with larger sets, the increase in time would grow exponentially larger. This was an unattractive option even with extremely fast computers.

The solution to the roadblock arose in the mid-1960s with the creation of the fast Fourier transform (FFT). Two brilliant IBM scientists named Cooley and Tukey published a paper showing the world how to compute a DFT using a clever scheme that reduced the number of mathematical operations from N^2 to $N \times \log2(N)$. This last notation should be read "N times the log of N to the base 2." For example, the number of operations for a 1024 sized data set goes from slightly over 1 million for the DFT to 10,240 for the FFT, a reduction of almost 100 to 1. Real-time FFT computations thus were made possible. Today, every application that converts sampled time data to frequency data uses the FFT. Amazingly, IBM never patented or protected this intellectual property and graciously made it available to everyone for the promotion of science and technology.

Receiving Aviation Data Signals

The SDR dongle is well suited to receiving aviation data signals, of which there are plenty constantly being broadcast. This next section describes how to monitor Mode-S transponders.

Monitor Mode-S Transponders

An aircraft Mode-S transponder consists of a radio system that transmits a data packet when it receives an incoming radar ping from an air traffic control (ATC) facility. It is important to realize that receiving these data packets will in no way harm or affect any aircraft operations.

Much of the following was based on a blog from Andrew Beck describing the `gr-air-modes` software package, which allows the SDR dongle to receive Mode-S airborne data signals. Enter the following command to start building this software:

```
sudo apt-get install sqlite pyqt4-dev-tools liblog4cpp5-dev swig libboost-dev
```

The source code has to be downloaded next by entering this command:

```
git clone https://github.com/bistromath/gr-air-modes.git
```

The next steps in this process involve building and installing the source code. Follow these steps to complete the software installation:

1. `cd gr-air-modes`
2. `mkdir build`
3. `cd build`
4. `cmake ../`
5. `make`
6. `sudo make install`
7. `sudo ldconfig`

The application is now run by entering the following command:

```
modes_rx -s osmocom
```

Figure 7-18 shows an output from this program. There are only a few entries because the location where I ran the program was a poor reception area for these types of signals.

Mode-S signals are not the only interesting aviation data packets that may be monitored using an SDR dongle. The next section describes another airborne data type that you may be interested in checking out and was also involved with an incident that made the worldwide news.

Monitoring ACARS

Have you ever wondered as you gazed at contrails high in the deep blue sky where those airplanes were coming from or going to? The Aircraft Communications Addressing and Reporting System (ACARS) is your answer. It is a digital data-link system for the transmission of short messages between aircraft and ground stations via airband radio or satellite. The ACARS Protocol was created and deployed in 1978 and follows the Telex format. It was designed by Aeronautical Radio, Inc. (ARINC) to promote safe and efficient airline operations. ACARS includes both Airline Operational Control (AOC) and ATC digital messages with an average message volume split of 80 percent

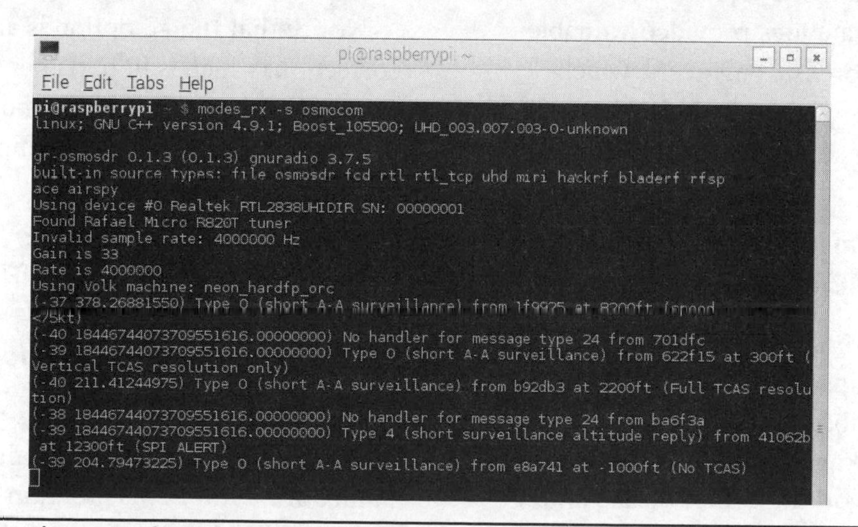

Figure 7-18 Example output from the modes_rx application.

for AOC and 20 percent for ATC. Table 7-3 shows the variety of information sent through ACARS.

TABLE 7-3 ACARS Message Types

Message Type	AOC	ATC
Takeoff and landing confirmation	x	
Airline weather update	x	
Gate information	x	
Engine data	x	
Navigation status		x
Position reports		x
Departure clearances		x
Oceanic clearances		x
Runway conditions		x

As you can see from the table, a great deal of information is constantly being exchanged between an aircraft and appropriate ground stations. The ACARS system is autonomous in operation and cannot be shut down by the pilots. This fact played a key role in the search for the Malaysia Airlines MH380 that disappeared several years ago over the southern Indian Ocean. For unknown reasons, all the pilot-controlled communications with the ground ceased as the aircraft deviated from its preplanned flight path. However, ACARS continued to broadcast from the plane for several hours until it suddenly stopped. ACARS transmissions to satellites provided valuable tracking data to forensic engineers, which allowed them to provide searchers with probable locations to begin searching for the aircraft.

Downloading and Installing a Multichannel ACARS Decoder

This procedure presumes that you already have downloaded and installed the rtl-sdr software package, as described earlier in this chapter. The ACARS package depends on having rtl-sdr available and in working order. Please follow these next steps in the order presented. Note that the version I downloaded was 3.1, but this might change when you download it because upgrades are constantly being done to this software. I would check the website http://sourceforge.net/projects/acarsdec/ to determine the current version. Then substitute that version number for 3.1 wherever it appears in these procedural steps:

1. `cd ~`
2. `mkdir acars`
3. `cd acars`
4. `wget http://sourceforge.net/projects/acarsdec/files/latest/download -O acarsdec-3.1.tar.gz`
5. `tar xvfz acarsdec-3.1.tar.gz`
6. `cd acarsdec-3.1`
7. `sudo apt-get install libsndfile1-dev`
8. `sudo apt-get install libasound2-dev`
9. `sudo apt-get install librtlsdr`
10. `sudo ldconfig`
11. `make`

I have also included Figure 7-19, which is a screenshot of the acarsdec help page to explain the options that are available when you run this application.

Notice that the `-p` option is a frequency correction, which is in parts per million. Correcting the receiver frequency should allow the tuner in the dongle to promptly lock onto the desired ACARS frequency. The next section details a calibration procedure to determining the appropriate frequency correction value.

Using the Kalibrate Application

Let me start by stating that the program used to determine the calibration value is actually named *Kalibrate*. This program uses any nearby cellular GSM base station as a frequency

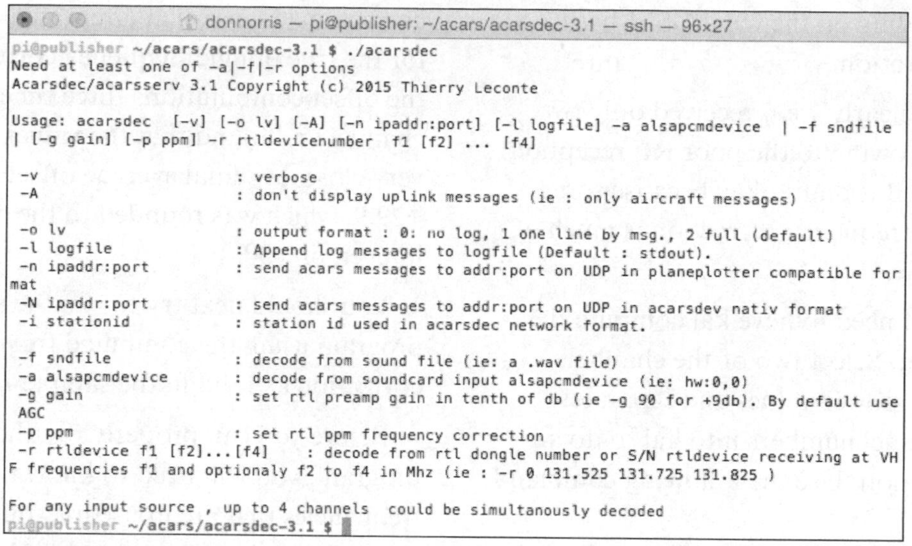

```
donnorris — pi@publisher: ~/acars/acarsdec-3.1 — ssh — 96×27
pi@publisher ~/acars/acarsdec-3.1 $ ./acarsdec
Need at least one of -a|-f|-r options
Acarsdec/acarsserv 3.1 Copyright (c) 2015 Thierry Leconte

Usage: acarsdec  [-v] [-o lv] [-A] [-n ipaddr:port] [-l logfile] -a alsapcmdevice  | -f sndfile
| [-g gain] [-p ppm] -r rtldevicenumber  f1 [f2] ... [f4]

 -v                     : verbose
 -A                     : don't display uplink messages (ie : only aircraft messages)
 -o lv                  : output format : 0: no log, 1 one line by msg., 2 full (default)
 -l logfile             : Append log messages to logfile (Default : stdout).
 -n ipaddr:port         : send acars messages to addr:port on UDP in planeplotter compatible for
mat
 -N ipaddr:port         : send acars messages to addr:port on UDP in acarsdev nativ format
 -i stationid           : station id used in acarsdec network format.

 -f sndfile             : decode from sound file (ie: a .wav file)
 -a alsapcmdevice       : decode from soundcard input alsapcmdevice (ie: hw:0,0)
 -g gain                : set rtl preamp gain in tenth of db (ie -g 90 for +9db). By default use
 AGC
 -p ppm                 : set rtl ppm frequency correction
 -r rtldevice f1 [f2]...[f4]      : decode from rtl dongle number or S/N rtldevice receiving at VH
F frequencies f1 and optionaly f2 to f4 in Mhz (ie : -r 0 131.525 131.725 131.825 )

For any input source , up to 4 channels  could be simultanously decoded
pi@publisher ~/acars/acarsdec-3.1 $
```

Figure 7-19 acarsdec help screen.

reference. In the United States, the AT&T cellular network uses GSM, which should mean that a base station will likely be able to be heard by the dongle. I believe that GSM is also widely used in Europe and likely as well in other parts of the world.

Please follow these steps to install and configure the Kalibrate program:

1. `mkdir ~/kal`
2. `cd ~/kal`
3. `sudo apt-get install libtool autoconf automake libfftw3-dev`
4. `git clone https://github.com/asdil12/kalibrate-rtl.git`
5. `cd kalibrate-rtl`
6. `git checkout arm_memory`
7. `./bootstrap`
8. `./configure`
9. `make`
10. `sudo make install`

Please be patient because some of these steps do take some time to complete.

Enter this next command to determine any local GSM channels that can be received:

```
kal -s GSM900 -d 0 -g 40
```

Figure 7-20 shows the result of my initial GSM frequency search. Your results will be

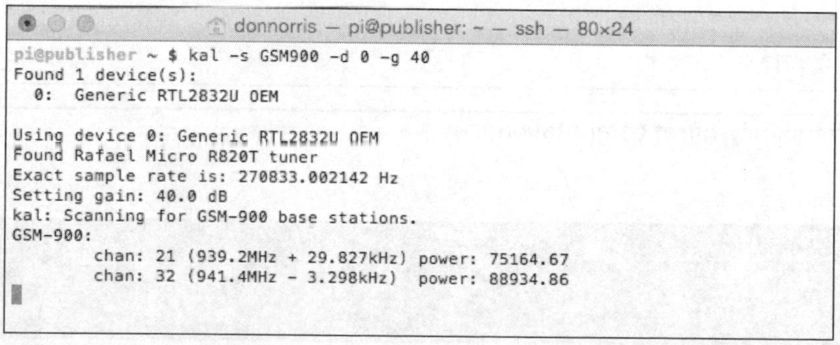

```
donnorris — pi@publisher: ~ — ssh — 80×24
pi@publisher ~ $ kal -s GSM900 -d 0 -g 40
Found 1 device(s):
  0:  Generic RTL2832U OEM

Using device 0: Generic RTL2832U OEM
Found Rafael Micro R820T tuner
Exact sample rate is: 270833.002142 Hz
Setting gain: 40.0 dB
kal: Scanning for GSM-900 base stations.
GSM-900:
        chan: 21 (939.2MHz + 29.827kHz) power: 75164.67
        chan: 32 (941.4MHz - 3.298kHz) power: 88934.86
```

Figure 7-20 Initial GSM frequency search.

different depending on the local GSM base station configuration.

As you may clearly see, I received only two GSM channels owing to the poor RF reception in my location. But that's okay because some data regarding frequency correction is much better than none.

You will next need to have kal compute the frequency offset. Select two of the channels with the largest power values and separately enter their channel numbers into kal to do an offset computation. Use the following command template:

```
kal -c <channel> -d 0 -g 40
```

Figure 7-21 shows the result of one of these frequency-offset computations for one of my two GSM channels.

The average absolute error shown in the figure is –32 ppm, so the correct offset would be +32 ppm because you are correcting for the error in absolute frequency. I would strongly suggest that you rerun the offset computations several times and then average all the resulting offset corrections to arrive at a representative value for the true dongle frequency correction. I reran the offset computations three times for each channel and found that the values returned were very close. My final average offset correction was +29.9, which was rounded to the nearest whole integer, or +30.

You should next try to run the acarsdec program using the computed frequency correction as I did in the sample-run section.

Before you can properly run the acarsdec program, you will need to know the ACARS frequencies that are in use for your region in order to monitor these data. Table 7-4 is a current ACARS frequency list.

CAUTION **The acarsdec program is a multichannel receiving program, meaning that it is capable of simultaneously receiving up to four separate ACARS radio channels. However, because of a sampling constraint, there cannot be more than 625-kHz bandwidth separating the highest and lowest channels. This means that monitoring 131.550 and 131.025 is permissible, but trying to monitor 131.550 and 129.125 is not and will generate the error shown in Figure 7-22.**

```
donnorris — pi@publisher: ~ — ssh — 80×24
pi@publisher ~ $ kal -c 21 -d 0 -g 40
Found 1 device(s):
  0:  Generic RTL2832U OEM

Using device 0: Generic RTL2832U OEM
Found Rafael Micro R820T tuner
Exact sample rate is: 270833.002142 Hz
Setting gain: 40.0 dB
kal: Calculating clock frequency offset.
Using GSM-900 channel 21 (939.2MHz)
average         [min, max]      (range, stddev)
+ 30.650kHz             [30608, 30707]  (98, 26.527403)
overruns: 0
not found: 0
average absolute error: -32.634 ppm
pi@publisher ~ $
```

Figure 7-21 `kal` frequency-offset computation.

```
pi@publisher ~/acars/acarsdec-3.1 $ ./acarsdec -p 30 -r 0 131.550 130.025 130.425 131.125
Found Rafael Micro R820T tuner
WARNING: too much frequencies, taking only the 4 firsts
Frequencies to far apart
[R82XX] PLL not locked!
Exact sample rate is: 1250000.002070 Hz
```

Figure 7-22 acarsdec frequency-separation error.

TABLE 7-4 ACARS Frequency List

Frequency	Region/Country
131.550	Primary channel worldwide
129.125	Additional channel for United States and Canada
130.025	Secondary channel for United States and Canada
130.425	Additional channel for United States
130.450	Additional channel for United States and Canada
131.125	Additional channel for United States
131.450	Primary channel for Japan
131.475	Air Canada company channel
131.525	European secondary
136.575	Additional channel for United States
136.650	Additional channel for United States
136.675	Delta and Jet Blue company channel
131.725	Primary channel in Europe
136.700	Additional channel for United States
136.725	Delta company channel
136.750	Additional channel for United States
136.775	Air Canada, AirTransat, American, Delta, Jet Blue, and United company channel
136.800	Additional channel for United States
136.900	European secondary
136.850	SITA North American frequency
136.750	New European frequency
131.850	New European frequency

The program will still appear to be operating, but it will not display any results.

Sample ACARS Results

I entered the following command to run the multichannel ACARS decoder application at my home. Note that I used two of the frequencies from Table 7-4 that were appropriate for my region and within 625 kHz of each other.

```
./acarsdec -p 30 -r 0 131.550 131.125
```

Figure 7-23 shows a sample of the ACARS messages received after I entered the preceding command.

This figure is a composite of some of the messages I received while I monitored the ACARS for approximately 1 hour. I do wish to further explain two messages that appear at the start of the figure because it will highlight why ACARS is so interesting. These two messages were sent 6 seconds apart from an aircraft that was flying over my home. The messages themselves do not contain anything of substance other than the aircraft registration and flight ID, which in this case were N822NW and NW0065, respectively. A quick Google search revealed that N822NW was an Airbus 242t A330-300 registered to Delta Airlines, which also meant that the flight ID referred to Delta Flight 0065 because NW is Delta's company designator. I next made a quick check on Flight Aware's website (www.flightaware.com) and entered Delta 65 as the flight number. Figure 7-24 shows the actual track of this aircraft, and you can see that it was over my home state of New Hampshire when I queried its status on the Flight Aware website.

Figure 7-25 shows much more information regarding the point of origin and destination for this flight.

Finally, I was interested in the aircraft itself. It turns out that this Airbus A330-300 is the newest addition to Delta's fleet and is used primarily for long-range trips such as this nonstop from Italy to Washington, DC. The 242t in the model descriptor refers to the gross weight, which is 242 metric tons, where 1 metric ton is equal to 1000 kg. Translated, this means that this type of aircraft carries almost 300 passengers. I believe Delta is replacing its 747 aircraft with this new type because it is so much more fuel efficient while carrying the same approximate payload of

```
pi@publisher ~/acars/acarsdec-3.1 $ ./acarsdec -p 30 -r 0 131.550 131.125
Found Rafael Micro R820T tuner
Exact sample rate is: 1250000.002070 Hz

[#1 (F:131.550 L:-31 E:0) 25/07/2015 19:14:20 ------------------------------
Aircraft reg: .N822NW Flight id: NW0065
Mode: 2 Msg. label: Q0
Block id: 3 Ack: !
Msg. no: S84A
Message :


[#1 (F:131.550 L:-31 E:0) 25/07/2015 19:14:26 ------------------------------
Aircraft reg: .N822NW Flight id: NW0065
Mode: 2 Msg. label: _d
Block id: 4 Ack: X
Msg. no: S85A
Message :


[#1 (F:131.550 L:-43 E:1) 25/07/2015 19:17:14 ------------------------------
Aircraft reg: .G-VGAL Flight id: VS073Q
Mode: 2 Msg. label: _d
Block id: 3 Ack: L
Msg. no: S46A
Message :


[#1 (F:131.550 L:-31 E:0) 25/07/2015 19:32:35 ------------------------------
Aircraft reg: .N116WJ Flight id: XA0000
Mode: 2 Msg. label: 15
Block id: 1 Ack: !
Msg. no: M51A
Message :
(2N43014W 68520--- 54429-55(Z

[#1 (F:131.550 L:-41 E:0) 25/07/2015 19:34:17 ------------------------------
Aircraft reg: .N171DN Flight id: DL0208
Mode: 2 Msg. label: H1
Block id: 7 Ack: !
Msg. no: D20A
Message :
#DFB239N171DN020807251519335881З 4437 -6909379-26-55317 4910-234 20400
 258 4690023245251030NXWLSZHKJFK

[#1 (F:131.550 L:-36 E:0) 25/07/2015 20:21:43 ------------------------------
Aircraft reg: .N720PS Flight id: US5222
Mode: 2 Msg. label: 32
Block id: 4 Ack: !
Msg. no: M13A
Message :
33,D,1,1,MHT,250398,-633089,-260,-708,10,32820,+03,669,15,38,289,288,7332,445FFAC

[#1 (F:131.550 L:-39 E:0) 25/07/2015 20:21:55 ------------------------------
Aircraft reg: .N720PS Flight id: US5222
Mode: 2 Msg. label: 32
Block id: 4 Ack: X
Msg. no: M13A
Message :
33,D,1,1,MHT,250398,-633089,-260,-708,10,32820,+03,669,15,38,289,288,7332,445FFAC

[#1 (F:131.550 L:-43 E:1) 25/07/2015 20:26:41 ------------------------------
Aircraft reg: .C-FEIX Flight id: RS7459
Mode: 2 Msg. label: 19
Block id: 1 Ack: !
Msg. no: M19A
Message :
2057,  28


[#1 (F:131.550 L:-26 E:0) 25/07/2015 20:26:47 ------------------------------
Aircraft reg: .N720PS Flight id: US5222
Mode: 2 Msg. label: 32
Block id: 5 Ack: !
Msg. no: M14A
Message :
33,D,2,1,MHT,249764,-635349,691,189,20,01615,-12,662,15,38,388,377,5990,6005703

[#1 (F:131.550 L:-28 E:0) 25/07/2015 20:26:53 ------------------------------
Aircraft reg: .N720PS Flight id: US5222
Mode: 2 Msg. label: _d
Block id: 6 Ack: Z
Msg. no: S19A
Message :
```

Figure 7-23 ACARS messages received at my home.

Figure 7-24 Delta flight 65's Flight Aware track.

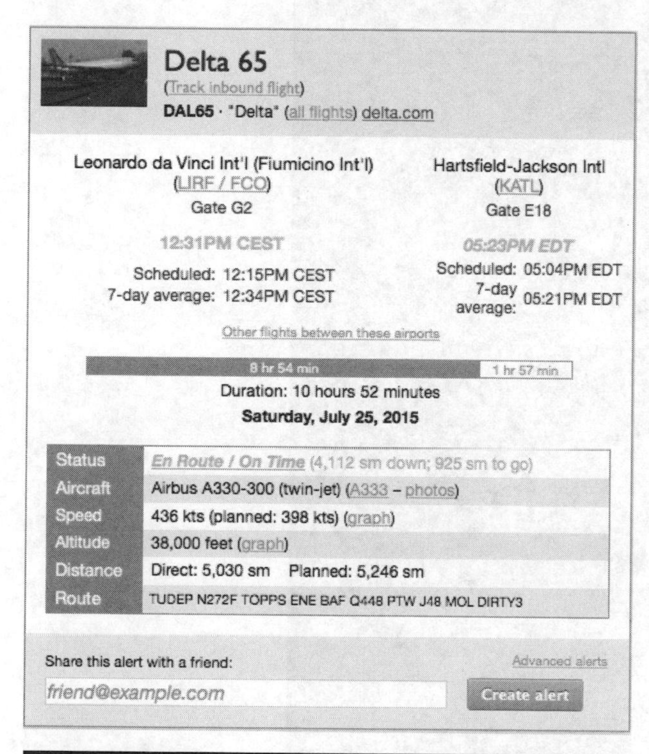

Figure 7-25 Flight itinerary for Delta flight 65.

Figure 7-26 Airbus 242t A330-300, Delta N822NW.

passengers and baggage. Figure 7-26 shows the aircraft that flew over my home that day.

I was also curious as to how the ACARS monitoring would function at a commercial airport, so I traveled to my closest airport, which was at Manchester, NH (KMHT is the IAOC designator). I needed to make my system completely portable, so I used a battery-operated 7-inch HDMI monitor for the display and ran the RasPi, SDR dongle, keyboard, and mouse using an external cell phone battery eliminator, which I discuss in a later section. Figure 7-27 shows a composite of some of the ACARS messages I monitored while being parked about a half mile from the airport. I only used the stubby antenna with this system, and it worked perfectly because the signals were quite strong.

Several messages in this figure refer to an aircraft with registration N339NG, but no flight ID is shown. A quick check on Google for this registration number revealed that it was

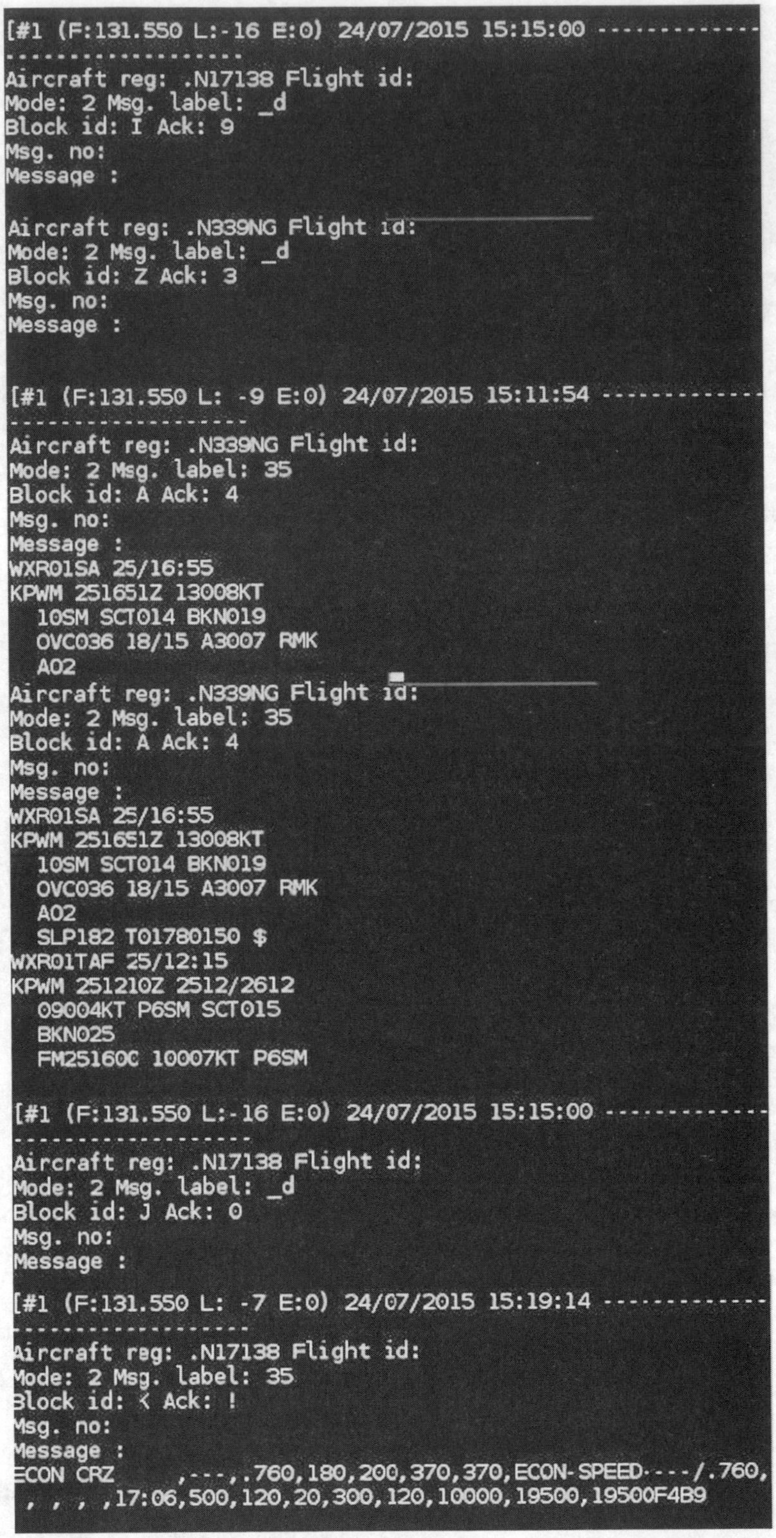

```
[#1 (F:131.550 L:-16 E:0) 24/07/2015 15:15:00 ------------
------------------
Aircraft reg: .N17138 Flight id:
Mode: 2 Msg. label: _d
Block id: I Ack: 9
Msg. no:
Message :

Aircraft reg: .N339NG Flight id:
Mode: 2 Msg. label: _d
Block id: Z Ack: 3
Msg. no:
Message :


[#1 (F:131.550 L: -9 E:0) 24/07/2015 15:11:54 ------------
------------------
Aircraft reg: .N339NG Flight id:
Mode: 2 Msg. label: 35
Block id: A Ack: 4
Msg. no:
Message :
WXR01SA 25/16:55
KPWM 251651Z 13008KT
  10SM SCT014 BKN019
  OVC036 18/15 A3007 RMK
  AO2
Aircraft reg: .N339NG Flight id:
Mode: 2 Msg. label: 35
Block id: A Ack: 4
Msg. no:
Message :
WXR01SA 25/16:55
KPWM 251651Z 13008KT
  10SM SCT014 BKN019
  OVC036 18/15 A3007 RMK
  AO2
  SLP182 T01780150 $
WXR01TAF 25/12:15
KPWM 251210Z 2512/2612
  09004KT P6SM SCT015
  BKN025
  FM251600 10007KT P6SM

[#1 (F:131.550 L:-16 E:0) 24/07/2015 15:15:00 ------------
------------------
Aircraft reg: .N17138 Flight id:
Mode: 2 Msg. label: _d
Block id: J Ack: 0
Msg. no:
Message :

[#1 (F:131.550 L: -7 E:0) 24/07/2015 15:19:14 ------------
------------------
Aircraft reg: .N17138 Flight id:
Mode: 2 Msg. label: 35
Block id: < Ack: !
Msg. no:
Message :
ECON CRZ      ,---,.760,180,200,370,370,ECON-SPEED---/.760,
 , , , ,17:06,500,120,20,300,120,10000,19500,19500F4B9
```

Figure 7-27 ACARS messages received at KMHT.

a United Express Dash8-400 twin turbo-prop commuter airliner that just happened to be passing overhead coming from Liberty Airport in Newark, NJ, and going to Portland, ME (KPWM). The messages associated with this flight contained some ATC content consisting of the current weather conditions at KPWM. These types of messages are known as *METARS* and are used extensively in aviation worldwide. In this case, I decode the METAR in the following sidebar for your information and to help you decipher any similar messages that you might monitor.

The ATC message content was

```
KPM 251651Z 13008KT 10SM SCT014 BKN019 OVC036 18/15 A3007 RMK A02
```

Decoded, it is

`KPWM`	**Destination (dest) is the Jetport, Portland, ME.**
`251651Z`	**Date/date in UTC or Zulu time. Local time was 12:51. The date was the 25th of the month, which happened to be July. The month and year are never stated, just assumed.**
`13008KT`	**Wind at dest reported at 130° at 8 knots.**
`10SM`	**Ten statue miles of visibility (at least).**
`SCT014`	**Beginning of cloud cover report. This one says that there are scattered clouds at 1400 ft above ground level (AGL).**
`BKN019`	**Broken clouds reported at 1900 ft AGL.**
`OVC036`	**Overcast reported at 3600 ft AGL.**
`18/15`	**Destination temperature is currently 18°C and dewpoint is 15°C.**
`A3007`	**Destination altimeter is 30.07 in Hg (inches of mercury).**
`RMK A02`	**Remark A02 shows that the weather originates from an automated weather station located on the grounds of the destination airport.**

The next section changes the pace a bit. I will show you how to build and use a portable spectrum analyzer.

Spectrum Analyzer

This section is based on a great tutorial by Tony DiCola that is available from learn.adafruit.com. In this SDR project, I used a dongle, a RasPi, and a PiTFT touchscreen display, which is the same one that was used in the Chapter 2 project. I mentioned back in Chapter 2 not to dismantle the project if you intended to complete this project. I will refer you back to the Chapter 2 discussion on how to install the PiTFT on a RasPi and use a custom case if you are so inclined. No other hardware installation is necessary because the SDR dongle simply plugs into one of the RasPi USB ports. Figure 7-28 shows the PiTFT mounted in the PiBow case along with the SDR dongle plugged into one of the RasPi's USB ports. I also used a USB extension adapter, which allowed me to plug additional USB cables into the RasPi's ports, which might otherwise be blocked by the dongle.

The RasPi also was connected to the Internet with an Ethernet patch cable. The whole assembly was powered by a wall-wart cube,

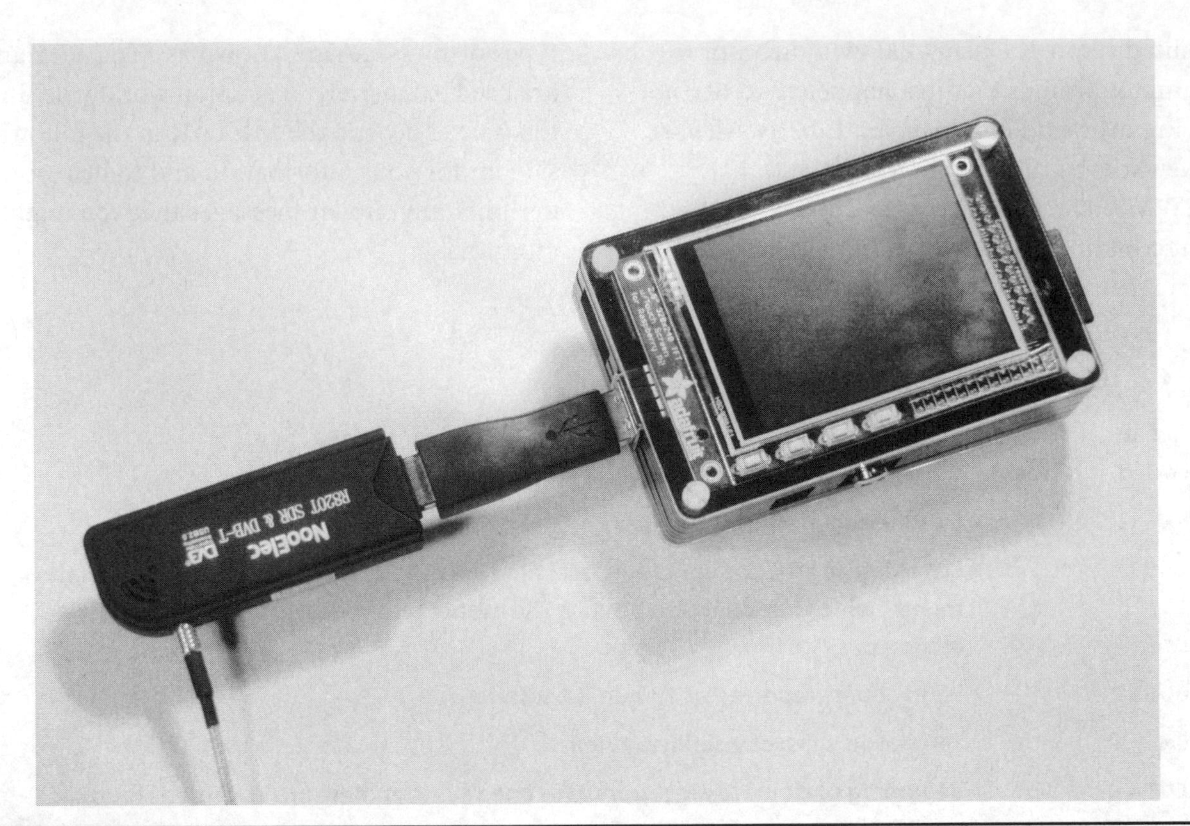

Figure 7-28 Spectrum analyzer assembly.

which I highly recommended during the software installation. Later on in this section I discuss how to make this project completely portable, which includes making it battery powered.

The next section discusses the software installation, which is the key to a successful project.

Software Installation

This project requires a fresh rtl-sdr installation in order to properly configure the RasPi with the PiTFT. I would not suggest trying to overlay this project's software with the previous rtl-sdr software because this software package uses Python and some other modules not found in the previous package. Just start with a fresh Wheezy distribution and you should be fine.

NOTE **I used an SSH connection from my MacBook Pro to the RasPi to accomplish all the following steps.**

Please follow these steps to start the installation and configuration:

1. `sudo apt-get update`
2. `sudo apt-get install cmake build-essential python-pip libusb-1.0-0-dev python-numpy git`
3. `cd ~`
4. `git clone git://git.osmocom.org/rtl-sdr.git`
5. `cd rtl-sdr`
6. `mkdir build`
7. `cd build`
8. `cmake ../ -DINSTALL_UDEV_RULES=ON -DDETACH_KERNEL_DRIVER=ON`
9. `make`
10. `sudo make install`
11. `sudo ldconfig`
12. `sudo pip install pyrtlsdr`

The preceding steps have downloaded, built, and installed all the dependencies required for the source code. The next steps are to download and run the application named *freqshow*. Follow these three steps to do this:

1. `cd ~`
2. `git clone https://github.com/adafruit/freqshow.git`
3. `cd freqshow`

All that is needed now is to run the program. Plug in the SDR dongle with an antenna attached and ensure that you are in the freqshow folder. Enter the following command:

```
sudo python freqshow.py
```

Figure 7-29 shows the PiTFT screen in the instantaneous frequency spectrum display mode.

The 97.5-MHz peak is the very strong radio signal from a local FM station. The 51-dB level shown on the left of the display indicates that it is indeed a very intense signal.

Touching the Switch mode button will change the screen display from the instaneous frequency to a waterfall display. This display is shown in Figure 7-30 for the same 97.5-MHz FM radio station signal.

The waterfall display plots the signal intensity versus time, with the latest traces starting at the

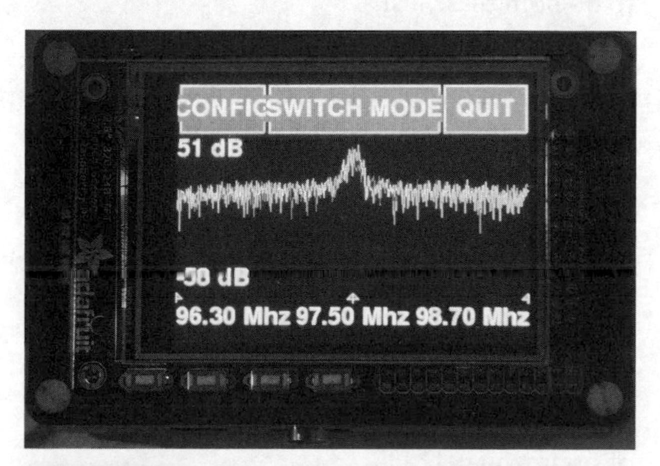

Figure 7-29 PiTFT screen displaying an instantaneous frequency spectrum.

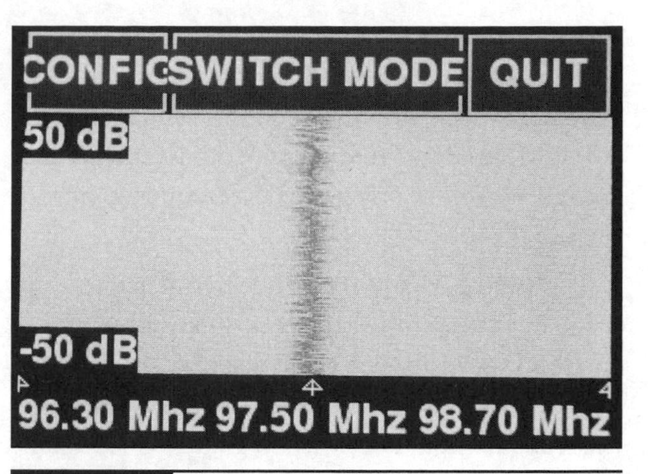

Figure 7-30 Waterfall display.

bottom and scrolling to the top. The traces are color-coded, with blue being a weak signal and red being the strongest. The 97.5-MHz signal in the center of the waterfall is quite red, indicating that it is the most intense signal shown within this particular spectrum bandwidth. Many experienced radio operators feel that the waterfall display is by far the best way to detect weak signals that are embedded in a noisy spectrum. The characteristic line is easy to discern. while a small peak that is bobbing around in an instantaneous display is almost impossible to detect.

Setting the center frequency is simply a matter of touching the Config button in the top left-hand corner. Once you do this, the screen shown in Figure 7-31 appears.

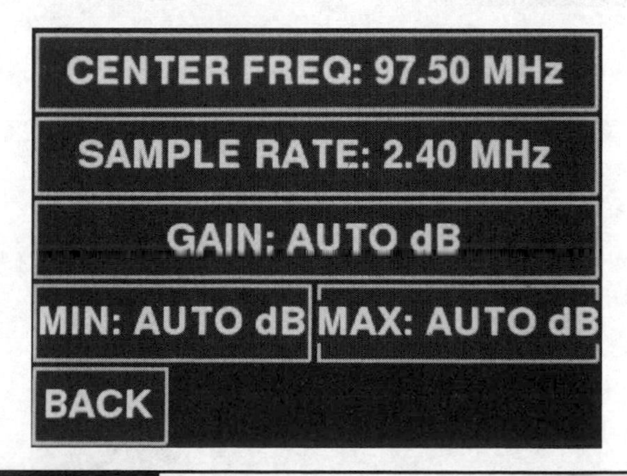

Figure 7-31 Config input screen.

Touch the Clear button to remove the existing entry, and then touch the buttons for the new center frequency. Figure 7-32 shows the entry for a 315-MHz center frequency. You must click the Accept button to enter the new frequency into the program.

I chose the 315-MHz center frequency to see if I could monitor the transmissions from a small RF key fob that I had used in a previous experiment. Sure enough, the analyzer detects the transmission, as shown in Figure 7-33. I did have to fiddle with the camera shutter and the key fob to capture the instant the data packet was sent.

My last test used a hand-held amateur radio transmitter set to the 2-m band at a frequency of 145.000 MHz. Please do not try this unless you are a licensed amateur radio operator, as

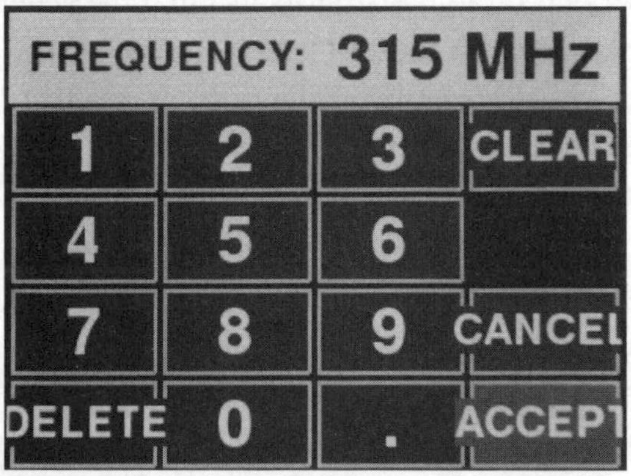

Figure 7-32 New center frequency entered.

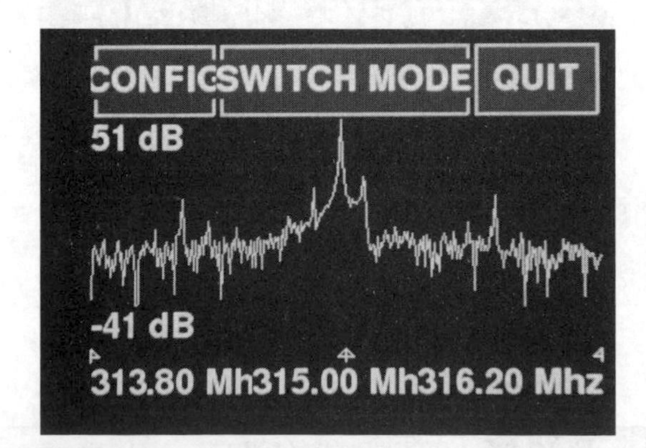

Figure 7-33 315-MHz key fob transmission.

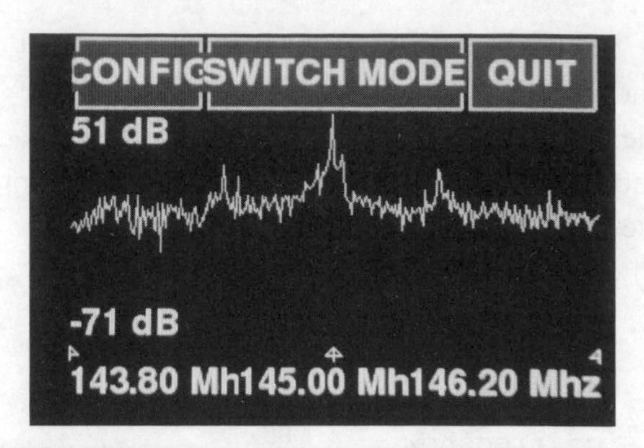

Figure 7-34 145.000-MHz CW transmission from an amateur radio transceiver.

am I. Figure 7-34 shows that the carrier wave was precisely at 145 MHz, as I expected.

In the next section, I will show you how to make this analyzer completely portable and boot right up into the `freqshow` program.

Making the Spectrum Analyzer Portable

It would certainly be advantageous to make the spectrum analyzer completely portable, meaning that it would be self-contained and battery powered. It would also need to be booted directly into the analyzer control program because it is not practical to carry around a keyboard to manually enter the login information. It turns out that it is relatively easy to do both these tasks.

The RasPi and PiTFT are already in a PiBow case, as described in Chapter 2 and shown in Figure 7-29. No more physical modifications are required with regard to the physical case. The battery-operation part was also very easy in that I used an external cell phone power pack, as shown in Figure 7-35.

This particular power pack is rated at 5600 mAh, which should power the analyzer for at least 5 hours, assuming an average current draw of approximately 1 A.

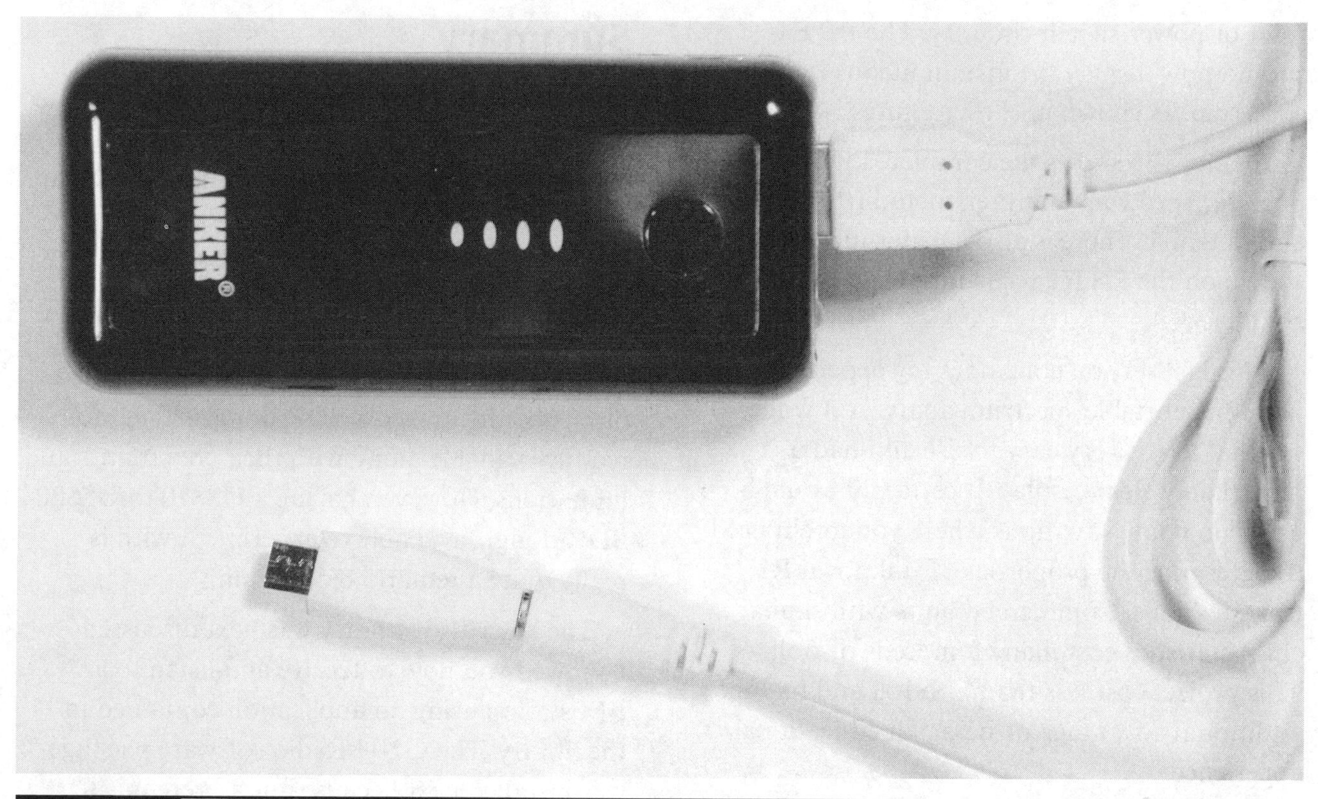

Figure 7-35 External power pack.

Having the RasPi boot directly into the Python program named freqshow.py does require a series of modifications to several files. All the changes are noted next. Just carefully follow the steps and you should be fine.

> **NOTE** I used the nano editor to make all these changes.

1. The file /etc/inittab needs to modified as follows: comment out the line that starts with:

```
1:2345:respawn:/sbin/getty ...
```

and replace it with:

```
1:2345:respawn:/bin/login -f pi
tty1</dev/tty1>/dev/tty1 2>&1
```

2. Add this line to the bottom of the file /etc/profile:

```
sudo python /home/pi/FreqShow/
freqshow.py
```

3. Make these two additions/changes to the freqshow.py code:

   a. Add this line to the very top of the program:

   ```
   #!/usr/bin/python
   ```

   b. Change the line:

   ```
   splash = pygame.image.
   load('freqshow_splash.png')
   ```

   to

   ```
   splash = pygame.image.load('/
   home/pi/freqshow_splash.png')
   ```

4. Do a reboot:

```
sudo reboot
```

Now you can try the auto login by disconnecting and reconnecting the RasPi power supply. I know that it is probably not the best way to reboot, but it is the only way I know of to reboot with the RasPi, which lacks any proper reset or power-switch circuitry. The PiTFT should now display the instantaneous frequency spectrum, as shown in earlier figures.

Figure 7-36 shows the complete kit for the portable spectrum analyzer, including the battery and antenna. There is also a waterfall display shown on the PiTFT receiving two local FM stations.

This last section concludes my experiments with this portable spectrum analyzer. I will say that it was very easy to use and had a remarkably clear display. It certainly would be useful in many situations where you might need to determine the properties of unknown RF transmissions. Commercial units with similar characteristics cost many hundreds of dollars. This project cost less than US$100 and had the additional advantage of being an educational experience.

Summary

I began this chapter with a brief review of some fundamental radiofrequency (RF) concepts because they should help you to understand the important ideas behind software-defined radio (SDR). I and Q signals were discussed next because they are key components in creating a functioning SDR system. SDR simply will not work without I and Q signals. The next section discussed the low-cost SDR dongles that are readily available along with their constraints and limitations. However, having a US$10 to US$20 SDR dongle available to experiment with is really quite a remarkable bargain.

The key rtl-sdr library was next discussed. I showed you how to receive and listen to an FM station using an application contained in the library. The GNU Radio software package was installed next because it is a prerequisite to monitoring many different types of aviation data signals, including Mode-S transponders

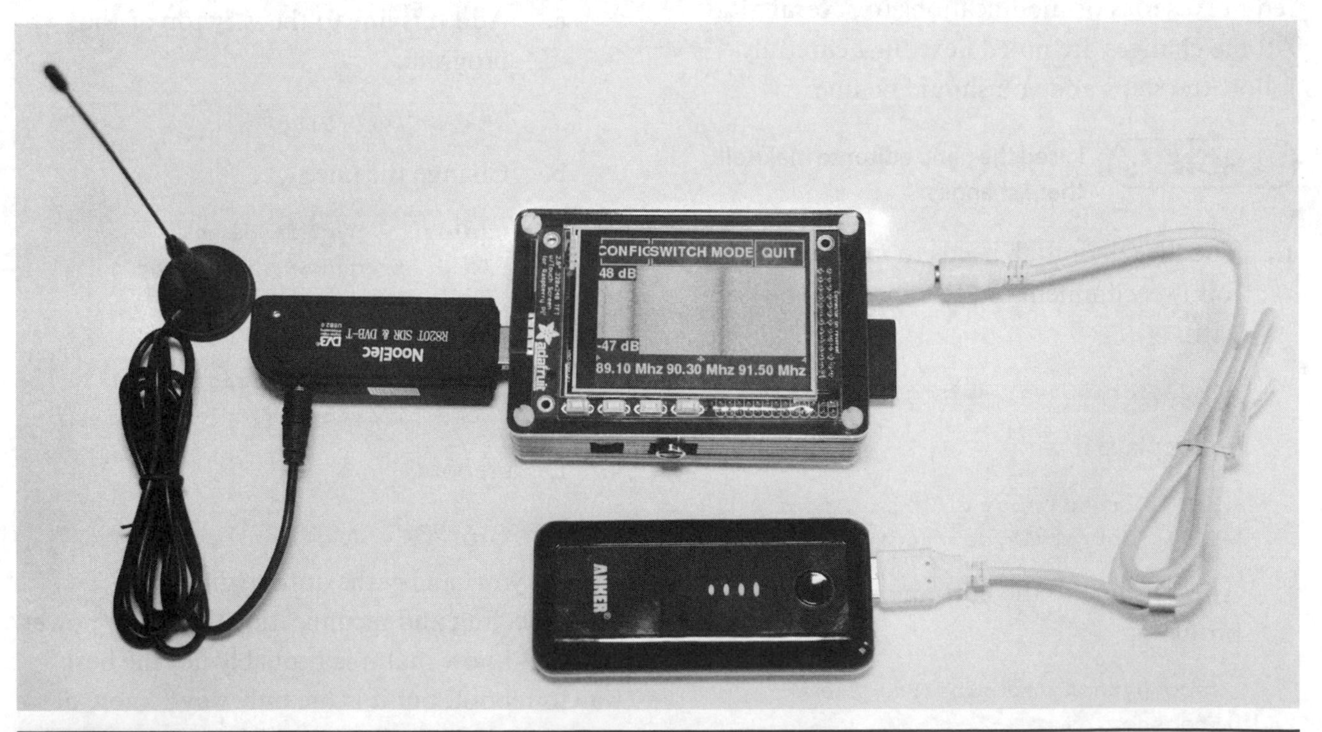

Figure 7-36 Complete portable spectrum analyzer kit.

and ACARS data packets. I went into several in-depth demonstrations of how to monitor and decode ACARS transmissions. I believe that you find this to be a very interesting subject once you start monitoring these aviation signals.

The chapter concluded with another demonstration project of a portable spectrum analyzer based on using the PiTFT-RasPi combination discussed in Chapter 2. This project used the SDR dongle to receive and display small sections of spectrum ranging from 24 MHz up to 1766 MHz.

CHAPTER 8

BrickPi Python Robot

IN THIS CHAPTER, I WILL BE showing you how to build a robotic car out of Lego building blocks with control provided by a RasPi. Using a Lego EV3 Mindstorms kit is normally the way such a project would be done with a control module nicknamed the "brick."

BrickPi

Figure 8-1 shows a current EV3 control module, which is powered by six AA batteries contained within the case.

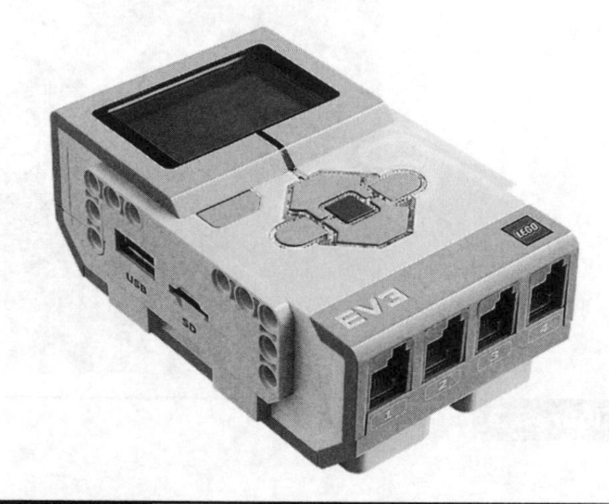

Figure 8-1 EV3 brick.

Instead of the EV3 brick, I will be using a BrickPi controller, which is a RasPi coupled with a special I/O board to control the Lego robotic car. I will also refer to the robotic car as the *CastorBot*. I made the controller substitution for the obvious reason that this project is in a book concerning RasPi control. However, another important reason was to demonstrate how an object-oriented language such as Python can be used to program this robot instead of the drag-and-drop graphical programming environment normally used with Mindstorms. I do wish to point out the drag-and-drop programming is perfectly fine for most users, but the Python approach allows for much more flexibility and utility in creating control programs for the CasterBot.

Figure 8-2 shows a BrickPi with a RasPi already mounted in its enclosure.

It is necessary to delve into how Mindstorms sensors and motors are controlled in order to understand how the BrickPi functions. I will first examine the sensor ports.

Chapter 8 Parts List

Item	Model	Quantity	Source
Lego EV3 Mindstorms kit	31313	1	amazon.com
BrickPi	Base kit	1	dexterindustries.com
Raspberry Pi Model B+	83-16530	1	mcmelectronics.com

Figure 8-2 BrickPi.

Sensor Ports

All Mindstorms sensors are controlled by using commands sent over the I2C bus. This bus was first introduced in Chapter 3, where it was used to establish a data communications link between the Arduino coprocessor and the Lidar module. In this project, the I2C Protocol will be used to create direct communications links between the RasPi and the Mindstorms sensors. You might want to go back to Chapter 3 and refresh yourself on how the I2C functions, but it is not really essential to build this project because the single sensor used in this project is pluggable and will play after the software is loaded. Table 8-1, which is courtesy of Wikipedia, shows the makeup of the connector used to interconnect all Mindstorms sensors with the BrickPi I/O board. All the electrical contacts are clearly identified in Figure 8-2 for your reference.

TABLE 8-1 Mindstorms Sensor Connector-Pin Layout

Pin	Name	Function	Color
1	ANALOG	Analog interface, +9-V supply	White
2	GND	Ground	Black
3	GND	Ground	Red
4	IPOWERA	+4.3-V supply	Green
5	DIGIAI0	I2C clock (SCL), RS-485 B	Yellow
6	DIGIAI1	I2C data (SDA), RS-485 A	Blue

You may have noticed that the I2C signal lines also have a corresponding RS-485 designation shown in the "Function" column. The reason for this is that an RS-485 serial link can communicate over much longer distances, up to 1 km, while an I2C is restricted to much closer distances, typically on the order of tens of meters. The RS-485 and I2C are very different from one another. The I2C is *synchronous*, meaning that it requires a clock signal to function, while RS-485 is *asynchronous*, relying instead on start and stop bits to form a data frame or packet. As far as I can determine, all the Mindstorms kits use an I2C for communications, but it is nice to know that the Mindstorms brick designers planned for optional longer-range communications.

You should also note that while the sensor/motor connector looks somewhat similar to an ordinary RJ11 telephone jack, it is not the same. For instance, the latching device is offset to the left as you look at the Mindstorms connector, with the pins facing down. The RJ11 has its latching device in the center top. Additionally, the pin spacing is slightly different, with the RJ11 spaced at 0.60 mm between pins and the Mindstorms spacing at 0.70 mm between pins. The bottom line in all this is that you should not try to jam a RJ11 connector into a Mindstorms jack; it will not work and likely will cause serious damage if you attempt to apply power. Lego Mindstorms, as well as a few other suppliers, have these connectors available for purchase if you want to build your own.

I will next discuss the Mindstorms motor ports, now that you have had a good introduction to the sensor ports.

Motor Ports

Mindstorms motor ports use the same connector as used with the sensor ports, but the pin designations differ significantly, and they do not use I2C as a communication protocol.

CAUTION **It is important to ensure that only motors are plugged into the motor jacks and only sensors into the sensor jacks. There is a potential for sensor electrical damage if you inadvertently plug a sensor into a motor jack because motor supply voltage may be present on pins 1 and 2, which might harm the sensor.**

Table 8-2 shows the makeup of the connector used to interconnect Mindstorms motors with the BrickPi I/O board.

TABLE 8-2 Mindstorms Motor Connector-Pin Layout

Pin	Name	Function	Color
1	MOTOR 1	Motor 1 power supply	White
2	MOTOR 2	Motor 2 power supply	Black
3	GND	Ground	Red
4	IPOWERA	+4.3-V supply	Green
5	TACH01	Phase A, quadrature encoder	Yellow
6	TACH02	Phase B, quadrature encoder	Blue

The MOTOR 1 and MOTOR 2 lines refer to the power supply for a single motor. Placing the positive power supply lead on MOTOR 1 and the negative lead on MOTOR 2 will cause the motor to rotate in a certain direction. Reversing the polarity will cause the motor to rotate in the opposite direction. The motor supply lines are connected to the AtMega pulse-width modulation (PWM) lines with an inline driver chip, which I further explain in the next section.

It is now time to examine the BrickPi's specialized I/O board in more detail because it is the reason why a RasPi can function as a substitute controller for the Lego brick.

BrickPi Specialized I/O Board

The block diagram in Figure 8-3 shows how the specialized I/O board functions. Incidentally, from now on I will refer to this board simply as the I/O board because you already know what I am talking about.

This block diagram is based on a schematic provided by the BrickPi manufacturer, Dexter Industries. One AtMega328 processer is shown in the figure, and it provides the I2C control signals for sensors S1, S2, and S5. The control signals for motors A and B are also provided

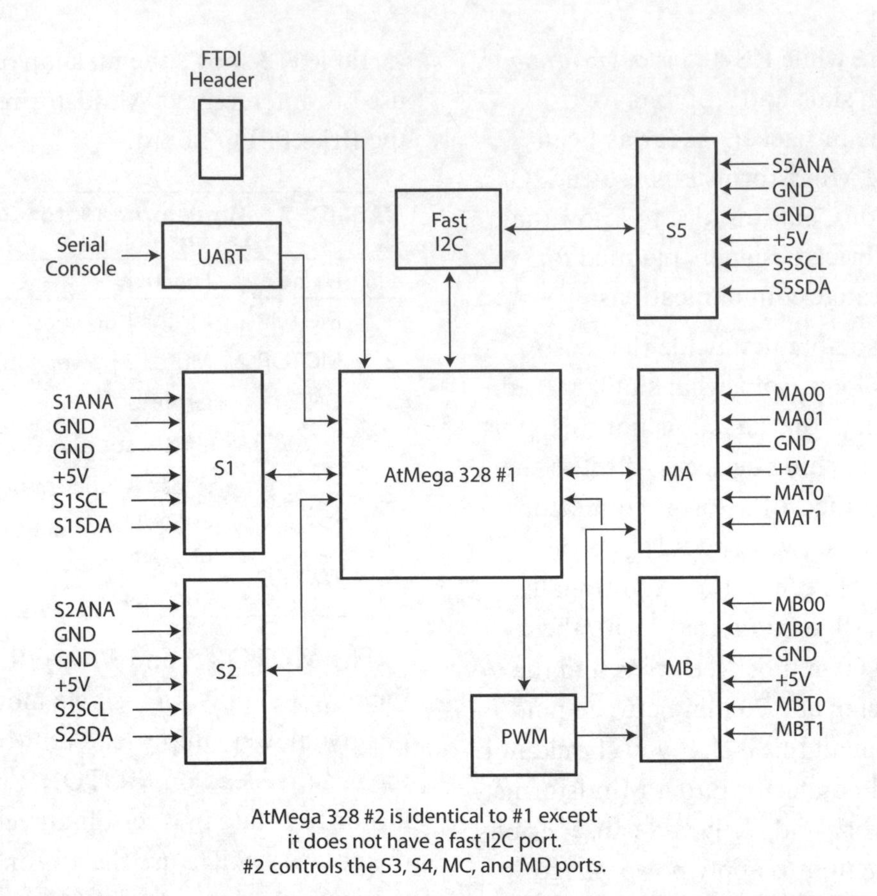

Figure 8-3 Block diagram for the I/O board.

by this chip in conjunction with a PWM driver chip. The I/O board also contains another identical AtMega328 processor that controls sensor ports S3 and S4 as well as motor ports C and D. These processors may be considered as coprocessors, as discussed in Chapter 2, because their primary functions are to process and manipulate data that otherwise would require the RasPi to handle. Without these coprocessors, it is highly likely that the RasPi would not be capable of processing all the data constantly flowing from the sensors and motors and thus be unable to control the robot as required. The data generated from the motors are encoder pulses, which come from quadrature encoders mounted in each of the motors. Hundreds of pulses are generated for every revolution of the motor shaft, and every pulse must be counted to track the motor's operation accurately.

Each AtMega328 controls the speed and rotation direction of two motors by sending low-level digital signals to a TI SN754410NE driver chip, which, in turn, converts these signals into high-level PWM pulses capable of driving the motors directly.

The sensor ports are configured in a parallel multidrop arrangement with external pull-up resistors to ensure that the I2C bus functions properly. The sensor input lines also are connected to the AtMega328 analog-to-converter (ADC) lines so that any compatible sensors with analog voltage outputs, such as a light-intensity sensor, also may be handled.

Of course, all the uncommitted RasPi GPIO pins are also available via the 26-pin header extension on the BrickPi board, as shown in Figure 8-4.

Figure 8-4 GPIO extension pins.

Another Python library named python-rpi.gpio enables I/O for these GPIO pins. I will demonstrate how to use one of the RasPi GPIO pins to control one of the BrickPi's LEDs in a later section.

At this point, I would recommend that you label all the sensor and motor ports, as shown in Figure 8-5. This will help you to identify the proper jacks to use when connecting the motors and sensors.

This concludes my introduction to the BrickPi. It is now time to introduce the robotic car and show you how to build it.

The CasterBot

The CasterBot shown in Figure 8-6 was built from the components contained in Lego EV3 Kit Number 31313. It is entirely possible to customize and build the CasterBot from other

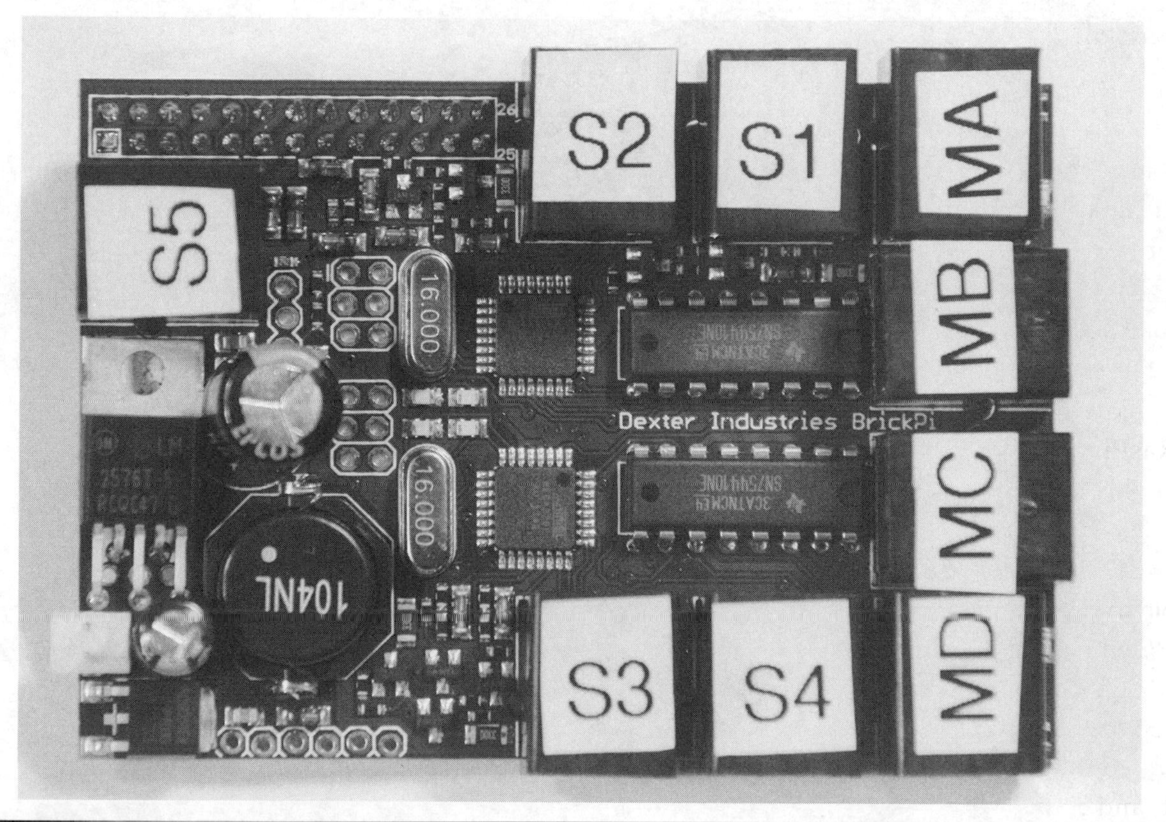

Figure 8-5 I/O board with labeled ports.

Lego kits, and I would urge you to do so, especially if you already have some components from other kits. The build plan and bill of material (BOM) are based on the 31313 kit, and you should use it as a template to follow and modify as needed to suit your situation. Of course, you will need motors and a sensor compatible with the Mindstorms connectors, which are either the NXT or EV3 models.

The build diagram, BOM, and build plan for the CasterBot are all on this book's companion website. All the instructions are quite detailed, and they should allow you to build the CasterBot without any problems. The only parts missing from the BOM are the BrickPi and four connector pegs used to mount the BrickPi to the two 13M Lego beams mounted on top of the CasterBot. In addition, you will need three small flexible cables to connect the two motors and the ultrasonic sensor to the BrickPi. The cable connections are as follows facing the CasterBot:

- Left motor connects to motor port B.
- Right motor connects to motor port C.
- Ultrasonic sensor connects to sensor port S1.

The CasterBot battery supply now warrants some additional discussion in the next section.

CasterBot Power Supply

The CasterBot has a substantial current draw when operating with two motors. I measured the average current drain to be approximately 0.5 A, which would quickly drain a six-pack of AA batteries, which typically powers the Lego brick. I chose instead to use a three-cell (3S) lithium ion polymer (LiPo) battery, which provides

Figure 8-6 The CasterBot.

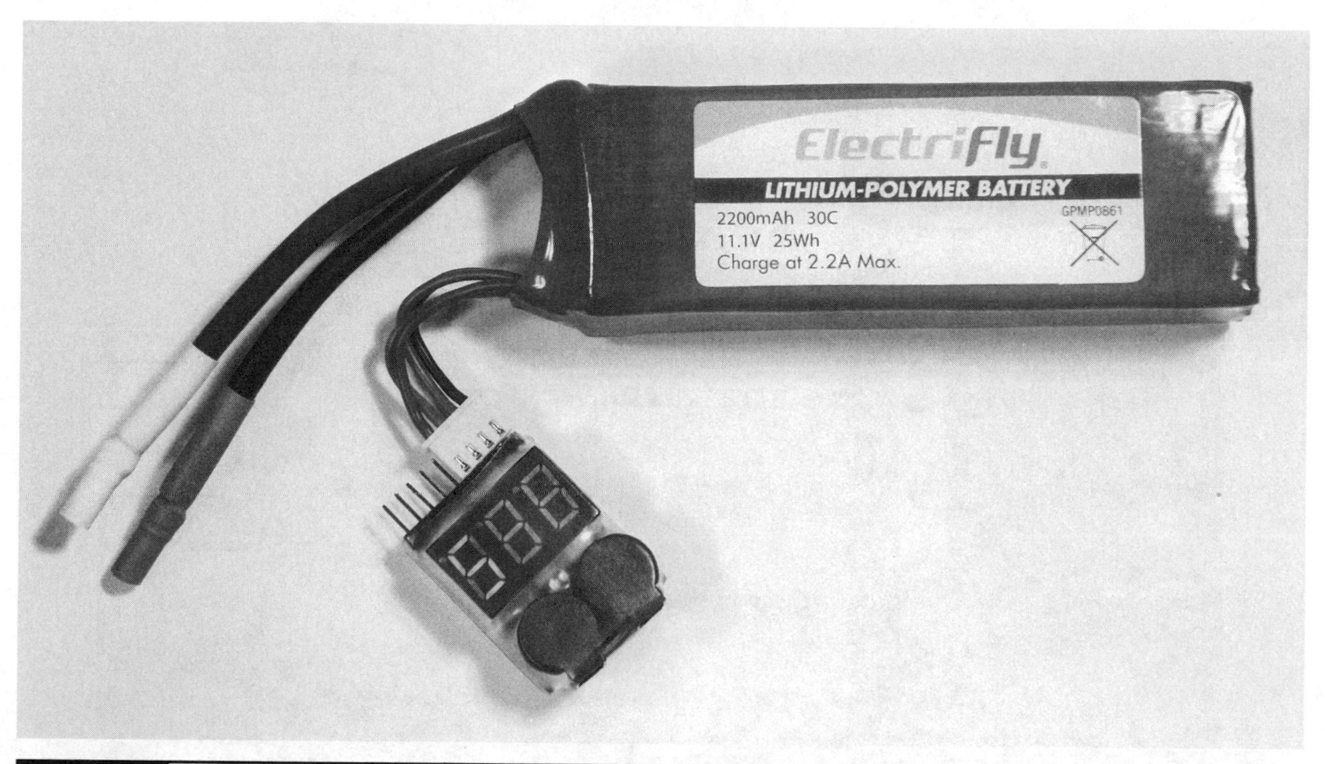

Figure 8-7 3S LiPo battery.

approximately 12 V when fully charged. Figure 8-7 shows the LiPo battery used in this project.

Also shown in the figure is a battery condition monitor attached to the LiPo's balancer connection. This device is very inexpensive and is well worth purchasing because it can indicate when the LiPo is nearing its discharged state. You do not want to discharge this particular LiPo battery below 10.2 V because it becomes quite difficult for the automatic charger to function correctly with a battery in this state.

This battery is also rated at 2200 mAh, which means that it should provide an operational time of about 2 hours for the CasterBot before it reaches its minimum voltage, as I described earlier, and needs a recharge. The operational time also may be shortened somewhat if the CasterBot is operated at maximum speed for an extended period of time.

CAUTION **This battery must be recharged using an approved LiPo charger. Attempting to use a charger designed only for nickel-cadmium (NiCad) or lead-acid batteries potentially could cause a mishap that might involve an exploding battery or even one that catches fire.**

Some readers also may know that Mindstorms motors are usually operated with 9 V and may wonder if 12 V would possibly overload them. You may be assured that 12 V will not harm them but simply cause them to rotate about 25 percent faster than with a 9-V supply. The BrickPi board also has a voltage regulator, which can easily handle the 12 V to supply regulated 5 V to the RasPi as well as to its own processors.

Figure 8-8 shows a bottom-side view of a CasterBot with the battery installed in a cradle that was designed especially to hold a battery of this particular size.

I used gold-plated pins and sockets to connect the battery to the wires leading to the BrickPi's power connector. This power connector comes with the BrickPi and has a 9-V battery clip

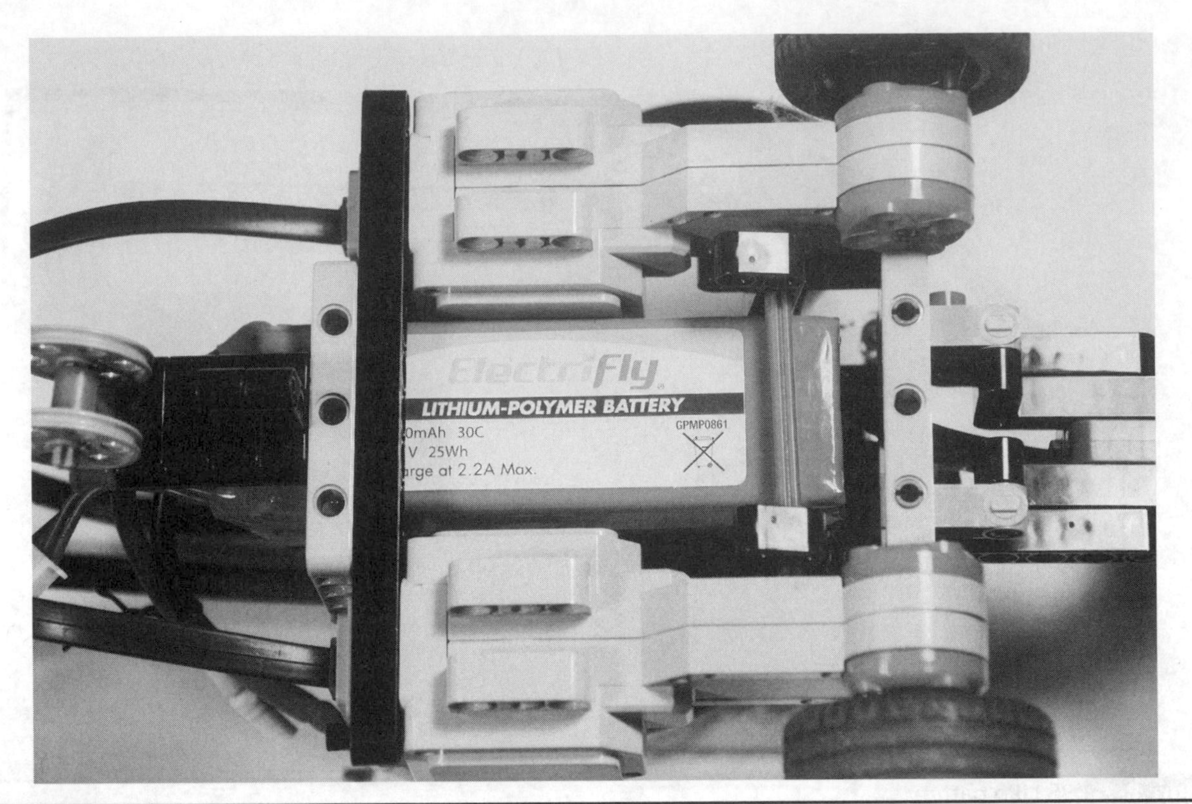

Figure 8-8 LiPo battery installed in a CasterBot.

attached. I cut this clip off and soldered the loose ends to two wires with the pin connectors already attached. The battery side likewise had two pin jacks soldered onto its wires. Be sure that you maintain the battery polarity, that is, positive to positive, or red, leads and negative to negative, or black, leads. Figure 8-9 shows a close-up of the battery connection configuration.

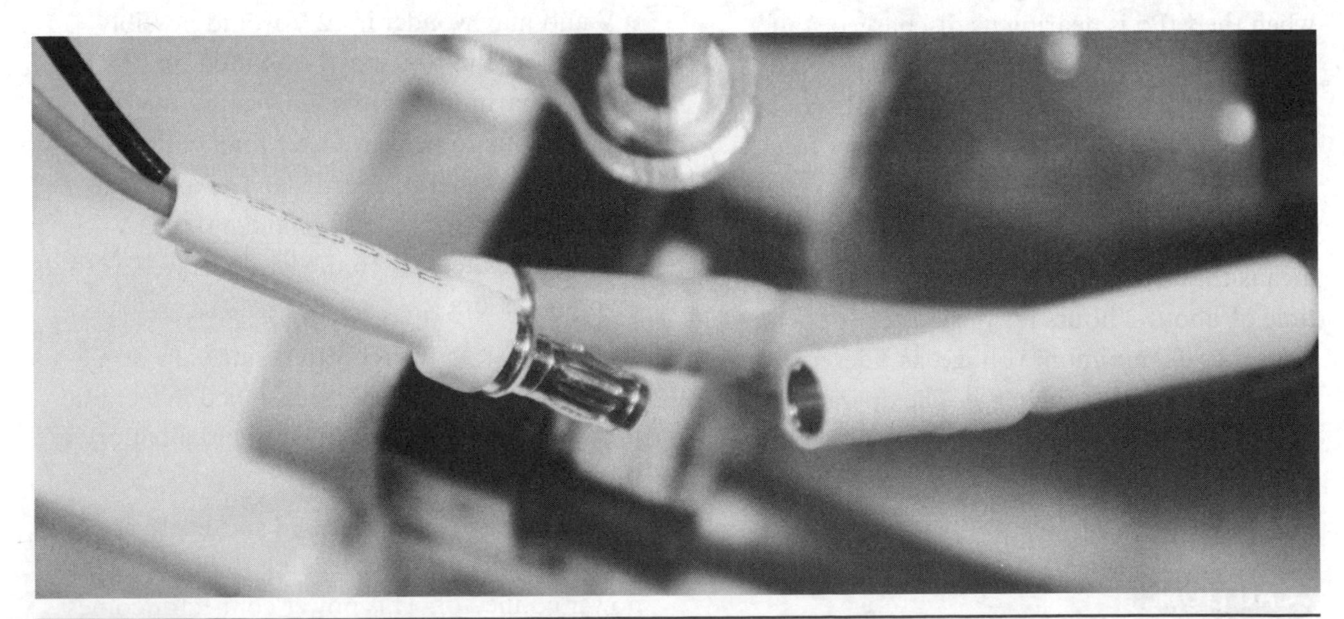

Figure 8-9 Battery connections.

The battery pins and jacks I used are readily available from most hobby stores because they are commonly used in radio-controlled (R/C) aircraft and cars. I also chose not to include a battery switch; I simply unplug the positive lead when I need to turn the CasterBot off.

WiFi Dongle

You also will need to install a WiFi dongle to enable the CasterBot's untethered movement. While it is possible to use a long USB cable to control the CasterBot from a laptop, it is not an optimal solution. A small WiFi dongle, as shown in Figure 8-10, will allow the bot to be completely portable while still maintaining a wireless connection to the laptop using SSH.

The WiFi dongle has a maximum range of 100 m, which should be more than adequate for all CasterBot movements.

You should also realize that you will need to do an initial setup on the BrickPi's RasPi using a normal workstation configuration, as described in Chapter 1. However, you first need to download and install a specially modified Wheezy Linux distribution, which is discussed in the next section.

Figure 8-10 WiFi dongle.

Software Installation and Configuration

The following instructions are based on an introductory tutorial available from the Dexter Industries website (www.dexterindustries.com). Please follow these next steps in order to successfully install the software required to program and control the CasterBot:

1. Download the Wheezy distribution created for the BrickPi at http://sourceforge.net/projects/dexterindustriesraspbianflavor/.
2. Unzip the download into an image file using WinZip or 7Zip.
3. Write the image to an SD card using the Win32DiskImager program.
4. Plug the SD card into the RasPi, which is set up in a stand-alone configuration with the WiFi dongle plugged in.
5. Power-on the RasPi, and enter the regular user name/password.
6. Run the raspi-config program by entering `sudo rasp-config`.
7. Expand the file system.
8. Set the appropriate location and time zone.
9. Check that SSH is enabled in the Advanced Options.
10. Close out of the rasp-config app, and start the GUI desktop by entering `startx`.
11. Click on the WiFi config icon, and scan for your access point.
12. Select your access point, enter the appropriate passphrase, and then click on Connect. Close the app.
13. Click on Accessories, and click on LX Terminal.

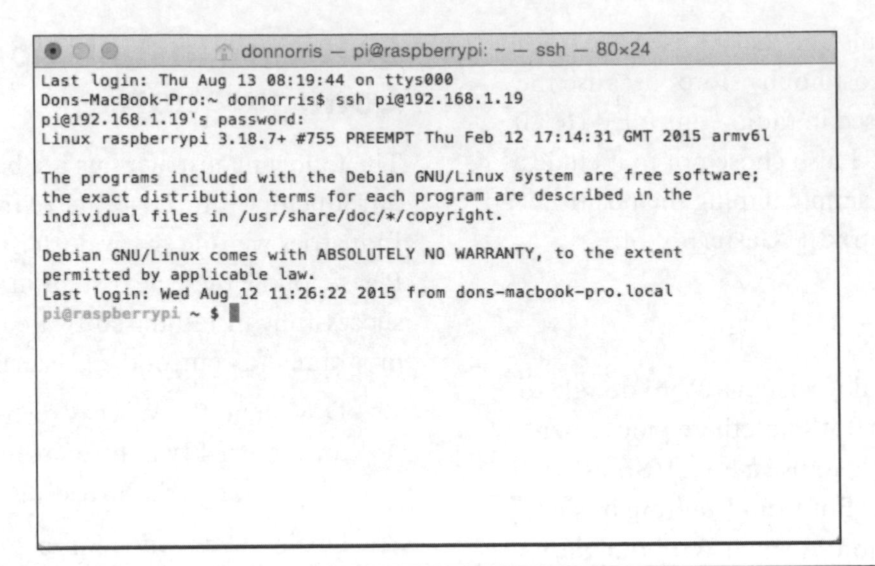

donnorris — pi@raspberrypi: ~ — ssh — 80×24
Last login: Thu Aug 13 08:19:44 on ttys000
Dons-MacBook-Pro:~ donnorris$ ssh pi@192.168.1.19
pi@192.168.1.19's password:
Linux raspberrypi 3.18.7+ #755 PREEMPT Thu Feb 12 17:14:31 GMT 2015 armv6l
The programs included with the Debian GNU/Linux system are free software;
the exact distribution terms for each program are described in the
individual files in /usr/share/doc/*/copyright.
Debian GNU/Linux comes with ABSOLUTELY NO WARRANTY, to the extent
permitted by applicable law.
Last login: Wed Aug 12 11:26:22 2015 from dons-macbook-pro.local
pi@raspberrypi ~ $

Figure 8-11 Initial RasPi SSH login on a MacBook Pro.

14. Enter `ifconfig`, and write down the IP address that was assigned by the access point. Close the Terminal window.
15. Close the desktop by clicking on the Reboot selection.

Next, connect to the RasPi using an SSH connection from your laptop with the IP address from step 10 and using an appropriate program suitable for the laptop, that is, PuTTY for a Windows machine and Terminal for a Mac. Figure 8-11 shows the Terminal screen for my initial login to the RasPi from my MacBook.

Now you are ready to run the following program, which will allow you to remote control the CasterBot. The program is named simplebot_speed.py and is located in the following directory:

```
/home/pi/BrickPi/Project_Examples/
simplebot/
```

You should change into this directory and enter the following to run the program:

```
sudo python simplebot_speed.py
```

Following are the single-letter commands that you may enter to instruct the CasterBot to perform various movements:

```
w - go forward
a - turn left
d - turn right
s - go back
x - stop
t - go faster
g - go slower
```

You will also need to press the ENTER key after typing a command to send it to the CasterBot. The source code for this program is listed next with annotations to assist you in understanding the various functions.

```
#!/usr/bin/env python

from BrickPi import * #import BrickPi.py file to use BrickPi ops

speed = 200 #initial speed (can lower it for 12V ops)

cmd = 'x'     #last used command (used when changing speed)

motor1 = PORT_B  # motor1 is on PORT_B
motor2 = PORT_C  # motor2 is on PORT_C

#Move Forward
def fwd():
   BrickPi.MotorSpeed[motor1] = speed
   BrickPi.MotorSpeed[motor2] = speed
#Move Left
def left():
   BrickPi.MotorSpeed[motor1] = speed
   BrickPi.MotorSpeed[motor2] = -speed
#Move Right
def right():
   BrickPi.MotorSpeed[motor1] = -speed
   BrickPi.MotorSpeed[motor2] = speed
#Move backward
def back():
   BrickPi.MotorSpeed[motor1] = -speed
   BrickPi.MotorSpeed[motor2] = -speed
#Stop
def stop():
   BrickPi.MotorSpeed[motor1] = 0
   BrickPi.MotorSpeed[motor2] = 0

#Move the simplebot depending on the command
def move_bot(val):
   global cmd
   if val=='w':
     cmd='w'
     fwd()
   elif val=='a' :
     cmd='a'
     left()
   elif val=='d':
     cmd='d'
     right()
   elif val=='s':
     cmd='s'
     back()
   elif val=='x':
     stop()
```

```
def main():
    global speed
    BrickPiSetup()                       # setup the serial port for communication

    BrickPi.MotorEnable[motor1] = 1 #Enable the Motor1
    BrickPi.MotorEnable[motor2] = 1 #Enable the Motor2

    BrickPiSetupSensors()                #Send sensors properties to BrickPi

    BrickPi.Timeout = 10000              #Set timeout value for motor ops
    BrickPiSetTimeout()                  #Set the timeout

    while True:
      inp=str(raw_input())               #Take input from the terminal
      move_bot(inp)                      #Send command to move the bot
      if inp=='t':                       #Increase the speed
          print "Speed: ",speed
          if speed > 234:
              speed = 255
          else:
              speed = speed + 10
          move_bot(cmd)                  #update motor values
      elif inp=='g':                     #Decrease the speed
          print "Speed: ",speed
          if speed < 11:
              speed = 0
          else:
              speed = speed - 10
          move_bot(cmd)

      BrickPiUpdateValues() #update motor values
      time.sleep(.01) #sleep for 10 ms

if __name__ == "__main__":
    main()
```

I will use this program as a base for additional functions, which will include an obstacle-avoidance algorithm. However, I will briefly discuss the ultrasonic sensor in the next section before adding the obstacle-avoidance functionality.

Ultrasonic Sensor

Figure 8-12 is a close-up of the ultrasonic sensor that is used to detect obstacles in the path of the CasterBot.

This particular sensor comes from an NXT Lego kit, but it is completely compatible with the BrickPi. The EV3 series also has a similar sensor, which may be used without any modifications to the software.

The ultrasonic sensor contains an embedded microprocessor as part of the encapsulated sensor hardware. This processor controls the ultrasonic transmitter and receiver transducers that physically measure distance by bouncing discrete sound wave pulses off objects and

Figure 8-12 Ultrasonic sensor.

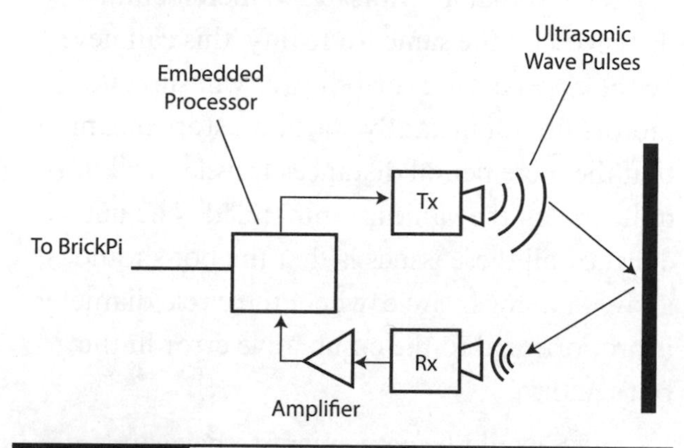

Figure 8-13 Ultrasonic sensor block diagram.

timing how long the sound takes to transit. The distance is easily calculated because the speed of sound in air is relatively constant. This is very similar to how bats navigate in caves and attics. Figure 8-13 is the sensor's block diagram, which shows how it functions.

The sensor uses an embedded processor, which off-loads any additional computational tasks from a RasPi in a similar fashion to the way the AtMega coprocessors function on the BrickPi. The sensor's processor measures sound pulse times to a resolution as fine as 1 ms, which a RasPi could not handle without introducing some timing errors.

The ultrasonic sensor measures distances from 3 to 250 cm with an accuracy of approximately ±2 cm. Distance measurements also depend on the size and texture of the object that reflects the sound pulses. A wall provides excellent reflections, while a stuffed teddy bear would be more problematic.

The ultrasonic sensor requires a few statements to integrate into a Python program. The first one associates the sensor with a specified port:

```
#Set the type of sensor connected to
PORT_1

BrickPi.SensorType[PORT_1] = TYPE_
SENSOR_ULTRASONIC_CONT
```

The next statement commands the BrickPi to update all sensor and motor data:

```
#Command the BrickPi to update values
for sensors/motors
result = BrickPiUpdateValues()
```

The final statement retrieves the most current ultrasonic distance measurement and assigns it to the variable `dest`:

```
dest = BrickPi.Sensor[PORT_1]
```

The issue now is to create software, which will be activated if an obstacle is detected within a preset threshold in the Casterbot's path. I discuss this software in the next section.

Obstacle-Avoidance Algorithm

Creating obstacle-avoidance software is an interesting exercise that involves thinking through all the various scenarios that the CasterBot may encounter in its travels from a start to a finish point. I had to severely constrain the types of obstacles to use because it would be literally impossible to account for all possible obstacle types. I chose to use a small cardboard box because it would be easy to place in the CasterBot's intended path, and if it were struck, it should not cause any damage to the robot. The obstacle-clearance software should be reasonably modifiable to accommodate other types of objects once this initial program is shown to be effective.

I don't believe that there is a formal procedure that will create an optimal obstacle-avoidance

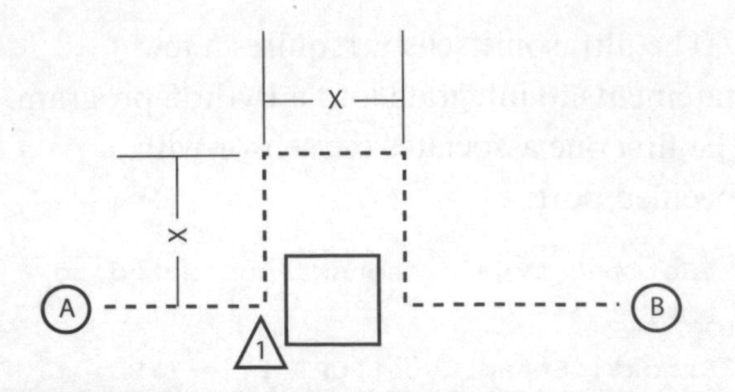

Figure 8-14 Obstacle playing field diagram.

algorithm. I have found that using a two-dimensional playing field with the start and stopping points as well as the obstacle in place helped me to visualize how the robot might maneuver between the start and stopping points while avoiding the obstacle. Figure 8-14 is a diagram of an example playing field with start point *A* and finish point *B*.

An obstacle detection point is shown as a numbered triangle. The CasterBot path is shown as a dotted line with the *X* predefined incremental path length. Table 8-3 is a pseudocode listing that describes the CasterBot behavior as it starts on a straight-line path from *A* to *B* and encounters the obstacle placed near point number 1. Note that I deliberately used only one obstacle in the direct path between *A* and *B* to simplify the example. Even so, the code to implement the avoidance behavior rapidly becomes complex.

All the steps to avoid an obstacle are contained in the method named `obstacleAvoid()`. This method is called when the ultrasonic sensor senses an obstacle within a threshold distance, which in this case is set at 25 cm. The whole process of tracking the bot is called *dead reckoning* and presumes that all 90° turns are perfect and that all transversed incremental distances are the same. In reality, this can never be the case because the bot tires will slip and motors do not instantly start and stop, meaning that the incremental distances traveled will not quite match the values commanded. The net effect of all these issues is that the bot should arrive at a finish circle, where the circle diameter is proportional to the cumulative error in the bot's path.

It also would be convenient to have an indicator on the BrickPi to show when the obstacle-avoidance algorithm is activated. This is easily accomplished because the BrickPi has two LEDs installed on the board edge that can be individually controlled by using a Python GPIO library named python-rpi.gpio. You will need to install the library on the RasPi by using this command:

TABLE 8-3 Pseudocode for a CasterBot Obstacle-Avoidance Algorithm

Playing Field Location	Obstacle Number	Behavior
Start point *A*	N/A	Straight-ahead motion on a direct path to finish point *B*
Obstacle detected	1	a. Turn left 90° b. Move *X* distance c. Turn right 90° d. Move *X* distance e. Turn right 90° f. Move *X* distance g. Turn left 90° h. Move *X* distance
On direct path	N/A	a. Straight-ahead motion in units of *X* distance b. Test for path completion (num*X* < 1) c. If at finish, exit program

```
sudo apt-get install python-rpi.gpio
```

You will need to import the library by using this command:

```
import RPi.GPIO as GPIO
```

The mode must next be set, followed by setting the GPIO that is connected to the LED as an output:

```
GPIO.setmode(GPIO.BOARD)
GPIO.setup(12, GPIO.OUT)
```

The `12` used in the setup method actually refers to `RasPi GPIO pin 18`. The other `BrickPi LED` is designated as `13` but really refers to `RasPi GPIO pin 27`. The pin numbering complies with the WiringPi designations that I discussed in Chapter 6. The python-rpi.gpio library is based, in large part, on the WiringPi library. I recommend that you review Figure 6-19, which clearly illustrates the relationships between the WiringPi and RasPi GPIO pin labels.

Turning on the LED is accomplished using this statement:

```
GPIO.output(12, True)
```

Of course, all you need to do is change `True` to `False` to turn off the LED.

I added the ultrasonic sensor, obstacle-avoidance algorithm, and indicator control to the preceding program and renamed it *obstacleAvoid.py*. I also modified the original program to remove the movement commands because they were unnecessary for this particular situation. The program is listed next and is also available on this book's companion website.

```
#!/usr/bin/env python

from BrickPi import *            #import BrickPi.py file to use BrickPi
import RPi.GPIO as GPIO          #import python-rpi.gpio for LED crtl

GPIO.setmode(GPIO.BOARD)
GPIO.setup(12, GPIO.OUT)
GPIO.output(12, False)           #reset the LED initially

#Set the type of sensor connected to PORT_1
BrickPi.SensorType[PORT_1] = TYPE_SENSOR_ULTRASONIC_CONT

speed = 80                       #initial speed
delay = 0.9                      # unit of delay timing in seconds

motor1 = PORT_B                  # motor1 is on PORT_B (Right)
motor2 = PORT_C                  # motor2 is on PORT_C (Left)
#Move Forward
def fwd():
    BrickPi.MotorSpeed[motor1] = speed
    BrickPi.MotorSpeed[motor2] = speed
    BrickPiUpdateValues()
    time.sleep(0.1*delay)

#Move Left
def left():
    BrickPi.MotorSpeed[motor1] = speed
```

```
    BrickPi.MotorSpeed[motor2] = -speed
    BrickPiUpdateValues()
    time.sleep(0.1*delay)

#Move Right
def right():
    BrickPi.MotorSpeed[motor1] = -speed
    BrickPi.MotorSpeed[motor2] = speed
    BrickPiUpdateValues()
    time.sleep(0.1*delay)

#Move backward
def back():
    BrickPi.MotorSpeed[motor1] = -speed
    BrickPi.MotorSpeed[motor2] = -speed
    BrickPiUpdateValues()
    time.sleep(0.1*delay)
#Stop
def stop():
    BrickPi.MotorSpeed[motor1] = 0
    BrickPi.MotorSpeed[motor2] = 0
    BrickPiUpdateValues()
    time.sleep(0.1*delay)

#Move forward several inches at a time
def moveX():
    fwd()
    time.sleep(1.5*delay)
    BrickPiUpdateValues()
    time.sleep(0.1*delay)

#Obstacle avoidance
def obstacleAvoid():
    left()
    time.sleep(0.85*delay)
    fwd()
    time.sleep(3.0*delay)
    right()
    time.sleep(0.85*delay)
    fwd()
    time.sleep(5.0*delay)
    right()
    time.sleep(0.85*delay)
    fwd()
    time.sleep(3.0*delay)
    left()
    time.sleep(0.85*delay)
```

```
def main():
    global speed
    global dest
    global totalDistance

    BrickPiSetup()#setup the serial port for communication

    BrickPi.MotorEnable[motor1] = 1 #Enable the Motor1
    BrickPi.MotorEnable[motor2] = 1 #Enable the Motor2

    BrickPiSetupSensors()#Send sensors properties to BrickPi

    BrickPi.Timeout = 10000          #Set timeout value for motor ops
    BrickPiSetTimeout()              #Set the timeout
    GPIO.output(12, True)            #turn on LED

    totalDistance = input("Distance (cm) between start and stop points: ")
    numX = totalDistance/10

    while True:
        moveX()
        numX -= 1
        print numX #number of path increments left
        if( numX < 1):
            print "At the finish area"
            stop()
            exit()
        BrickPiUpdateValues()
        time.sleep(.01*delay)
        dest = BrickPi.Sensor[PORT_1]   #get distance
        if( dest < 25): #about 10 inches to obstacle
            obstacleAvoid()

if __name__ == "__main__":
    main()
```

In the next section, I discuss how the obstacle-avoidance algorithm worked, pointing out the successes and where some improvements should be made.

Obstacle-Avoidance Demonstration

Figure 8-15 shows the CasterBot setup on the real obstacle course. The total distance between the start and stop points is 120 cm, or approximately 5 ft, so you need a moderate-sized area to set up the course.

I also set up an SSH session between the CastorBot and my MacBook Pro to initiate the program and start the robot on its path. Run the program by entering:

```
sudo python obstacleAvoid.py
```

You will be prompted to enter the direct path length between the start and finish points in

Figure 8-15 Obstacle course.

centimeters. This path length is divided by 10 to create the number of *X* increments that the bot uses to transverse the path. The program will display the number of path increments left to be completed and will halt when it completes the last increment.

I initially observed that the 90° turns were more like 120° turns because the timing was off a bit. I subsequently reduced the time spent in a turn to about 0.8 second, which corrected the overturning problem. I also noticed that the playing surface on which I ran the CasterBot seriously affected its performance. Running on a carpet would slow the bot down, which then would cause the turns to be much less than 90° and subsequently cause the obstacle avoidance to fail. In addition, the bot would not complete the commanded path length because completing the total path depends on time duration, not actual distance traveled. I tried to use the motor encoder pulses to overcome this timing problem, but that led to a whole series of programming issues with the BrickPi library. After much struggling with the library, I decided that it was just more practical to select a proper playing surface, such as a wood floor, than trying to adapt the library software.

This last demonstration concludes this chapter on how to build and operate a BrickPi-controlled robotic car.

Summary

I began this chapter with a brief introduction to the BrickPi, which is a substitute controller for an EV3 Lego Mindstorms brick control module. Using a BrickPi in place of the regular Mindstorms brick allows you to program in whatever Raspian-compatible language you choose instead of being restricted to the Mindstorms drag-and-drop environment. I chose to use Python as the language to control a robotic car, named the CasterBot.

I next discussed how you could build this car out of Lego components using a complete set of referenced instructions. A LiPo battery also was recommended as a power source because the BrickPi and Mindstorms motors consume quite a bit of power.

A simple Python program was next shown that allowed a user to remotely control the car using an SSH connection on a laptop. The car could be commanded to go forward, backward, turn right or left, and stop. It also could be sped up or slowed down, all by using single-character commands from the laptop.

An ultrasonic sensor was discussed next in regard to its internal operation and how it would be used to avoid obstacles when the car was operated autonomously. I developed a program that drove the car in a straight line between two points but could maneuver around an obstacle it detected in its path.

The chapter concluded with my observations on how well the bot performed on the obstacle course and where improvements could be made.

CHAPTER 9

Python-Controlled Robotic Arm

IN THIS CHAPTER, I WILL SHOW YOU how to build two robotic arms, each with its own features and capabilities. You can choose to build one or both depending on your desires and your interest in this topic. However, I will begin the chapter with a reasonably detailed discussion on how robotic arms are defined and designed because that will help you to establish a useful comprehension of this interesting topic.

Background for Robotic Arms

Robotic arms are used extensively in many industrial applications in manufacturing operations. Knowing the basic principles underlying how these arms work will assist you in evaluating when it is appropriate to use this type of robot. Figure 9-1 shows an advanced Denso robotic arm that has 6 degrees of freedom

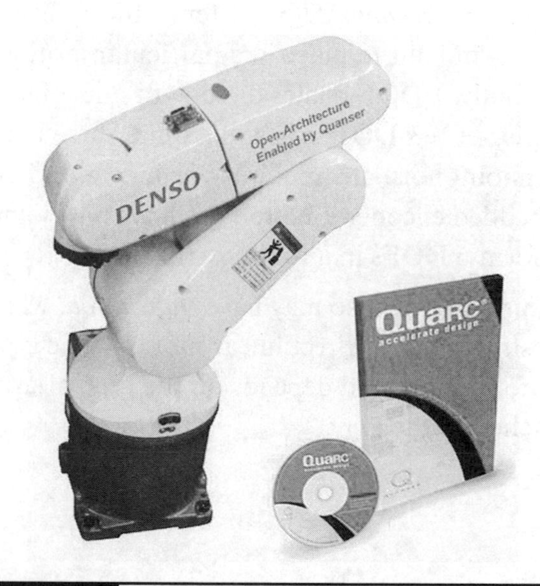

Figure 9-1 Advanced Denso robotic arm.

Chapter 9 Parts List

Item	Model	Quantity	Source
3-DOF 3-Axis Robotic Arm	20-014-305	1	sainsmart.com
DIY 6-Axis Servos Control Palletizing Robot Arm Model	20-014-304-US-KS	1	sainsmart.com
Servo/PWM Pi HAT board	2327	1	adafruit.com
Raspberry Pi Model B+	83-16530	1	mcmelectronics.com
Aluminum block measuring 6 × 12 × 1.5 inches	Commodity	1	amazon.com
Machine screws and nuts	Commodity	Various	Local home-improvement store
6-V, 4-A power supply with 2.1-mm connector	Commodity	1	amazon.com

(DOFs), enabling it to perform many intricate operations.

Understanding robotic arms requires you to know the basic terms and definitions used in this field. The following terms and definitions were sourced from the Society of Robotics website (www.roboticssociety.org).

Degrees of Freedom

Degrees of freedom (DOF) refer to the ability of an arm joint to move in a particular motion. Normally, 1 DOF relates directly to one joint. Therefore, a 3-DOF arm will require three joints. Each joint normally requires one motor and one embedded encoder. Figure 9-2 shows how joints perform 6 DOFs in a robotic arm.

Joint motion also may have *translation*, which is a straight-line or rectilinear motion. The type of motion involved depends on the type of arm attached to the joint.

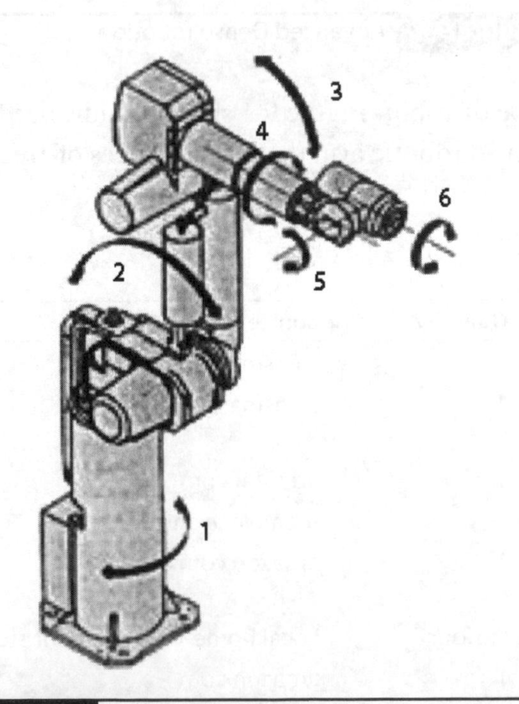

Figure 9-2 6-DOF robotic arm.

Free-Body Diagram

A *free-body diagram* (FBD) is often used with the Denavit-Hartenberg (DH) Convention to depict joint motion in both translation and rotation. Figure 9-3 shows a Denso 4-DOF robotic arm with a DH overlay.

Please note that links are also shown on the FBD. Link lengths are important in that they form a moment arm, which will have a significant effect on motor loading and performance. Also, joints may have multiple DOFs with essentially zero-length links. Consider your shoulder as an example: it has a multiple range of movement without noticeable links. The human shoulder could be described as a joint with a 3-DOF motion range.

Robot designers often create industrial arm systems use the DH Convention along with FBDs. Such systems typically involve significant forces and are intended for heavy-duty industrial applications. This chapter's project has no such requirement and thus will not require a DH analysis.

Workspace

Workspace is the volume of space circumscribed by the robot arm's maximum range of motion.

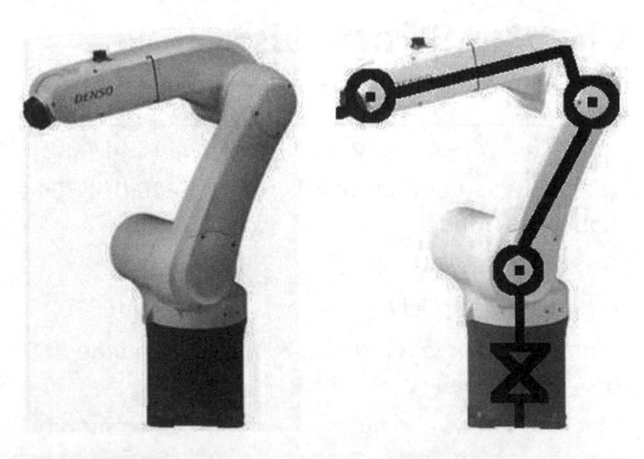

Figure 9-3 Free-body robotic arm with DH joint designations.

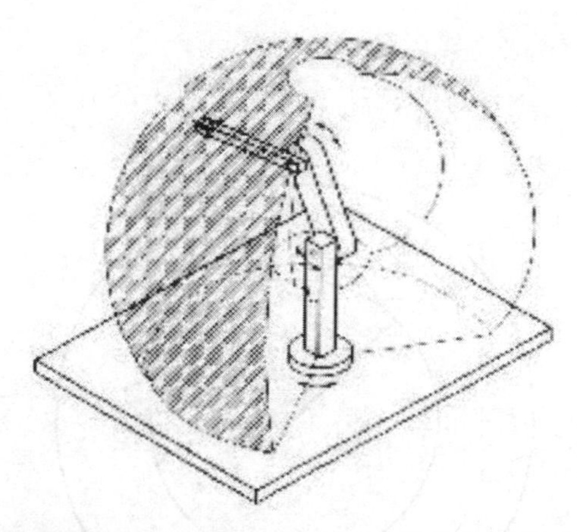

Figure 9-4 Workspace volume.

Figure 9-4 shows the workspace for the robot in Figure 9-3.

Workspace volume typically is closely controlled and posted because personnel entering anywhere in this space could be struck and seriously injured by an operating robotic arm. It is also important to ensure that a robotic arm can extend to all points that need to be reached for the particular manufacturing operation that is being automated. It wouldn't make much sense to have a painting robotic arm that couldn't quite completely apply paint to the product being manufactured. Ensuring adequate coverage and reach is the responsibility of the project manager in charge of installation of the robotic system.

Robotic Arm Classifications

There are also three broad categories for classifying robotic arms.

SCARA

A *selectively compliant articulated robot arm* (SCARA) is a type of robotic arm that has a very wide range of rotary joint motion. There is a maximum of 620° of joint rotation for the SCARA-compliant robotic arm shown in Figure 9-5. This wide range of motion is quite remarkable when you consider that typical robotic arm joint motion is 180° or less.

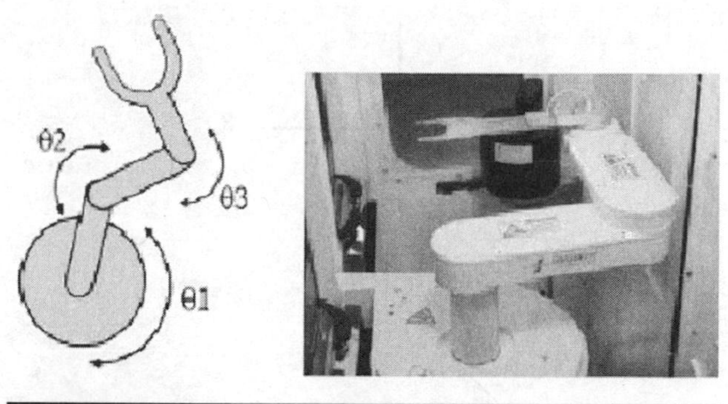

Figure 9-5 SCARA robotic arm.

R-Theta

The *radial-theta angle* (*R-Theta*) robotic arm is a more traditional design with more limited joint rotary motion than a SCARA type. The R-Theta also has a linear (radial) placement parameter. Any required linear motion for the end-effector position is automatically computed by the robot controller and is translated to appropriate joint rotary motion. Figure 9-6 illustrates an R-Theta robotic arm.

X-Y-Z Coordinate (Cartesian Coordinate)

The end-effector position for the *X-Y-Z* coordinate (Cartesian Coordinate) robotic arm is specified as a set of *x*, *y*, and *z* coordinate

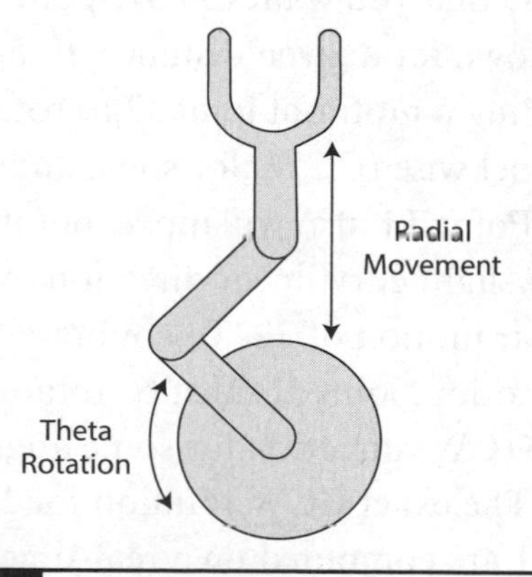

Figure 9-6 R-Theta robotic arm.

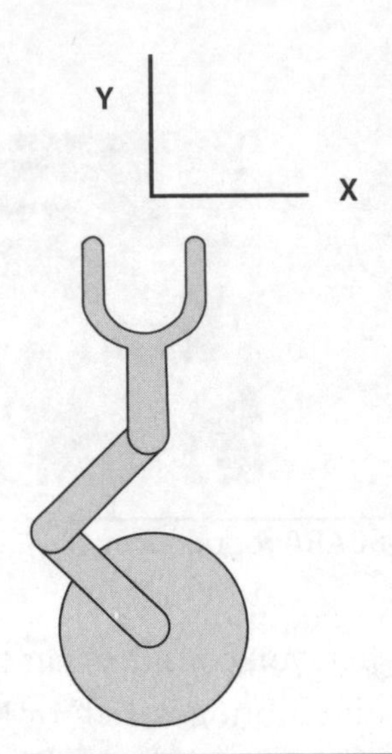

Figure 9-7 *X-Y-Z* Coordinate robotic arm.

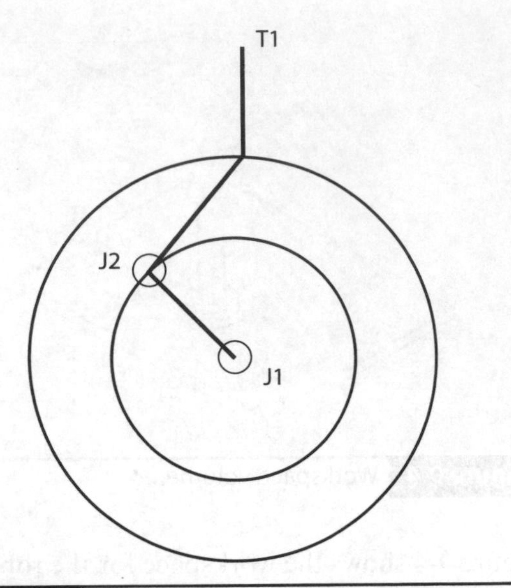

Figure 9-8 Rotary-to-linear motion conversion.

positions. The robot controller automatically translates an *x-y-z* coordinate pair to appropriate joint motion to move the end-effector to the desired position. Figure 9-7 shows a two-dimensional *x-y* coordinate robot arm. The *z* direction would be into or out of the plane of the page.

Linear-to-Rotary Translation

Linear motion was mentioned several times in the preceding descriptions of the various robotic arms. A question you might ask yourself is how linear motion can be accomplished when only rotary joints are being used? The answer is quite simple, and Figure 9-8 shows how it is done.

Suppose that you want to have point T1 travel straight down for a given distance. Imagine commanding a motor at joint J2 to rotate counterclockwise (CCW) for some angular rotation. Point T1 also will move, but it will move in a slight curvilinear direction owing to the angular motion of J2. To counteract this curving motion, joint J1 must be rotated in a clockwise (CW) direction for some angular rotation. The exact CCW rotation for J2 and CW for J1 are computed on a real-time basis by the robotic controller integrated into the arm. The mathematics involved consists of purely trigonometric functions and is a bit tedious—something ideal for a computer to handle.

Arm Positioning

Positioning a robotic arm effector to perform whatever action it has to do, such as welding an auto body seam or painting a body panel, is critical to the overall success of the robot. The process of positioning is called *training*, and it may be accomplished in several ways. One common way is to manually position the effector and then press a button to record the spatial location as *x*, *y*, and *z* coordinates. Figure 9-9 shows a typical control pendant used to manually position a robotic arm as well as record the incremental positions. The complete path is eventually recorded as a series of points that are then used by the arm controller to smoothly control the robotic arm in its autonomous operation.

The robot programmer also will use the controller to make small adjustments to refine the path and to modify the speed of the effector for both safety and product quality assurance. It is very important to have the correct effector

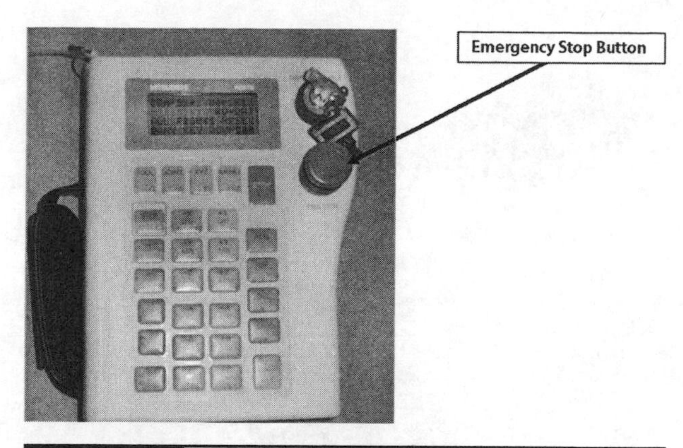

Figure 9-9 Teaching pendent.

speed as a weld is being placed or paint is applied. The robotic arms used in this chapter do not use teaching pendants because such devices are used only in expensive industrial robotic arms.

This completes the introductory discussion. It is now time to show you this chapter's robotic arms.

SainSmart Robotic Arms

I decided to use two robotic arms to demonstrate the concepts presented in this chapter. Both are distributed by SainSmart, with one being a 3-DOF system with a gripper mechanism and the other a 6-DOF system without a gripper. I chose these two systems to allow you the opportunity to experiment with an actual robotic arm, but one that you could afford. The 3-DOF arm costs about one-fourth as much as the 6-DOF arm. Obviously, the 6-DOF arm offers more flexibility in positioning, but the fundamental concepts underlying these arms are the same. I strongly recommend that you purchase the 3-DOF system and play around with it to see whether your results warrant the additional expense of purchasing the 6-DOF arm. I discuss each of these arms in the next two sections.

Sainsmart 3-DOF Robotic Arm

Figure 9-10 shows the simpler of the two SainSmart robotic arms I chose as project demonstrators. It is a 3-DOF arm with a gripper assembly, which accounts for one of the DOFs. SainSmart calls its DOFs *axes*, but they are synonymous.

The precise name of this robotic arm is the SainSmart DIY 3-Axis Servos Control Palletizing Robot Arm Model for Arduino UNO MEGA2560, Item # 20-014-305. Some of the key technical specifications are listed in Table 9-1.

TABLE 9-1 SainSmart 3-Axis Robotic Arm Key Technical Specifications

Motion/Rotation/Reach/Lift	Value
Elbow rotation	120°
Base rotation	120°
Vertical reach	290 mm (gripper closed)
Horizontal reach	240 mm
Lifting capacity	TBD
Finger motion (gripper mechanism)	Open and close 55 mm

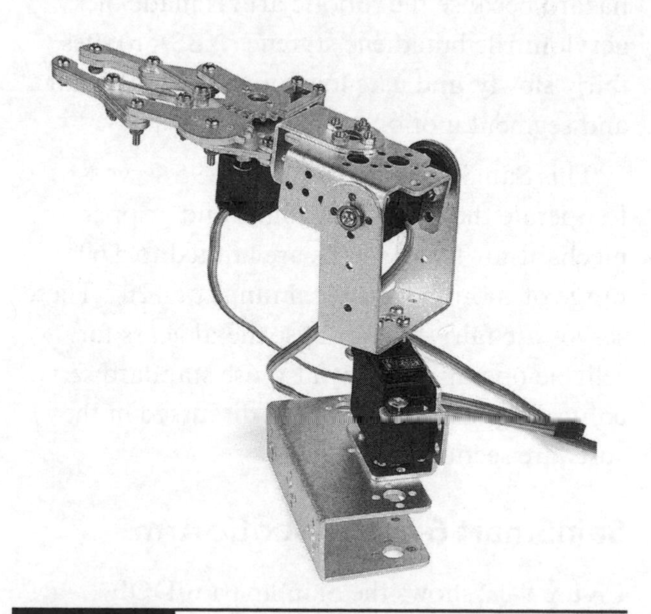

Figure 9-10 3-DOF SainSmart 3-Axis Robotic Arm.

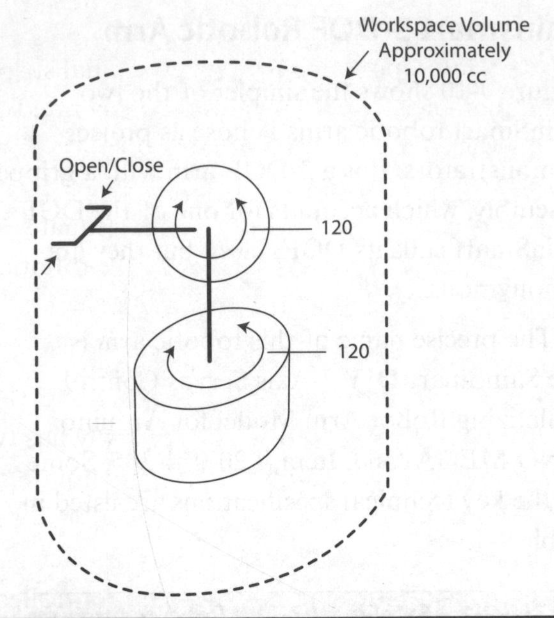

Figure 9-11 SainSmart 3-DOF arm joint rotary range of motion and volumetric workspace diagram.

Figure 9-11 shows the rotary-motion range for each joint as well as the total volumetric workspace for the arm.

It was a bit surprising to see that there was a fairly large volume of 9692 cm^3 (0.009692 m3) swept through 120° by the arm when fully extended. This really doesn't present a safety hazard because the robotic arm is made of acrylonitrile butadiene styrene (ABS), rotates fairly slowly, and uses low forces for both joint and segment motion.

This SainSmart arm uses MG995 servos to operate the two rotary joints and gripper mechanism. These servos are limited to ±60° range of motion for a total range of 120°. These servos are fairly fast and use metal gears for reliable operation. They also use standard servo control signals, which will be discussed in the software section.

SainSmart 6-DOF Robotic Arm

Figure 9-12 shows the SainSmart 6-DOF robotic arm. It is not equipped with a gripper mechanism, as is the 3-DOF arm.

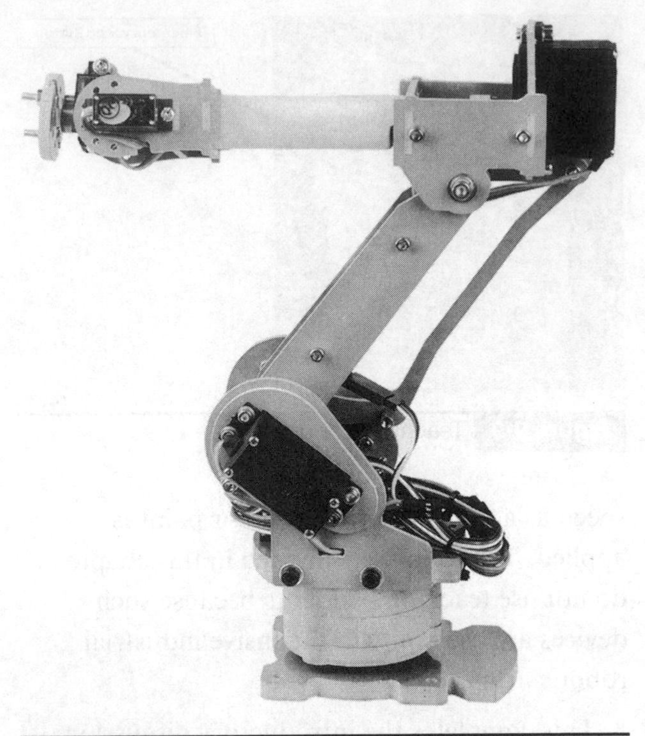

Figure 9-12 6-DOF SainSmart robotic arm.

This arm has more than twice the number of joints as the 3-DOF arm and thus is much more flexible in positioning within its workspace. The arm can lean, which is not possible with the 3-DOF arm, whose main vertical segment is fixed in a single vertical plane. The precise name of this robotic arm is the SainSmart DIY 6-Axis Servos Control Palletizing Robot Arm Model for Arduino UNO MEGA2560, Item # 20-014-304-US-KS. Some of the key technical specifications are listed in Table 9-2.

TABLE 9-2 SainSmart 6-Axis Robotic Arm Key Technical Specifications

Motion/Rotation/Reach/Lift	Value
Finger rotation (end effector)	120°
Fingers up/down	120°
Forearm rotation	120°
Elbow rotation	120°
Shoulder rotation	120°
Base rotation	120°
Vertical reach (from base)	414 mm
Horizontal reach (from shoulder joint)	275 mm

The volumetric workspace is substantially more than that of the 3-DOF arm. I estimated the volume to be approximately 49,748 cm3. It is very difficult to compute the workspace volume because of the many joints in the arm, which means that arm positioning is quite variable. I settled on using a volume that is one-sixth of a sphere with a 414-mm radius. This is very much an overestimate, but it is better to be conservative than compromise this important safety criterion.

In the next section, I discuss a servo control board, which is an essential component between the RasPi and the robotic arms.

Servo Control Interface Board

I used the Adafruit 16-Channel PWM/Servo HAT for Raspberry Pi, Minikit # 2327, as a servo control interface board connected between the RasPi and the robotic arm. This control board is shown in Figure 9-13 mounted on a RasPi Model B+.

Each of the three robotic arm servos plugs directly into one of the three pin terminal strips mounted on the board. A power supply rated to handle all the attached servos also must be plugged into the control board. I used a 5-V, 4-A supply, which should be more than adequate to power all the servos for either of the robotic arms used in this chapter's projects.

One distinct advantage of using this particular interface board is its ability to control up to 16 channels of PWM/servo. The RasPi only has two built-in PWM channels, which is quite limiting and would not be sufficient to control either one of the two robotic arms used in this chapter. The interface board makes use of an I2C-controlled PWM controller chip to drive the onboard 16 channels. It is even possible to connect up to 62 additional interface boards in a parallel multidrop I2C network to control an astounding 992 PWM/servo channels. What is even more impressive is that the RasPi's computational requirement is minimal and totally independent of the number of servos being controlled.

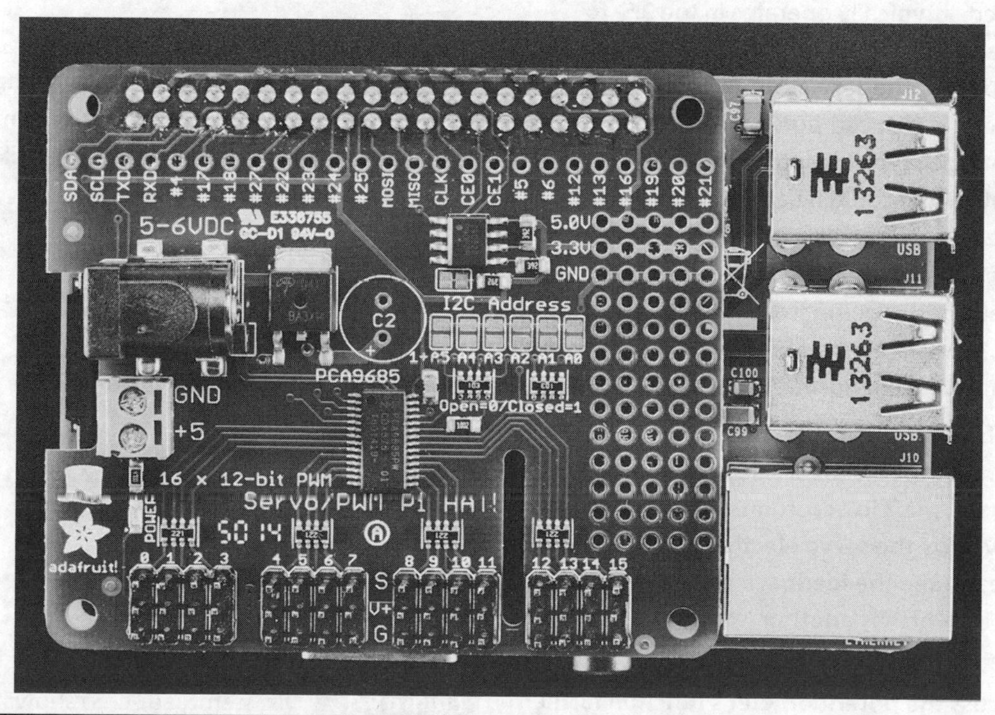

Figure 9-13 16-Channel PWM/Servo HAT board.

Each interface board continually controls the servos plugged into it based on the last I2C commands sent to the board by the RasPi. The RasPi does not have to continually refresh any digital servo control signals, which minimizes its computational burden and frees up the RasPi to handle other important real-time tasks.

At this point in the discussion, I would like to offer you an opportunity to gain some knowledge regarding how the servos function in this robotic arm. I do this in the form of the following sidebar, which is purely optional and will not matter for successful project completion if you choose to skip reading it.

Figure 9-14 is a somewhat transparent view of the inner workings of a standard analog radio-controlled (R/C) servo motor.

I would like to point out five components in this figure:

- **Brushed electric motor (left side).**
- **Gear set (just below the case top).**
- **Servo horn (attached to a shaft protruding above the case top).**
- **Feedback potentiometer (at the bottom end of the same shaft with the horn).**
- **Control PCB (bottom on the case to the motor's right).**

Figure 9-14 Inner view of a standard R/C servo motor.

The electric motor is just an inexpensive ordinary motor that probably runs at approximately 12,000 rpm unloaded. It typically operates in the 2.5- to 5-VDC range and likely uses less than 200 mA even when fully loaded. The servo torque advantage results from the motor spinning the gear set such that the resulting speed is reduced significantly, producing a very large torque increase compared with the motor's ungeared rating. A typical motor used in this servo class might have a 0.1 oz-inch torque rating, while the servo output torque could be about 42 oz-inches, which is a 420 times increase in torque production. Of course, the speed would be reduced by the same proportional amount going from 12,000 rpm to about 30 rpm. This slow speed is still sufficiently fast to move the servo shaft to meet normal R/C requirements.

The feedback potentiometer attached to the bottom of the output shaft is a key element in positioning the shaft in accordance with the pulses being received by the servo electronic control board. You may clearly see the feedback potentiometer in Figure 9-15, which is another image of a disassembled servo.

Figure 9-15 Disassembled servo showing the feedback potentiometer.

I will discuss the potentiometer's function further during the control board discussion. The electronics

board is the heart of the servo and controls how the servo functions. I will describe an analog control version because that is by far the most popular type used in low-cost servo motors. Figure 9-16 shows a Hitec control board that is in place for its Model HS-311, which is a very common and inexpensive analog servo.

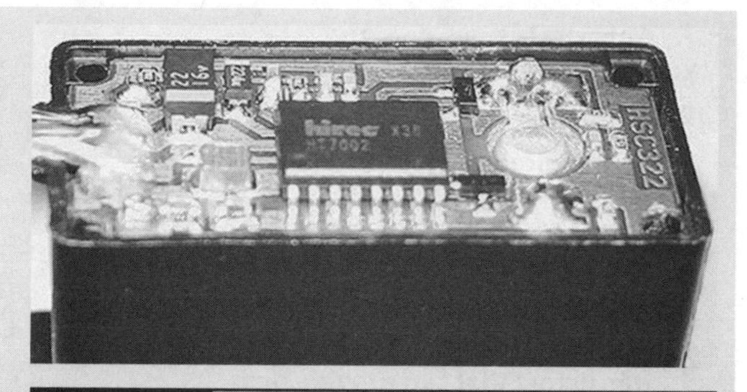

Figure 9-16 Hitec HS-311 electronics board.

The main chip is labeled HT7002, which is a Hitec private model number as well as I could determine. I believe that this chip functions the same as a commercially available chip manufactured by Mitsubishi with Model # M51660L. I will use the M51660L as the basis of my discussion because it is used in a number of other manufacturers' servo motors and is representative of any chip used in this situation. The Mitsubishi chip is called a Servo Motor Controller for Radio Control, and its pin configuration is shown in Figure 9-17.

Don't be put off by the different physical configuration between the HT7002 in Figure 9-16 and the chip outline in Figure 9-17 because it is often the case that identical chip dies are placed in different physical packages for any number of reasons. The M51660L block diagram shown in Figure 9-18 illustrates the key functional circuits incorporated into this chip.

Next, I will provide an analysis that will go hand in hand with the demonstration circuit in Figure 9-19 that was provided in the manufacturer's datasheet (as were the preceding two figures).

This analysis should help you to understand how an analog servo functions and why there are certain limitations inherent in its design:

1. The start of a positive pulse appearing on the input line (pin 5) turns on the R-S flip-flop and also starts the one-shot multivibrator running.
2. The R-S flip-flop works in conjunction with the one-shot multivibrator to form a linear one-shot or monostable multivibrator circuit whose on time is proportional to the voltage appearing from the tap on the feedback potentiometer and the charging voltage from the timing capacitor attached to pin 2.
3. The control logic starts comparing the input pulse to the pulse being generated by the one-shot multivibrator.
4. This ongoing comparison results in a new pulse called the *error pulse* that is than fed to the pulse stretcher and deadband and trigger circuits.
5. The pulse stretcher output ultimately drives the motor control circuit that works in combination with the directional control inputs that originate

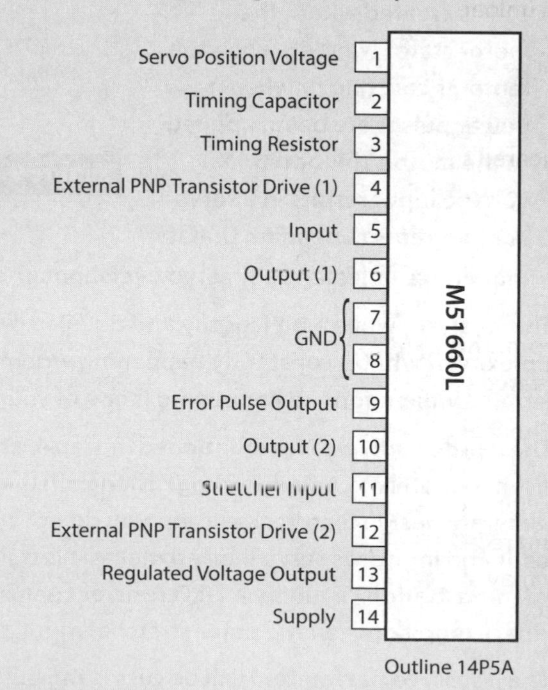

Figure 9-17 Mitsubishi M51660L pin configuration.

from the R-S flip-flop. The trigger circuits enable the PNP transistor driver gates for a time period directly proportional to the error pulse.

6. The PNP transistor drive gate outputs are pins 4 and 12, which control two external PNP power transistors that can provide over 200 mA to power the motor. The M51660L chip can only provide up to 20 mA without using these external transistors. This is too little current flow to power the motor in the servo. The corresponding current sinks (return paths) for the external transistors are pins 6 and 10.

7. The 560-kΩ resistor (R_f) connected between pin 2 and the junction of one of the motor leads and pin 6 feeds the motor's back electromotive force (EMF) voltage into the one-shot multivibrator. Back EMF is created within the motor stator winding when the motor is coasting or when no power pulses are being applied to the motor. This additional voltage input results in a servo *damping effect*, meaning that it moderates or lessens any servo overshoot or in-place dithering.

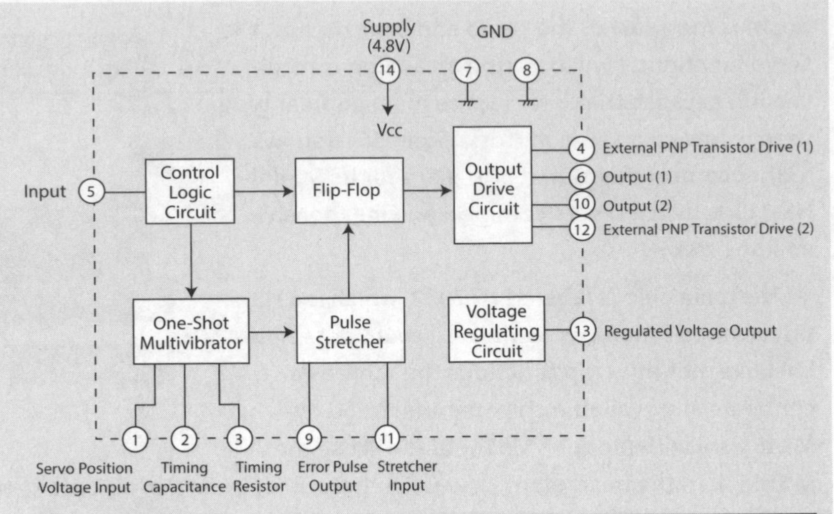

Figure 9-18 M51660L block diagram.

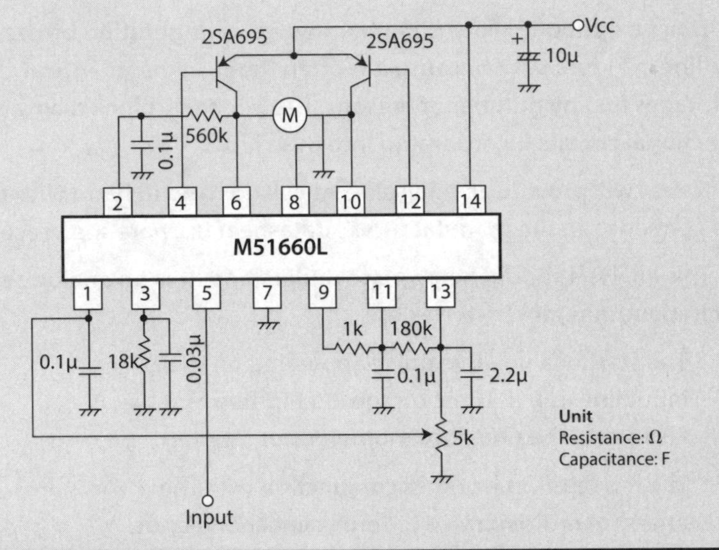

Figure 9-19 Demonstration M51660L schematic.

This analysis, while a bit lengthy and detailed, was provided to give you an understanding of the complexity of what is constantly happening within the servo case. This knowledge should help you to determine what might be happening if one of your servos starts operating in an erratic manner.

The word *deadband* was mentioned in step 4, and it is worth some more explanation. Deadband used in this context refers to a slight voltage change in the control input, which should not elicit an output. This is a deliberate design feature: basically, you do not want the servo to react to any slight input changes. Using a deadband improves servo life and makes it less jittery during normal operations. The deadband is fixed in the demonstration circuit by a 1-kΩ resistor connected between pins 9 and 11. This resistor forms another feedback loop between the pulse stretcher input and output.

The last servo parameter I will discuss is the pulse stretcher gain, which largely controls the error pulse length. This gain in the demonstration circuit is set by the values of the capacitor from pin 11 to ground and

the resistor connected between pins 11 and 13. This gain would also be referred to as the *proportional gain* (*Kp*) in closed-loop control theory. It is important to have the gain set to what is sometimes jokingly called the "Goldie Locks region"—not too high nor too low, but just right. Too much gain makes the servo much too sensitive and possibly could lead to unstable oscillations. Too little gain makes it too insensitive with a very poor time response. Sometimes experimenters will tweak the resistor and capacitor values in an effort to squeeze out a bit more performance from a servo, but I believe that the manufacturers already have set the component values for a good compromise between performance and stability.

Servo Control Pulses

Understanding how digital pulses control an analog servo is the key to knowing how to program it. A series of pulses on the servo signal line with widths of 1.5 ms and a frequency of 50 Hz will cause the servo to remain stationary at its center rotation point. Figure 9-20 illustrates this pulse train.

A 50-Hz frequency is commonly used for these kinds of analog servos, and the pulse amplitude is typically 5 V. The servo will rotate to its full CW rotation position when the pulse width is increased to 2.0 ms. Similarly, the servo will rotate to its full CCW rotation position when the pulse width is decreased to 1.0 ms. Consequently, changing the pulse width from 1.0 to 2.0 ms will cause the servo to move its shaft through its entire range of motion. You also should note that the pulse widths do vary with different servos. In some very inexpensive servos, the range can be from 0.5 to 2.4 ms, while in some pricier ones the range is 0.9 to 2.1 ms. My recommendation is that you check the technical specifications for the servo you will use in your application and adjust the pulse widths appropriately. I would recommend using the website www.servodatabase.com to retrieve the data on your particular servo.

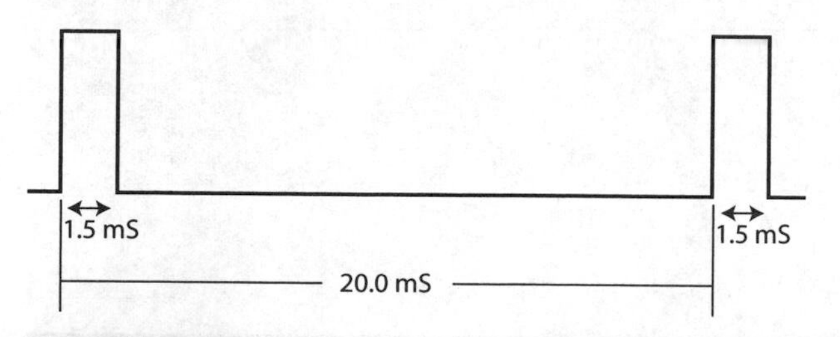

Figure 9-20 50-Hz, 1.5-ms pulse train.

The software controlling the servo is required to generate the pulse width corresponding to the commanded position for the servo rotation angle. I will examine this requirement in further detail in the next section, which discusses the software.

Robotic Arm Software

The servo interface control board is controlled by a RasPi using the I2C bus, as was mentioned in my introduction to the interface board. The Wheezy Raspian Linux distribution must be configured to use the I2C bus. The configuration starts with enabling I2C using the raspi-config application. Enter the following to start this config process:

```
sudo raspi-config
```

Select the Advanced Options and then select the Enable I2C option. Next enter

```
sudo reboot
```

You next have to install the Python library, which allows you to program the I2C bus. Follow these two steps to install and configure the required software:

1. `sudo apt-get install python-smbus`
2. `sudo apt-get install i2c-tools`

Using the nano editor, add these two lines to the file /etc/modules:

```
i2c-bcm2708
i2c-dev
```

Use the nano editor to see whether the file /etc/modprobe.d/raspi-blacklist.conf has the following line, and if so, comment it out by placing a number symbol (#) at the front of the line:

```
blacklist i2c-bcm2708
```

Using the nano editor, add these two lines to the file /boot/config.txt:

```
dtparam=i2c1=on
dtparam=i2c_arm=on
```

Then:

```
sudo reboot
```

Check for all installed I2C devices by entering:

```
sudo i2cdetect -y 1
```

or:

```
sudo i2cdetect -y 0
```

(for very early RasPi models).

Figure 9-21 shows the result of the `i2cdetect` command, which I ran on the RasPi with the 16-channel PWM/Servo HAT module installed.

The 0x40 address shown comes from the PCA9685 chip on the HAT board. This means that the RasPi is in proper communication with the HAT board using the I2C bus and ready to be programmed, as discussed in the next section.

```
pi@raspberrypi ~ $ sudo i2cdetect -y 1
     0  1  2  3  4  5  6  7  8  9  a  b  c  d  e  f
00:          -- -- -- -- -- -- -- -- -- -- -- -- --
10: -- -- -- -- -- -- -- -- -- -- -- -- -- -- -- --
20: -- -- -- -- -- -- -- -- -- -- -- -- -- -- -- --
30: -- -- -- -- -- -- -- -- -- -- -- -- -- -- -- --
40: 40 -- -- -- -- -- -- -- -- -- -- -- -- -- -- --
50: -- -- -- -- -- -- -- -- -- -- -- -- -- -- -- --
60: -- -- -- -- -- -- -- -- -- -- -- -- -- -- -- --
70: 70 -- -- -- -- -- -- --
pi@raspberrypi ~ $
```

Figure 9-21 `i2cdetect` command result.

Initial Test Program

I created a very simple test program named *ServoTest.py* that exercises a standard Hitec HS-311 servo connected to channel 0 on the HAT board. Figure 9-22 shows the test setup with the servo being powered from a supply connected to the 2.1-mm barrel connector on the HAT board.

The commented ServoTest.py program is listed next and is also available from this book's companion website.

```
#!/usr/bin/python

# D. J. Norris 8/2015
# ServoTest.py
# This code is in the public domain and
is freely available
#

from Adafruit_PWM_Servo_Driver import PWM
import time
```

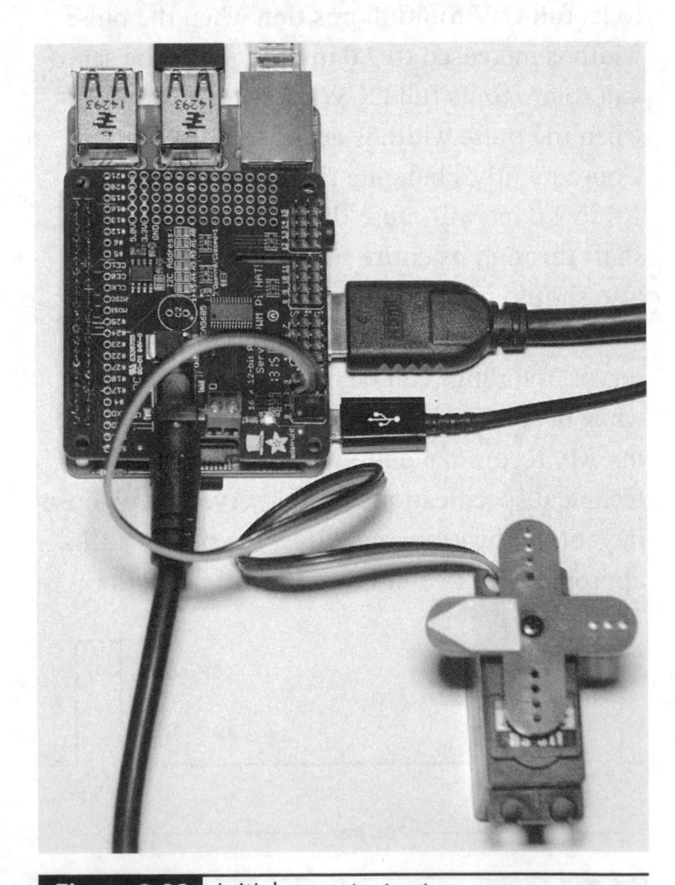

Figure 9-22 Initial servo test setup.

```
# Initialize the PWM device using the
default address
pwm = PWM(0x40)
# For debug output run:
#pwm = PWM(0x40, debug=True)

servoMin = 350 # Min pulse length out of
4096
servoMax = 1100 # Max pulse length out
of 4096
servoMid = 725 # Mid-point

pwm.setPWMFreq(100) # Set frequency to
100 Hz

while (True):
    #rotate between 0, 180 and 90 degrees
       on a standard servo
    pwm.setPWM(0, 0, servoMin) # 0 deg
    time.sleep(1)
    pwm.setPWM(0, 0, servoMax) # 180 deg
    time.sleep(1)
    pwm.setPWM(0, 0, servoMid) # 90 deg
    time.sleep(1)
```

Please note that I set the PWM frequency to 100 Hz versus the normal 50 Hz I mentioned in my servo discussion. This was done purely for convenience sake to ease the pulse-length calculations. The increased frequency has absolutely no effect on the servo's performance.

I observed that when I ran the program, the servo smoothly rotated between the 0, 180, and 90° positions without any issues. This test confirmed that the RasPi/HAT board was properly configured and ready to handle the robotic arms. Satisfied that I could successfully control a servo, I started the next task of creating a program designed to control all the servos in a robotic arm. It is now time to discuss the software that I created to control the robotic arms.

Robotic Arm Software

I decided to take a measured approach to developing the robotic arm control software and first try it out with a stand-alone servo as I did with the initial test. In this way, I could resolve any problems that appeared without damaging an expensive robotic arm. I also chose to use a graphical user interface (GUI) for the robotic arm program because such a program naturally lends itself to the way a user would interact with the arm. The GUI was created using the Tkinter library, which is part of the normal Python distributions, including versions 2.7 and 3.0. The arm program also must be run in the X Windows environment because it is a GUI and requires X Windows to correctly display the GUI window and widgets.

3-DOF Robotic Arm Servo Control Program

The first arm control program named 3DOF_Robot_Arm.py was created for the 3-DOF arm because that arm uses only three servos, and naturally, such a program would be easier to develop and test. The listing is shown next with a heavy dose of comments included within the code. The source code is also available on this book's companion website.

```
from Tkinter import *
from Adafruit_PWM_Servo_Driver import PWM
import time

#Default I2C address for the HAT board
pwm = PWM(0x40)
#Used 100Hz for the servo freq as the algorithm requires it
pwm.setPWMFreq(100)
```

```
#all the servo functions are now in this class
class Servo:
    def __init__(self, top1):
      win = Frame(top1) #container for the GUI components
      win.pack()
      label1 = Label(win,text='Gripper') #Gripper label
      sc1 = Scale(win, from_=-60, to=60, orient=HORIZONTAL, command=self.update0)
#Slider component for gripper angle
      sc1.grid(row=0) #Layout manager arranges components
      label1.grid(row=1)
      label2 = Label(win, text='Elbow')
      sc2 = Scale(win, from_=-60, to=60, orient=HORIZONTAL, command=self.update1)
        sc2.grid(row=2)
      label2.grid(row=3)
        label3 = Label(win, text='Base')
      sc3 = Scale(win, from_=-60, to=60, orient=HORIZONTAL, command=self.update2)
      sc3.grid(row=4)
      label3.grid(row=5)

    def update0(self, angle): #Called when value is changed
      num = float(angle)
       #Algorithm to convert angle to pulse width for 100Hz
      pulseWidth = int((750*(num+60))/120 + 350)
      pwm.setPWM(0, 0, pulseWidth) #Servo repositions

    def update1(self, angle):
      num = float(angle)
      pulseWidth = int((750*(num+60))/120 + 350)
      pwm.setPWM(1, 0, pulseWidth)

    def update2(self, angle):
      num = float(angle)
      pulseWidth = int((750*(num+60))/120 + 350)
      pwm.setPWM(2, 0, pulseWidth)

top = Tk()  #GUI desktop window object
top.wm_title("3 DOF Robotic Arm Servo Control")
app = Servo(top) #instantiate a Servo object
top.update_idletasks() #Tkinter likes this before displaying
top.geometry("400x200+0+0") #Create a good sized window
top.mainloop() #This kicks off everything
```

The program is started by first going to the directory, which contains both the 3DOF_Robot_Arm.py and the Adafruit_PWM_Servo_Driver library. You will get an error if you attempt to run the arm program without the library being present in the same directory. The command to cd into the directory containing the 3DOF_Robot_Arm.py and the Adafruit driver library from the home directory is shown next:

```
pi@raspberrypi ~ $ cd Adafruit
-Raspberry-Pi-Python-Code/Adafruit_PWM
_Servo_Driver
```

Once in the directory, enter these two commands to start X Windows and run the arm program:

```
startx
sudo python Robot_Arm.py
```

Figure 9-23 is a screenshot of the GUI showing the three labeled slider controls.

Changing the slider position on each of the slider GUI widgets changes the digital pulse width being sent to the HAT servo channel corresponding to the slider widget being changed. The pulse width is 1.0 ms when the slider is at the –60 position and is 2.0 ms when the slider is at the +60 position. The pulse width is 1.5 ms when the slider is at 0, or midway between –60 and +60.

Before running this program with the arm, you must ensure that all the servo leads from the arm have been labeled properly. I identified the lead connections by running the program and then connecting one lead at a time to channel 0. I then moved the slider for the gripper and observed which servo moved. This procedure quickly identified the leads, and I used a label maker to tag the leads. This procedure is not really critical for the 3-DOF robotic arm because the servo leads are easy to identify by visually inspecting the arm. However, the 6-DOF arm has its leads woven all through the structure, making physical lead identification impossible.

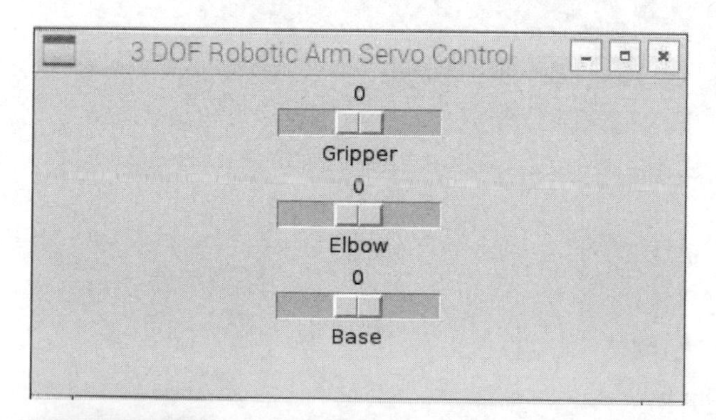

Figure 9-23 Robot_Arm program GUI.

Besides, identifying the leads using the program also checks the physical connections early in the testing phase and will quickly identify any electrical connection issues. Once all the leads are labeled, you should connect them to the HAT board, as shown in Table 9-3.

TABLE 9-3 3-DOF Servo Lead Connections

Servo Lead	Channel Number
Fingers open/close (gripper)	0
Elbow rotate	1
Base rotate	2

Figure 9-24 shows all the servo leads connected to the HAT board, which, in turn, is mounted on a RasPi.

Testing the 3-DOF Robotic Arm

You must clamp the robotic arm to a tabletop before using it. I used a small bar clamp to secure the arm to my worktable. I then tested the 3-DOF robotic arm by running a common

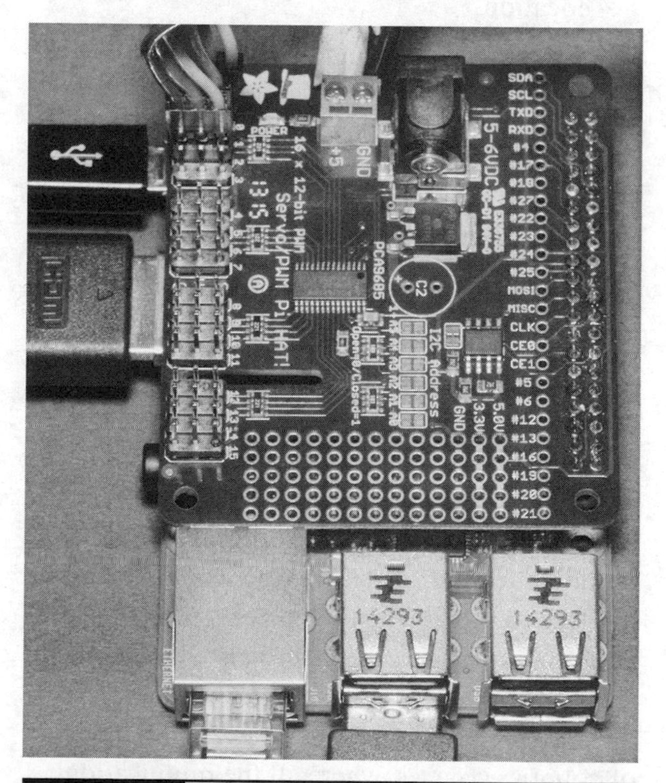

Figure 9-24 3-DOF servo leads connected to HAT board.

industrial robot operation known as *pick-and-place*. This operation requires the arm to pick up an object and transport it to a different location all within the robot's volumetric workspace. I used a small wooden block as the transported object. The pick-and-place operation is detailed in the following step sequence:

1. Place the object to be picked up within the reach of the 3-DOF arm.
2. Ensure that the gripper is open and the elbow is rotated to clear the top of the object.
3. Rotate the base such that the gripper opening is centered on the object.
4. Rotate the elbow until the gripper is at the midpoint of the object.
5. Close the gripper to the point where it securely holds the object.
6. Rotate the elbow to lift the object above the mount plane.
7. Rotate the base until it is at the drop-off location.
8. Rotate the elbow until the object rests on the mount surface.
9. Open the gripper to free the object.
10. Rotate the elbow until the gripper is clear of the object.
11. Repeat steps 1 to 10 for additional objects to be moved.

Conveyers also could be required to both deliver objects to be picked as well as to whisk away the objects placed (dropped off). Of course, in a real-world scenario, the robotic arm would be controlled by a real-time program without human intervention. There also would be sensors in place to signal when the object to be picked up was properly arranged and also to signal when the object was dropped off. Development of these real-time programs is normally the topic of robotics courses, where all the ramifications of robotic arm control can be explored.

Figure 9-25 shows the 3-DOF robotic arm with an object in its gripper midway between the pick-and-place locations.

It took me about 20 seconds to manually complete the pick-and-place operation. I would estimate that an automated pick-and-place likely would take no more than 2 seconds, which is a 10 times improvement over manual mode.

Now it is time to examine the 6-DOF arm. I will try a slightly different operational approach with this arm because it significantly more flexible in its positioning owing to the 6 DOFs.

6-DOF Robotic Arm Servo Control Program

The control program for the 6-DOF arm is named *6DOF_Robot_Arm.py* and is listed next. I have not included all the same comments as were in the 3 DOF program because they are all identical. Essentially, the 6-DOF program is simply an extension of the 3-DOF program with three servo control channels added to accommodate the extra DOFs.

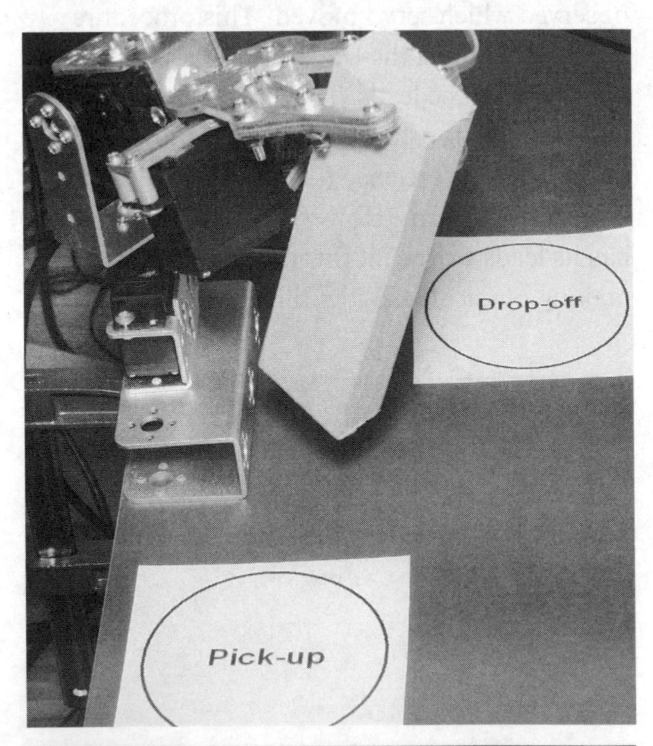

Figure 9-25 3-DOF robotic arm in a pick-and-place configuration.

```
#!/usr/bin/python

from Tkinter import *
from Adafruit_PWM_Servo_Driver import PWM
import time

pwm = PWM(0x40)
pwm.setPWMFreq(100)

# Channel assignments
# 0 = Wrist Rotate
# 1 = Wrist Up/Down
# 2 = Forearm Rotate
# 3 = Elbow Rotate
# 4 = Shoulder Rotate
# 5 = Base Rotate

class Servo:

    def __init__(self, top1):
      win = Frame(top1)
      win.pack()
      label1 = Label(win,text='Wrist Rotate')
      sc1 = Scale(win, from_=-60, to=60, orient=HORIZONTAL, command=self.update0)
      sc1.grid(row=0)
      label1.grid(row=1)
      label2 = Label(win, text='Wrist Up/Down')
      sc2 = Scale(win, from_=-60, to=60, orient=HORIZONTAL, command=self.update1)
            sc2.grid(row=2)
      label2.grid(row=3)
            label3 = Label(win, text='Forearm Rotate')
      sc3 = Scale(win, from_=-60, to=60, orient=HORIZONTAL, command=self.update2)
      sc3.grid(row=4)
      label3.grid(row=5)
      label1 = Label(win,text='Elbow Rotate')
        #Adjusted range from (-60 to +60) to (-75 to -30)
      sc1 = Scale(win, from_=-75, to=-30, orient=HORIZONTAL, command=self.update3)
      sc1.grid(row=6)
      label1.grid(row=7)
      label1 = Label(win,text='Shoulder Rotate')
        #Adjusted range from (-60 to + 60) to (-6- to +20)
      sc1 = Scale(win, from_=-60, to=20, orient=HORIZONTAL, command=self.update4)
      sc1.grid(row=8)
      label1.grid(row=9)
      label1 = Label(win,text='Base Rotate')
      sc1 = Scale(win, from_=-60, to=60, orient=HORIZONTAL, command=self.update5)
      sc1.grid(row=10)
      label1.grid(row=11)
```

```
    def update0(self, angle):
      num = float(angle)
      pulseWidth = int((750*(num+60))/120 + 350)
      pwm.setPWM(0, 0, pulseWidth)

    def update1(self, angle):
      num = float(angle)
      pulseWidth = int((750*(num+60))/120 + 350)
      pwm.setPWM(1, 0, pulseWidth)

    def update2(self, angle):
      num = float(angle)
      pulseWidth = int((750*(num+60))/120 + 350)
      pwm.setPWM(2, 0, pulseWidth)

    def update3(self, angle):
      num = float(angle)
      pulseWidth = int((750*(num+60))/120 + 350)
      pwm.setPWM(3, 0, pulseWidth)

    def update4(self, angle):
      num = float(angle)
      pulseWidth = int((750*(num+60))/120 + 350)
      pwm.setPWM(4, 0, pulseWidth)

    def update5(self, angle):
      num = float(angle)
      pulseWidth = int((750*(num+60))/120 + 350)
      pwm.setPWM(5, 0, pulseWidth)

top = Tk()
top.wm_title("6 DOF Robotic Arm Servo Control")
app = Servo(top)
top.update_idletasks()
top.geometry("400x400+0+0")
top.mainloop()
```

Mounting the RasPi/HAT on the 6-DOF Robotic Arm

A microcontroller mounting plate is included on this arm because of the large number of servo leads, which potentially could limit the base rotation of the arm. This mounting plate comes with four standoffs placed to support an Arduino Mega 2560 board. I removed these standoffs and drilled three holes to accommodate a RasPi 2 Model B. I also reused one of the existing holes, as shown in Figure 9-26.

I next mounted the RasPi onto the four repositioned standoffs. I would also recommend that you defer connecting the HAT to the RasPi until you complete the servo lead identification procedure.

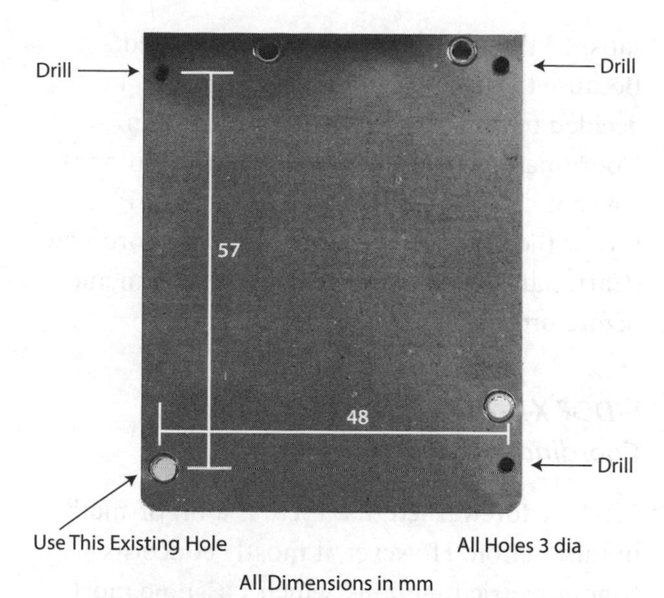

Figure 9-26 Hole placement for a RasPi on the 6-DOF robotic arm mount plate.

You must next ensure that all the servo leads from the arm have been labeled properly. Please use the procedure that I detailed in the preceding section to label all the leads. Once all the leads are labeled, you should connect them to the HAT board, as detailed in Table 9-4.

TABLE 9-4 6-DOF Servo Lead Connections

Servo Lead	Channel Number
Wrist rotate	0
Wrist up/down	1
Forearm rotate	2
Elbow rotate	3
Shoulder rotate	4
Base rotate	5

Connect the whole assembly to the RasPi once all the servo leads are connected to the HAT board. You are now almost ready to test this arm. The arm must be attached to a secure base or platform before operating it because it will fall over if it is extended beyond its stable center of gravity (CG). I will show you the base I used in the next section.

Robotic Arm Base

The platform I used was a solid block of aluminum measuring 6 × 12 × 1.5 inches. I drilled and tapped out four mounting holes in the block that matched the mounting holes already drilled in the robotic arm's mounting plate. The aluminum block weighs a bit over 9 pounds, which is more than adequate to provide a stable platform for the arm. Figure 9-27 shows the 6-DOF robotic arm mounted on the platform.

You do not need to use an aluminum block as I did, but you will need some sort of support. A medium-sized piece of ¾-inch plywood also would serve nicely for a mounting platform. I would recommend a piece at least 18 inches on a side to provide sufficient stability for the arm.

Figure 9-27 Aluminum mounting platform with arm attached.

You also can clamp the robotic arm to a table in the same manner as I did for the 3-DOF robotic arm test. All that matters is that you secure the arm before proceeding with the initial testing, as I discuss in the next section.

Testing the 6-DOF Robotic Arm

The initial test for this arm starts by running the program. The program must be in the same directory I showed you in the 3-DOF arm discussion because it uses the exact same Adafruit library. Enter the following:

```
startx
sudo python 6DOF_Robot_Arm.py
```

Figure 9-28 is a screenshot of the GUI showing the six labeled slider controls.

I tested all the servos on the arm by moving the slider controls starting from the top to the bottom. The corresponding servo should move throughout its total range of motion. Just be careful to slowly move each slider because there is some mass to the arm, which, in turn, can create a significant force on the arm's structure if moved rapidly. Remember that Newton's second law, $F = MA$, still applies, and moving the slider rapidly creates a large acceleration A.

It is time to explore an interesting application for this 6-DOF robotic arm once you are satisfied that the arm operates as expected. Because the arm does not have a gripper, I decided to pursue a generalized Cartesian Coordinate positioning application based on the concept discussed earlier in this chapter. Please review the very brief section "*X-Y-Z* Coordinate (Cartesian Coordinate)" to refresh your memory before proceeding to the next section.

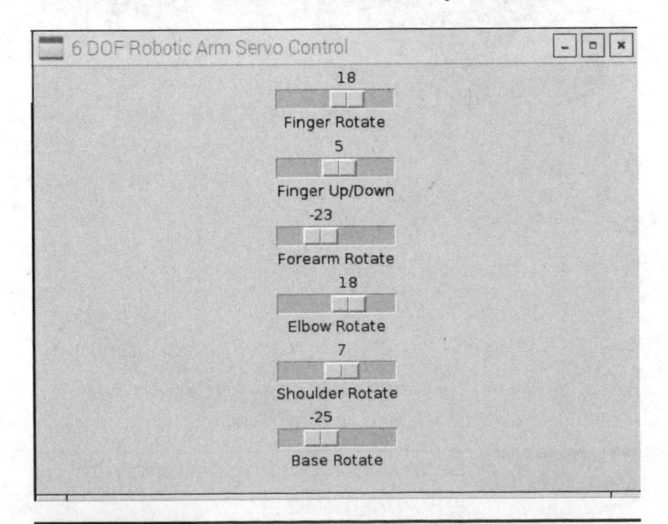

Figure 9-28 6-DOF_Robot_Arm program GUI.

6-DOF X-Y-Z Coordinate (Cartesian Coordinate) Application

First, be forewarned that there is a bit of math in this section. However, it mostly concerns trigonometric functions, which I assume most of my readers (except for the younger ones) have been exposed to at some point in their educational experience. The primary purpose of this *X-Y-Z* Coordinate application (referred to from this point on as the *app*) is to translate a given coordinate located in the 6-DOF arm's volumetric workspace into servo angle commands for the base, shoulder, and elbow joints. You will see that just changing these three joint angles will position the fingers at any point in the volume corresponding to the given *x*, *y*, and *z* coordinates. Figure 9-29 is a vector diagram illustrating this configuration.

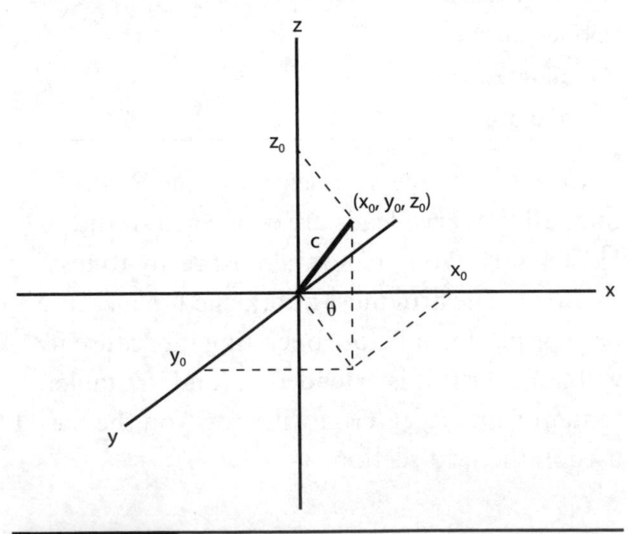

Figure 9-29 Vector diagram.

The given coordinate is shown in the diagram as a point located at x_0, y_0, and z_0, which are points on the respective *x*, *y*, and *z* axes. The vector from the point of origin (0, 0, 0) is labeled c, and its length is determined by this formula:

$$c_2 = x_{02} + y_{02} + z_{02}$$

The angle θ shown on the diagram is the angle of the vector *c* projected onto the *x-y* plane and the *x* axis. This angle would be used to rotate the base such that the robotic arm points precisely in the direction of the given coordinate point. Determining the angles for both the shoulder and elbow joints is more complex but still doable. Figure 9-30 is a simplified diagram representing both the shoulder and forearm segments. The shoulder segment is labeled *a* and the forearm *b*.

I used the law of cosines to compute the angles *A*, *B*, and *C* shown in Figure 9-30. The basic formula is

$$c_2 = a_2 + b_2 - 2 \times a \times b \times \cos(C)$$

The values of segments *a* and *b* came from a SainSmart website diagram, shown in Figure 9-31. The shoulder segment is 12 cm, and the forearm segment is 15.5 cm.

This next equation computes the *C* angle, which is part of an elbow joint servo angle:

$$\cos(C) = (a_2 + b_2 - c_2)/(2 \times a \times b)$$

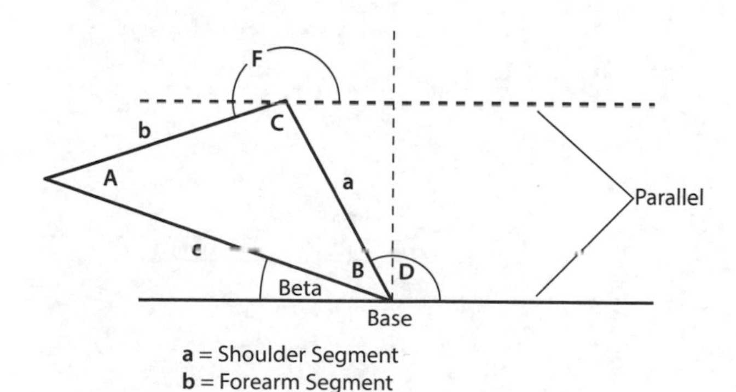

Figure 9-30 Free-body diagram showing the shoulder, forearm, and vector segments.

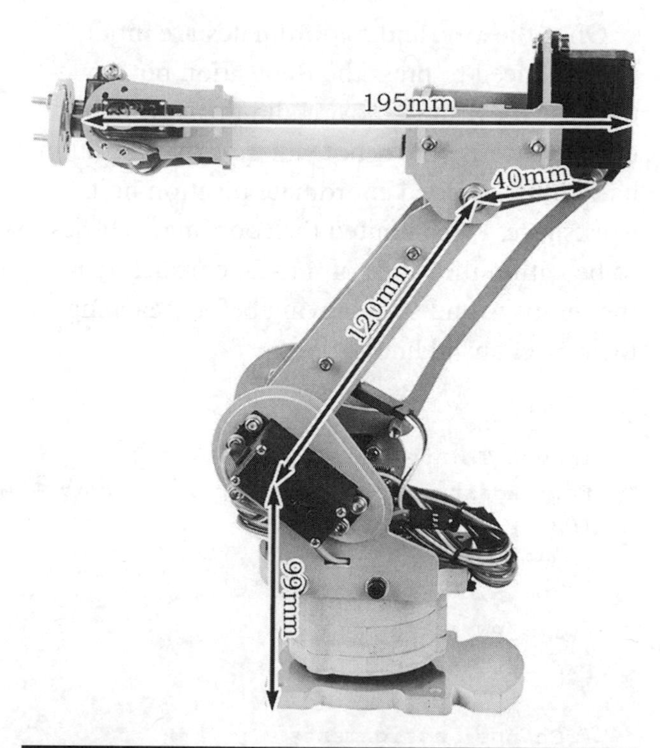

Figure 9-31 Segment lengths for *a* and *b*.

I used this next formula to compute the *B* angle, which is part of the shoulder joint:

$$\cos(B) = (a_2 - b_2 + c_2)/(2 \times a \times c)$$

Of course, all the angle computations are done in radians and must be converted back into degrees using the relationship that 2π radians is equivalent to 360°.

There is one more angle that must be accounted for, and this is the angle β between the vector and the *x-y* plane. Looking at Figure 9-30, it is straightforward to determine that this angle is computed as follows:

$$\sin^{-1}(\beta) = z_0/c$$

I incorporated all the mathematics I just went through into a function named `coordinateTransform()` and added it to the 6-DOF_Robot_Arm.py program. I then rewrote 6-DOF_Robot_Arm.py to query the user for the *x*, *y*, and *z* coordinates in centimeters. I also removed all the sliders because the purpose of this new app is to implement arm positioning based on coordinate inputs.

Once the *x*, *y*, and *z* coordinates are input, you will need to press the Reposition button to cause the arm to move. I did this as a safety feature because the robot will quickly reposition itself to the desired coordinate position in its workspace. I also limited the coordinate entries to be within the range of 4 to 24 cm such that the servos would stop moving before reaching their mechanical limits.

CAUTION **Be very careful because the robotic arm will swing quite rapidly if you enter new coordinates that are widely separated from its resting position. Keep small children, pets, and limbs away from the robotic arm while you run this program.**

The changed program 6-DOF_Robot_Arm.py was renamed *XYZ_Robot_Arm.py* and is available on this book's website and also listed next:

```
import Tkinter as tk
from Adafruit_PWM_Servo_Driver import PWM
import time
import math

pwm = PWM(0x40)
pwm.setPWMFreq(100)

# Channel assignments
# 0 = Wrist Rotate
# 1 = Wrist Up/Down
# 2 = Forearm Rotate
# 3 = Elbow Rotate
# 4 = Shoulder Rotate
# 5 = Base Rotate

root = tk.Tk()
root.title("XYZ Coordinate 6 DOF Robotic Arm")

win = tk.Frame(root)
win.grid()

x = tk.StringVar()
y = tk.StringVar()
z = tk.StringVar()

def showValues():
    print("x: " + x.get())
    print("y: " + y.get())
    print("z: " + z.get())

def coordinateTransform():
    x0 = float(x.get())
    if(x0 < 4):
      x0 = 4
    if(x0 > 24):
      x0 = 24
```

```
    y0 = float(y.get())
    if(yo < 4):
      y0 = 4
    if(y0 > 24)
      y0 = 24

    z0 = float(z.get())
    z0 = 30 - z0

    a = 12.0
    b = 15.5
    c = math.sqrt(x0*x0 + y0*y0 + z0*z0)
    C = math.degrees(math.acos((a*a + b*b - c*c)/(2*a*b)))
    B = math.degrees(math.acos((a*a - b*b + c*c)/(2*a*c)))

    Beta = math.degrees(math.asin(z0/c))
    Theta = math.degrees(math.atan(y0/x0))
    D = 180 - (B + Beta)
    E = 180 - (D + 90)
    F = 180 - (D - C)
    T = Theta

    #Scale and offsets for the servo commands
    SCmd = -.707*D + 57.5
    ECmd = .55*F -170
    BCmd = 1.0*T

    update3(ECmd)
    time.sleep(1)
    update4(SCmd)
    time.sleep(1)
    update5(BCmd)
    time.sleep(1)

    #these print statements are for debugging
    print("c: ", c)
    print("C: ", C)
    print("B: ", B)
    print("Beta: ", Beta)
    print("Theta: ", Theta)
    print("D: ", D)
    print("F: ", F)
    print("SCmd: ", SCmd)
    print("ECmd: ", ECmd)
    print("BCmd: ", BCmd)

label1 = tk.Label(win, text='X coordinate')
xCor = tk.Entry(win, textvariable=x, bd=5)
label1.grid(row=0, column = 0)
```

```
xCor.grid(row=0, column = 16)
label2 = tk.Label(win, text='Y coordinate')
yCor = tk.Entry(win, textvariable=y, bd=5)
label2.grid(row=1, column = 0)
yCor.grid(row=1, column=16)
label3 = tk.Label(win, text='Z coordinate")
zCor = tk.Entry(win, textvariable=z, bd=5)
label3.grid(row=2, column=0)
zCor.grid(row=2, column=16)

button = tk.Button(win, text='Show', command = showValues)
button.grid(row=3, column=16)

button1 = tk.Button(win, text='Reposition', command = coordinateTransform)
button1.grid(row=4, column=16)

#Elbow
def update3(angle):
    num = float(angle)
    pulseWidth = int((750*(num+60))/120 + 350)
    pwm.setPWM(3, 0, pulseWidth)
#Shoulder
def update4(angle):
    num = float(angle)
    pulseWidth = int((750*(num+60))/120 + 350)
    pwm.setPWM(4, 0, pulseWidth)
#Base
def update5(angle):
    num = float(angle)
    pulseWidth = int((750*(num+60))/120 + 350)
    pwm.setPWM(5, 0, pulseWidth)

root.update_idletasks()
root.geometry("400x200+0+0")

root.mainloop()
```

You should run the program in the same directory as all the other arm control programs. Enter the following to run this program:

```
sudo python XYZ_Robot_Arm.py
```

Figure 9-32 is a screenshot of the program running with a set of *x*, *y*, and *z* coordinates entered. The arm started to reposition immediately after I clicked the Reposition button.

You also may notice that I had the debugging print statements displayed because they helped me immensely to correctly modify the program such that it functioned as I wanted it to. I also included 1-second delays so that the individual segment positioning could be observed separately. You can easily remove them if you want all the segments and base to move together.

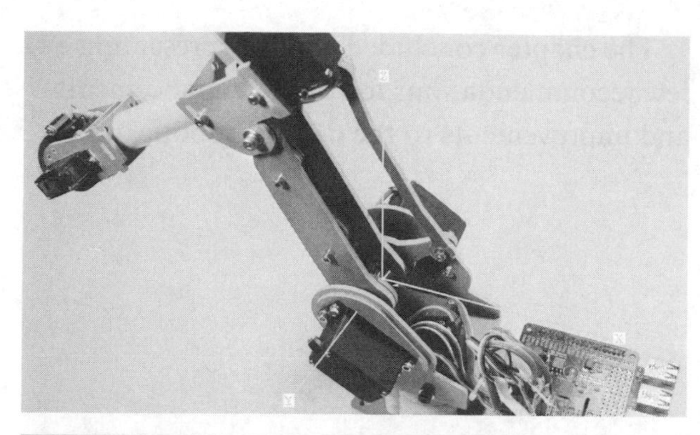

Figure 9-32 Screenshot of the XYZ_Robot_Arm program running.

Before concluding this section, I want to mention some constraints and limitations within this program. The arm positioning can only be approximated because the point of origin I used in the geometric calculations differs from the real robotic arm. By this, I mean that the pivot where the shoulder segment rotates at the base is displaced approximately 4.5 cm from the vertical base rotation axis. This displacement will cause a constant error in the resulting angle calculations owing to the *z*-axis offset. I chose to leave the error in place because the robotic arm was designed for an educational/fun experience and not for a precision setup. There is also another slight deviation in that there is a small offset between the shoulder and elbow pivot points, again introducing some error in the angle calculations. Readers are encouraged to modify the program to reduce or eliminate these errors or simply use the program as is.

I have listed below several improvements and enhancements to this program that interested readers may explore further:

1. Incorporate a file I/O operation where a series of coordinates may be read in and the robotic arm would be continuously repositioned in accordance with the coordinates stored in the file.
2. Attach a felt-tipped pen to the end effector (fingers), and draw text and figures on a white sheet using only *x* and *y* coordinates to move the effector.
3. Attach a video camera to the end effector and feed the video stream to an image-processing program capable of recognizing simple objects, such as a ball or block. These programs usually have center-of-gravity capture algorithm that can provide control signals to the robotic arm to keep the observed object centered in the frame of the video stream.

Presenting these enhancements and improvements to the 6-DOF robotic arm concludes this section and the chapter.

Summary

The chapter began with a comprehensive introduction to robotic arms that explained the concepts and methodologies used in designing these devices. Two robotic arms were next introduced, and they were used as demonstrators to show how robotic arms actually function.

I next discussed how servos function as they are key components of all robotic arms. Software was discussed next, and I showed you how to configure a RasPi to control a robotic arm using the I2C bus. I also went through an initial test using a stand-alone servo to test the software installation and configuration process.

I next demonstrated a relatively simple 3-DOF arm used in a pick-and-place application that was programmed using Python. A 6-DOF arm was used next to demonstrate how a much more complex robotic arm worked. I wrote two Python programs to control this arm. The first simply allowed you to control each of the six servos using a GUI widget. The second program allowed you to enter *x*, *y*, and *z* coordinates and have the robotic arm reposition its end effector to the corresponding point in its volumetric workspace.

The chapter concluded with my presenting a few recommendations for future enhancements and improvements to the 6-DOF robotic arm.

CHAPTER 10

Gigapixel Camera System

IN THIS CHAPTER, I WILL SHOW YOU how to build two digital camera systems capable of composing gigapixel photographs. I will not be building digital cameras with billion-plus pixel resolution because only a few of those exist today, and they are prohibitively expensive. Instead, I will use moderate cost digital cameras, each with millions of pixel resolution, and then take multiple photographs of a subject and use post-processing techniques to form a final gigapixel image. The multiple-image approach is really the only practical way to capture huge-resolution photographs using consumer-grade digital cameras.

The first system will use a small, relatively inexpensive point-and-shoot digital camera mounted on an inexpensive pan-and-tilt servomechanism. The second system will use a more expensive digital single-lens reflex camera (DSLR) mounted on a "prosumer"-grade pan-and-tilt mechanism. I took this two-system approach to accommodate readers who would like to experiment with this technology but do not have the resources to invest in the expensive photographic equipment that would be needed to build the second system.

Chapter 10 Parts List

No cameras or tripod components are listed. Please refer to the text regarding suitable cameras/tripods to use for these projects.

Item	Model	Quantity	Source
Servo pan/tilt mechanism for point-and-shoot camera	Generic	1	amazon.com
Servo pan/tilt mechanism for DSLR camera	PT785-S System	1	servocity.com
Servo/PWM Pi HAT board	2327	1	adafruit.com
Push button	1479	2	adafruit.com
Raspberry Pi Model B+	83-16530	1	mcmelectronics.com
5-V cell phone external battery	Generic	1	amazon.com
6-V sealed lead-acid battery	Commodity	1	amazon.com
0.125 inch of Lexan sheet material	Commodity	See build drawings	Local home-improvement store
Machine screws and nuts	Commodity	Various	Local home-improvement store

I will start the chapter by discussing two post-processing techniques, both of which allow a gigapixel image to be composed.

Stack and Stitch

Stacking and stitching are both required for image post-processing, which allows the gigapixel camera system to be implemented. I will discuss stacking first, followed by the stitching process.

Stacking

The *stacking* in the section title refers to a technique more formally called *focus stacking* but also known as *image stacking*. It is necessary to review some fundamental photography basics to understand what focus stacking encompasses. Figure 10-1 is a diagram showing a simple camera lens and sensor that helps to illustrate the key concept of how a camera focuses.

In this figure, the lens has been set either manually or automatically to converge the majority of incoming light rays reflecting off the subject object onto the plane of the sensor. However, some light rays reflecting off objects that are separate from the subject object will converge before or after the sensor plane and are said to be *out of focus*. This loss of focus is also referred to as *loss of sharpness* and is the result of an optical principle called *diffraction*. Diffraction is a fundamental physics principle and cannot be overcome simply by using a better lens. However, using focus stacking can mitigate it.

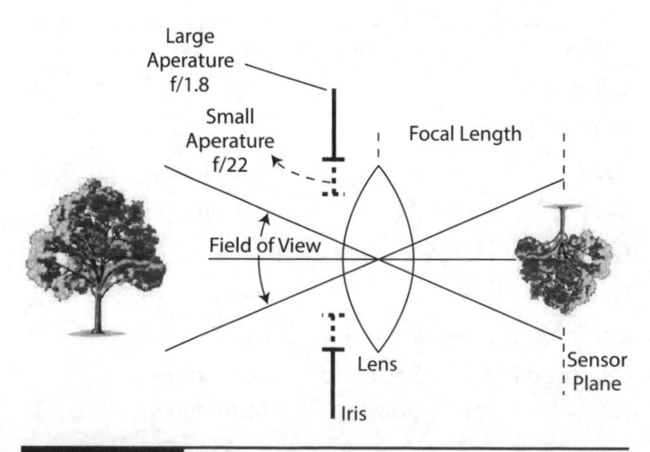

Figure 10-1 Diagram of camera focus system.

The distance between the optical center of the lens and the sensor plane is known as the *focal length*. This attribute does not affect the lens focus directly but is related more to image size. An iris also can be seen in the figure situated just in front of the lens, and it controls the amount of light passing through the lens. The opening in the iris is referred to as the *aperture*, and it also plays an important role in determining the image focus. Figure 10-2 shows an iris set to various aperture settings known as *f-stops*. The f-stop is defined as the ratio between the lens diameter and the open-aperture diameter.

This figure should convey the concept that a smaller aperture setting means that a greater depth of field exists, and hence objects being photographed will tend to remain in focus on the sensor plane. The underlying optical physics determining why this happens is complex and really does not need to be discussed. Simply accept the fact that smaller apertures mean greater depth of field for an image. Of course, there is a distinct disadvantage to using small apertures in that less light strikes the image sensor, and there is a good chance that the image will be underexposed depending on the total intensity of light reflecting off the object to be photographed. The main way photographers overcome this issue is to increase the time during which the shutter remains open, exposing light from the lens onto the sensor. This then leads to the issue of image blurriness owing to camera shake, especially if the camera is hand held. Use of a tripod normally mitigates this situation quite well, which is why you see professional photographers using tripods during night shots.

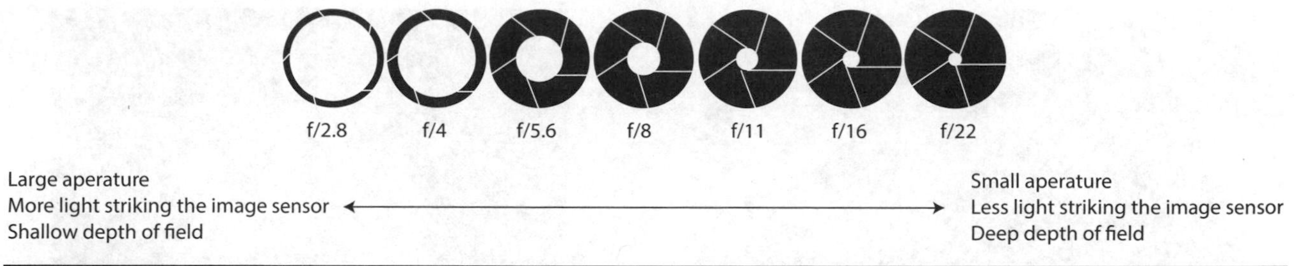

Figure 10-2 Aperture settings.

Figure 10-3 is an excellent representation of how focus, aperture, and depth of field are related. This image is courtesy of Dave Shaker and is available as part of his "The Camera Lens" blog at www.thatfish.com.

I also wanted to mention that the issue of depth of field becomes very problematic when you attempt to take extreme close-up photographs. This type of image taking is known as *macrophotography*, and it has greatly benefited from the focus-stacking technique.

The type of lens being used also plays a role in this whole interrelationship of focus, aperture, and depth of field. You are likely to have a small zoom lens if you are using a modern point-and-shoot camera such as a Canon SX160 IS, as shown in Figure 10-4.

This is the camera I used for the point-and-shoot part of this chapter's project. It has a 5.0- to 80-mm zoom lens permanently attached, as shown in the figure. This lens is well suited for wide-angle to long-range shots because it incorporates a 16× optical zoom in the lens. Some point-and-shoot cameras do not have a telescopic capability but instead rely on electronic processing to achieve telescopic range. Electronic image magnification usually has less quality than a quality optical telescopic lens. But this is an acceptable tradeoff that most point-and-shoot camera users readily accept for the convenience of using a compact camera at a

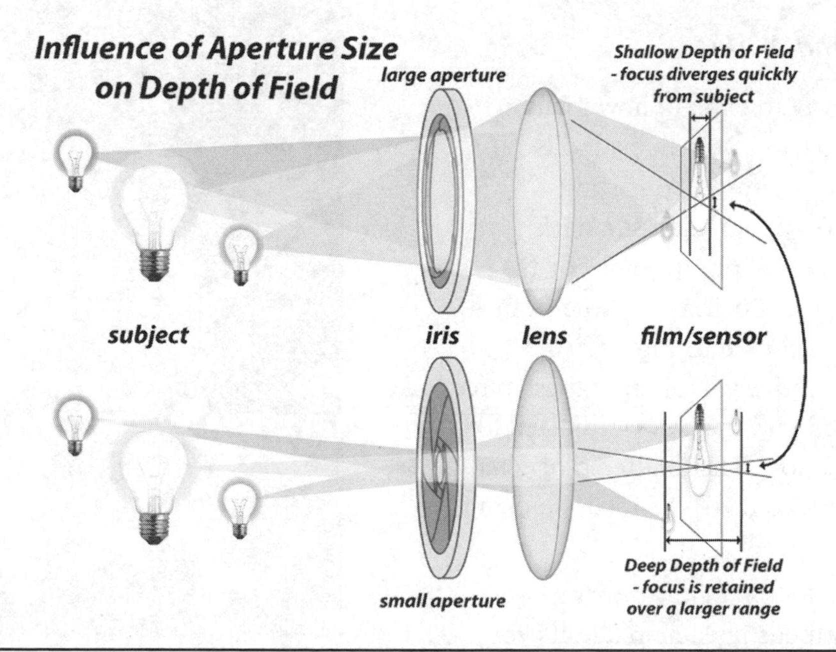

Figure 10-3 Camera focus, aperture, and depth-of-field relationships. (www.thatfish.com)

Figure 10-4 Canon SX160 IS point-and-shoot camera.

good price point. In the second portion of this chapter project, I do use a DSLR, which accepts a wide range of lens, including wide-angle and telescopic lenses.

So far, I have presented a brief background on how a camera focuses on a subject object and the related issue of why out-of-focus objects are also typically seen in the frame of view. It is now time to discuss how focus stacking overcomes this problem.

How Focus Stacking Works

Probably the best way to explain how focus stacking works is to show you an example. This example is pretty simple, consisting of a photograph of a tie set against a dark background. I took three photographs of the tie, with the focus points being set at the front, middle, and back, as shown in Figures 10-5, 10-6, and 10-7, respectively. Note that these three figures were cropped and the colors adjusted for publication in this book. The unaltered original images were used as inputs to the focus-stacking application.

You should be able to see how only one area of the tie is in focus in each image. These three images were next used as inputs to Zerene

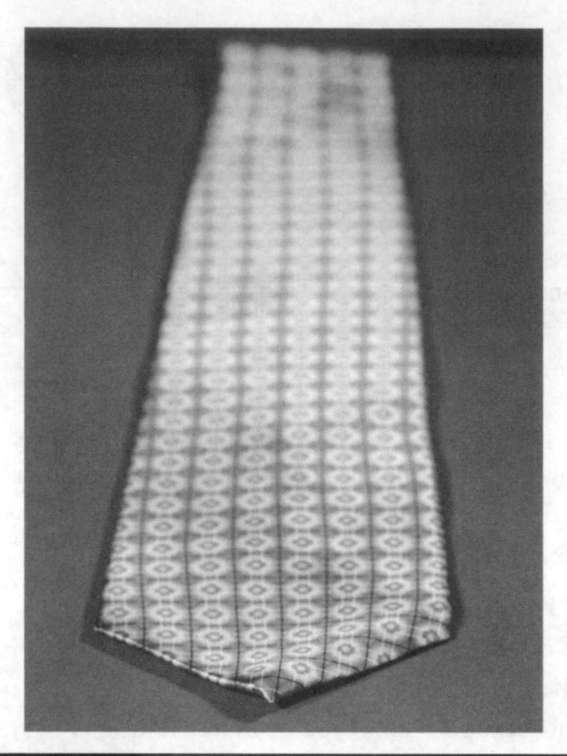

Figure 10-5 Near focus.

Systems, focus-stacking software, ZereneStacker. Figure 10-8 is a screen shot of the ZereneStacker application running on my MacBook Pro.

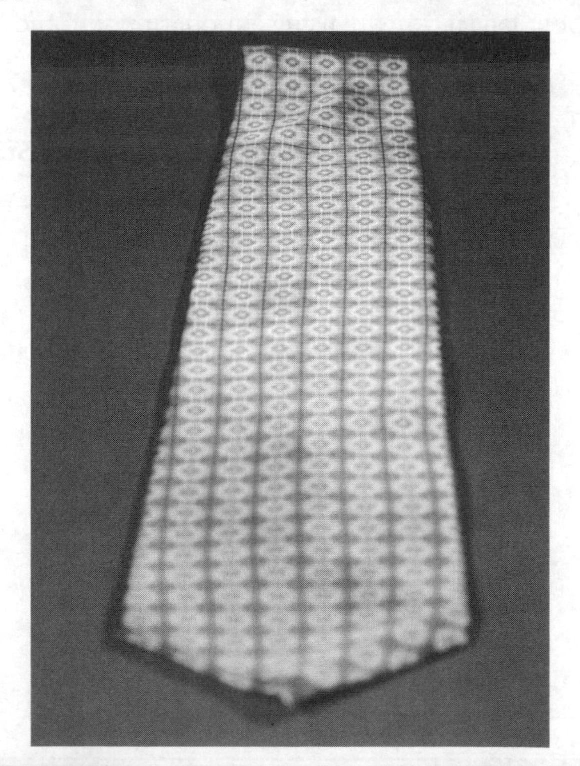

Figure 10-6 Middle focus.

Figure 10-7 Far focus.

You can generate an output image using PMax or DMap methods. PMax uses a *pyramid method*, which is very good at finding and retaining details in an image. It is good at processing images containing matrix-like structures such as hair mats and crisscrossed bristles. It also avoids creating halos, which can mask details and is common in other focus-stacking applications.

DMap uses a *depth-map method*, which tends to retain more of the original image's smoothness and colors. It does not retain as much detail as the PMax method.

Zerene Systems recommends processing stacked images using both methods and selecting the best result. Often, selecting the best output image and then retouching it will yield the best results. Figure 10-9 shows the final tie output image, which I cropped and retouched for optimal intensity level, contrast, and color.

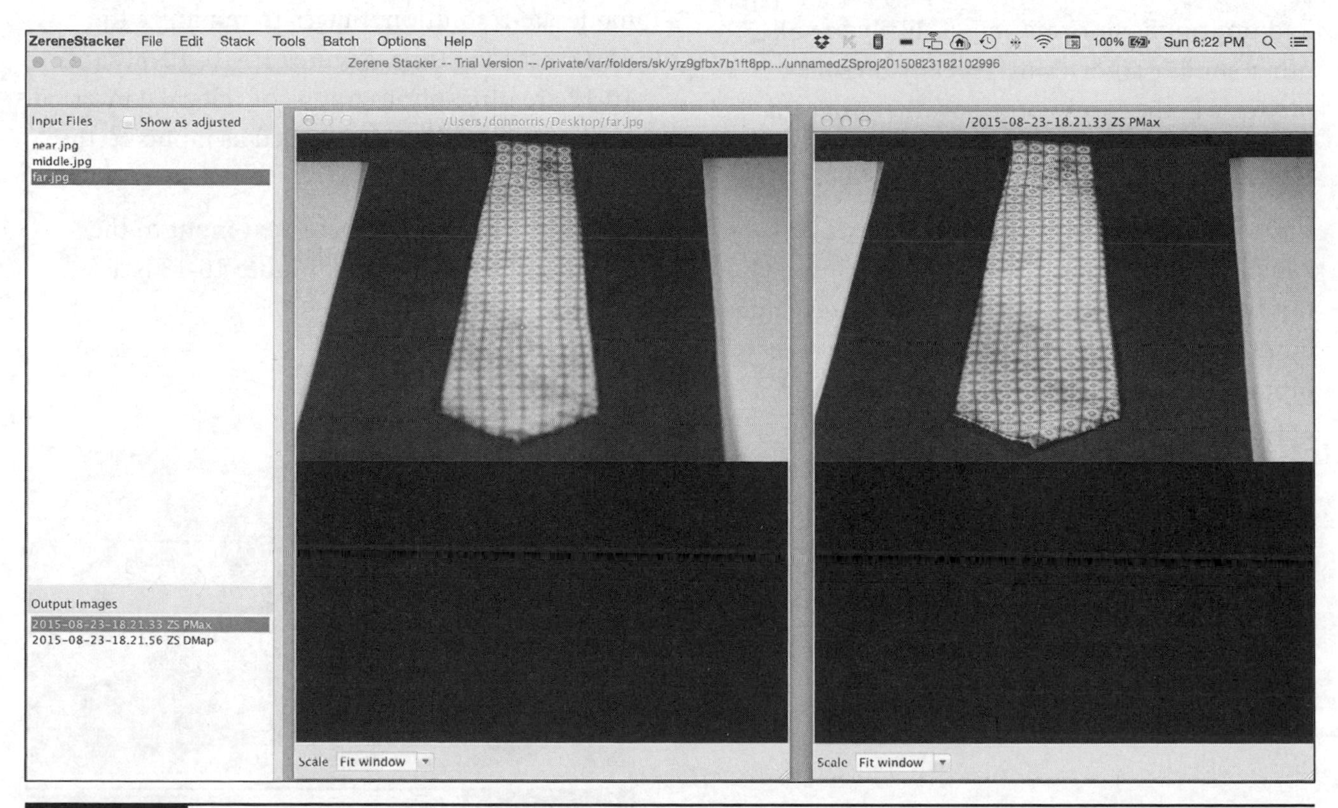

Figure 10-8 Zerene focus-stacking application.

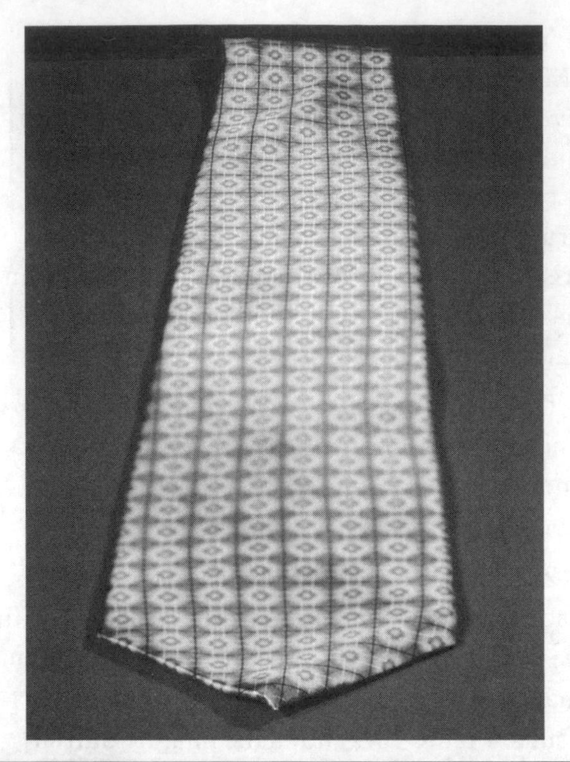

Figure 10-9 Focused-stacked tie image.

Of course, the actual algorithms that Zerene uses to focus stack images are proprietary and not public information, as likely are all other similar algorithms. It really doesn't matter whether you know how the program functions other than to know how to use it effectively. Some readers may know that Adobe's Photoshop also includes a photo *merge function*, which is Adobe's term for focus stacking. I chose not to use Photoshop because it requires many more steps than the ZereneStacker to process input images. In addition, a focus-stacking application is optimized for that particular process as opposed to Photoshop, which has to perform a multitude of different and often unrelated image-processing functions.

Stitching is the next part of this project's image postprocessing, which I discuss in the following section.

Stitching

Stitching is the imaging process of seamlessly joining adjacent photographs such that they form one panoramic view. Sometimes panoramic photographs are simply referred to as *panos*, which I will do from now on for brevity's sake. Specialized pano image-processing applications are available as well as functions contained in Photoshop that can perform image stitching. I chose to use an easy pano specialized program named *Panoweaver v9.1 Standard Edition* to create my project panos using similar reasoning as to why I chose a specialized stacking software over the Photoshop merge image function.

Taking Some Sample Pictures

For best results, I took a series of images with a digital camera mounted on a tripod. The camera was leveled on the tripod, and I overlapped each image by 40 to 50 percent with the preceding one. Also, I set the camera to take moderate-resolution images to minimize the postprocessing time. Figures 10-10, 10-11, and 10-12 are three photographs of a local converted textile mill building that I used as inputs to the stitching application.

These three images were next input to the image-stitching software. Figure 10-13 is a

Figure 10-10 Left.

Figure 10-11 Center.

Figure 10-12 Right.

screenshot of the Panoweaver application running on my MacBook Pro.

You generate an output image by first importing the source images. The panoramic stitch function will not even appear on the ribbon menu bar without the source images being selected. You will next need to select the type of pano desired from one of the radio-button selections appearing on the right-hand side of the application. I list these selections next with a few brief comments about each one.

- **Spherical.** This is the default selection and was the one I used.
- **Cubic.** There is very little perceived difference between this and spherical selection.
- **Cylindrical.** The mill roofline was convex with a "bowed out" appearance.
- **Littleplanet.** Extremely concave; not recommended except for a special effect.

The generated pano has an uneven border because of the image manipulations necessary to create the pano. I cropped and slightly retouched

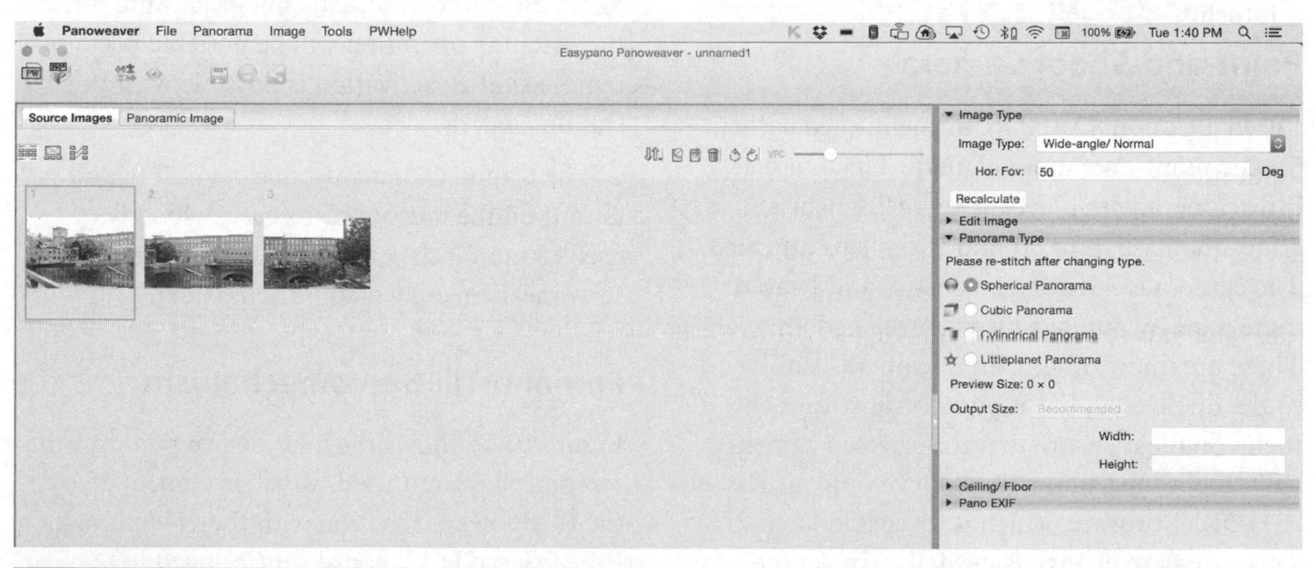

Figure 10-13 Panoweaver image-stitching application.

Figure 10-14 Mill building pano.

the pano image using Photoshop to achieve a pleasing image, which is shown in Figure 10-14.

A lot of information is available regarding the stitching algorithms, which was not the case for the stacking algorithms. However, the algorithms are quite complex, and not much can be gained in a book of this type by going through them.

At this point, I have covered the entire pertinent image postprocessing that needs to be done to achieve the goals of a gigapixel camera system. It is time to discuss in some detail the point-and-shoot camera, which is the critical part of the first gigapixel camera system.

Point-and-Shoot Camera

I used a Canon SX160 IS, as mentioned earlier, for the point-and-shoot camera. This is a nice little camera with a decent 16-MB sensor and a good-quality 5.0- to 80-mm zoom lens attached. The camera is not very expensive, and I saved quite a bit by purchasing a refurbished unit. There are many other Canon cameras similar to the one I used, but if you do substitute one, then you must ensure that the selected camera will accept the Canon Hack Development Kit (CHDK) firmware, which is discussed later. This firmware change is needed to make the camera suitable to accept RasPi commands for shutter activation. There are no other required modifications to the camera other than the firmware change. Note that even changing the original firmware will not impair normal camera operations because the CHDK firmware is easily bypassed to place the camera in its normal operating condition.

You should be aware that the CHDK cannot control the camera focus operation. It thus will be impossible to implement automated focus stacking using the CHDK firmware. In fact, most modern point-and-shoot cameras do not have any provision for manual lens focus. However, automatic image capture for the stitching operation will be possible because remote shutter activation is possible with this Canon camera.

The Canon camera also has a ¼-20 tripod mount on the bottom of its case, which is used to mount the camera to the pan-and-tilt servomechanism, which I discuss next.

Pan-and-Tilt Servomechanism

Figure 10-15 shows the inexpensive pan-and-tilt servomechanism I used, which is controlled by the RasPi in conjunction with the 16-channel PWM/Servo HAT board I introduced in Chapter 9.

Figure 10-15 Pan-and-tilt servomechanism.

This mechanism has 2 DOFs and will allow the camera to be tilted in excess of 120°. The same is true of the panning servo. In reality, I expected to do very few tilt adjustments but to heavily use the panning to create the pano. This is why I made the tilt operation manual but automated the panning.

You also should note that this pan-and-tilt mechanism handles the lightweight point-and-shoot camera just fine but will not be able to move the much heavier and massive DSLR camera, which is the reason I separated the project into two parts.

You will need to remove the C-bracket from the top of the tilt servo and drill out the center hole to provide clearance for the ¼-20 screw that secures the camera to the C-bracket. Figure 10-16 shows this bracket with the center hole enlarged using a ¼-inch drill bit.

I do wish to point out that these inexpensive pan-and-tilt mechanisms are a commodity product built in China by various suppliers. Other units may not have the same mounting

Figure 10-16 C-bracket with mounting hole drilled out.

hole as this one does, so you will have to be flexible and adapt the mounting arrangement to suit what you have in hand. However, I strongly recommend that you attach the camera to the C-bracket while it is not attached to the servo. It will be impossible to install the mounting screw once the C-bracket is reattached to the servo. Figure 10-17 shows the complete assembly of the camera attached to the pan-and-tilt mechanism.

Figure 10-17 Pan-and-tilt servomechanism with camera mounted.

Note that I also attached the mechanism to a temporary Lexan stand for testing purposes. You don't have to do this as long as you secure the bottom servo so that the whole assembly does not fall over. I eventually removed the stand and attached the bottom servo to a tripod mount adapter, which I describe later in this chapter.

You also must ensure that the servo cables are free and clear of any pinch points and obviously not in the way of the lens. This will be easy to do once the RasPi has been mounted on the tripod adapter along with the battery supply. Again, all this will be discussed a bit later.

Now it is time to discuss the CHDK firmware, which must be installed on the camera to automate the pano process.

CHDK Firmware

The Canon Hack Development Kit (CHDK) is a unique software application that runs on microprocessors contained in Canon point-and-shoot cameras. CHDK software makes no permanent changes to the camera, which means that the original Canon firmware can be restored, if desired.

CHDK allows control over many camera features, some of which are detailed in Table 10-1.

TABLE 10-1 CHDK Features

Feature	Details
Camera control	RAW files, bracketing, exposure, zebra mode, grids
Detection	Trigger exposure in response to motion detection
USB	Remote camera control over USB
Programming	Create scripts with uBasic and/or Lua

I do wish to make one item very clear, and that is that the CHDK firmware strictly uses digital pulses to control the camera. It does not use digital text sent over the USB data lines for camera control. This means that camera control is created by both counting the number of pulses and the durations of those pulses. Of course, all the appropriate pulses and durations will be done for you using the RasPi.

A key component for controlling the camera is to modify a USB cable such that it can transfer digital pulses from the RasPi to the camera. This is the topic of the next section.

USB Control Cable

You will need to modify an existing USB cable that has a mini-A socket as shown in Figure 10-18.

The mini-A socket matches the point-and-shoot camera I used, but if you substituted a different Canon camera, just ensure that you match the cable to the existing USB socket. You next need to cut off the standard USB connector and strip back the cable insulation about 1.5 inches to expose the enclosed wires. The good news is that only two wires will be used for the connections, and these are shown in Figure 10-19.

Typically, the two wires are colored red and black, with the red one connected to pin 1 and the black one connected to pin 4. Note that it is not guaranteed that all USB cable manufacturers follow this color-coding convention, so you might have to use a volt-ohm meter (VOM) to confirm the wire-to-pin connections. You also

Figure 10-18 Mini-A USB socket.

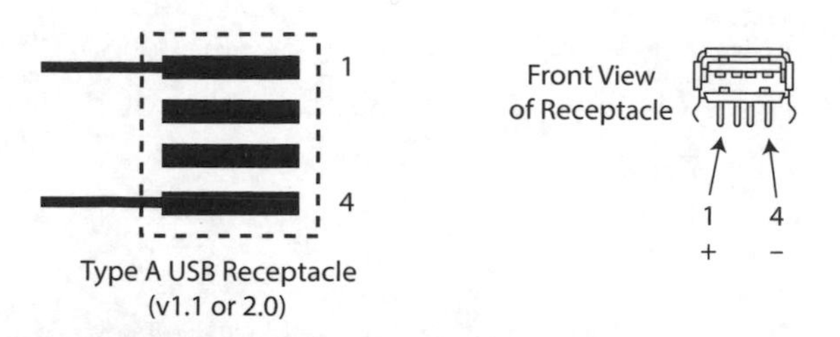

Figure 10-19 Two USB control wires.

should tin the wire ends to strengthen them a bit so that they can be easily soldered onto the HAT board. By *tin*, I mean to apply a little solder to the wire ends. Just be careful because the wires are very small gauge and easily damaged by overheating.

I next set up for an initial test using the same RasPi that was used to control the pan-and-tilt mechanism. You cannot use a Pi Cobbler for the USB cable connections because the servo HAT covers all 40 GPIO pins. Instead, you will need to solder the black USB wire to an open ground point and the red wire to the pin labeled pin 4 on the HAT board. Figure 10-20 shows these wires soldered to the HAT board.

I wrote a simple Python test program that checked out the USB cable connection and ensured that the camera could be controlled by the RasPi. This program is a modified version of the 3-DOF robotic arm program I used in Chapter 9. There are only two sliders in this GUI, one for the pan and the other for the tilt. I also included a GUI button that may be clicked to take a picture. The program is named *PnS_Camera1.py* and is available on this book's companion website. It is listed next with comments.

Figure 10-20 USB wires attached to HAT board.

```
#!/usr/bin/python

#PnS_Camera1.py
#D. J. Norris 2015
#This code is in the public domain.

from Tkinter import *
from Adafruit_PWM_Servo_Driver import PWM
import RPi.GPIO as GPIO
import time

GPIO.setmode(GPIO.BOARD)
#7 is the actual pin number used by the RasPi GPIO pin 4
GPIO.setup(7, GPIO.OUT)
GPIO.output(7, False)

pwm = PWM(0x40)
pwm.setPWMFreq(100)

def takePicture():
    GPIO.output(7, True)
    time.sleep(1.0)
    GPIO.output(7, False)

class Servo:
    def __init__(self, topl):
      win = Frame(topl)
      win.pack()
      label1 = Label(win, text='Pan')
      sc1 = Scale(win, from_=-90, to=30, orient=HORIZONTAL, command=self.update0)
      #Centers the slider tab
      sc1.set(-35)
      sc1.grid(row=0)
      label1.grid(row=1)
      label2 = Label(win, text='Tilt')
      sc2 = Scale(win, from_=-90, to=30, orient=HORIZONTAL, command=self.update1)
      #centers the slider tab
      sc2.set(-32)
      label2.grid(row=3)
      sc2.grid(row=2)
      button = Button(win, text='Take Picture', command=takePicture)
      button.grid(row=4)

    def update0(self, angle):
      num = float(angle)
      pulseWidth = int((750*(num+60))/120 + 350)
      pwm.setPWM(0, 0, pulseWidth)

    def update1(self, angle):
```

```
        num = float(angle)
        pulseWidth = int((750*(num+60))/120 + 350)
        pwm.setPWM(1, 0, pulseWidth)

top = Tk()
top.wm_title("Point-n-Shoot Pan and Tilt Servo Control")
app = Servo(top)
top.update_idletasks()
top.geometry("400x200+0+0")
top.mainloop()
```

You must be in X Windows to run this program, as was the case for the robotic arm programs. You also must be in the same directory, which I specified in Chapter 9, which contain all the robotic arm programs. This program uses the same Adafruit PWM/Servo library as was used in the robotic arm programs.

The PWM/Servo HAT board must be attached to the RasPi with a connected 5-V servo power supply, as it was set up with the robotic arm tests. In this case, the pan servo is connected to channel 0, and the tilt servo is connected to channel 1.

Enter the following commands to run the camera program:

```
sudo startx
sudo python PnS_Camera1.py
```

Figure 10-21 is a screenshot of this program running on a RasPi.

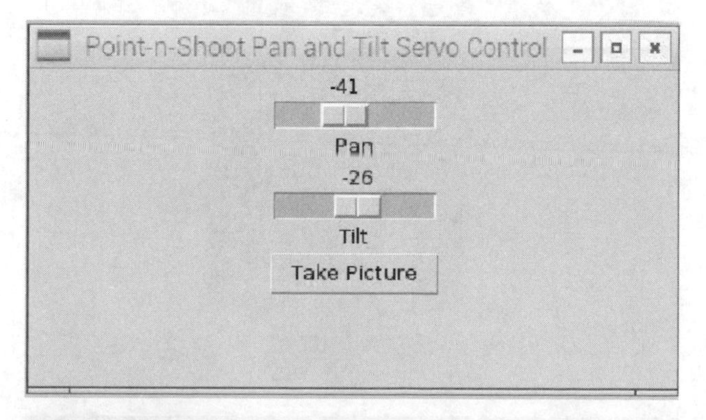

Figure 10-21 PnS_Camera1 program GUI.

You should be able to easily tilt and pan the camera using the sliders. Just be careful because the camera is sufficiently heavy to the point of causing the stand to tip over if it is tilted too far forward or backward.

The next step in completing the point-and-shoot camera system is to build a tripod adapter to hold all the components, including a hefty battery supply for the RasPi and the servomechanism.

Building the Tripod Adapter

I will start this section by stating that my design is not the only way to build a frame that will safely hold all the components and yet allow you to mount them on a sturdy tripod for the picture taking. After some experimenting, I settled on an O-frame structure made out of ⅛- × 1.5-inch aluminum flat stock, which allowed me to mount all the items without too much difficulty in machining yet was strong enough and light enough to be mounted on a normal camera tripod. I used a typical tripod connect adapter called a *shoe plate* that matched my Vanguard tripod. These adapters are inexpensive and are really the only proper way to mount devices on tripods. Just be sure that you purchase a matching adapter that fits your tripod because they come in several sizes.

The shoe plate then was mounted to the bottom of the O-frame using a ¼-20 machine screw, thumbscrew, and washer to spread the

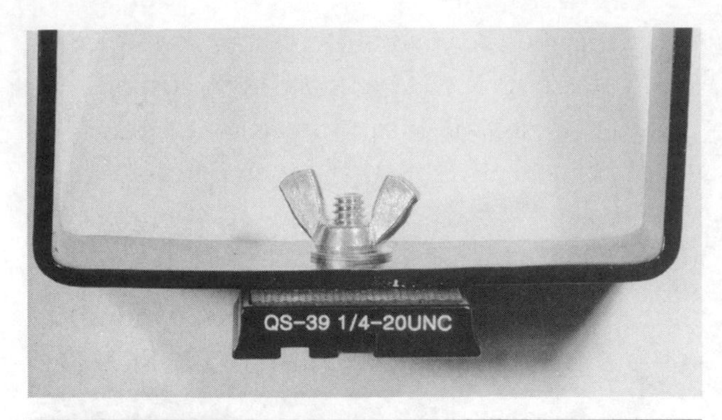

Figure 10-22 O-frame with shoe plate attached.

load out onto the frame. Figure 10-22 shows the O-frame with the shoe plate attached.

I next had to fashion a small L-bracket, which I attached between the existing bottom servo C-frame and the top portion of the aluminum O-frame. This bracket also serves as the connecting plate between the ends of the O-frame. I found this to be the most convenient way to join the O-frame ends unless you happen to have an aluminum welding rig. Figure 10-23 shows this small L-frame, which I made of a strip of aluminum flat stock.

The RasPi-HAT combination is attached to the vertical portion of the C-frame using an inexpensive plastic case designed for a Model B+. I first attached the case using 4-40 machine screws/nuts and then inserted the RasPi into the case. Just be sure that there is sufficient slack in the servo wires connecting to the HAT board. Figure 10-24 shows the completed assembly mounted on a tripod.

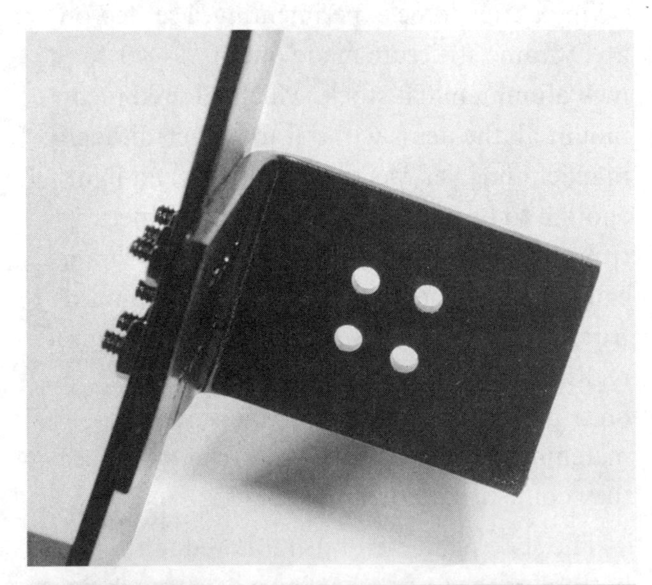

Figure 10-23 Aluminum L-frame holds the servos to the top of the steel C-frame.

Notice that I simply placed two battery packs on the bottom portion of the O-frame. One pack consists of four AA batteries, which power the servos. The other is a commercial cell phone battery extender, which powers the RasPi. Both these packs should provide sufficient capacity for a least one photographic expedition. This arrangement allows you to carry along some fresh AA batteries as well as an extra cell phone battery extender. Having the battery packs

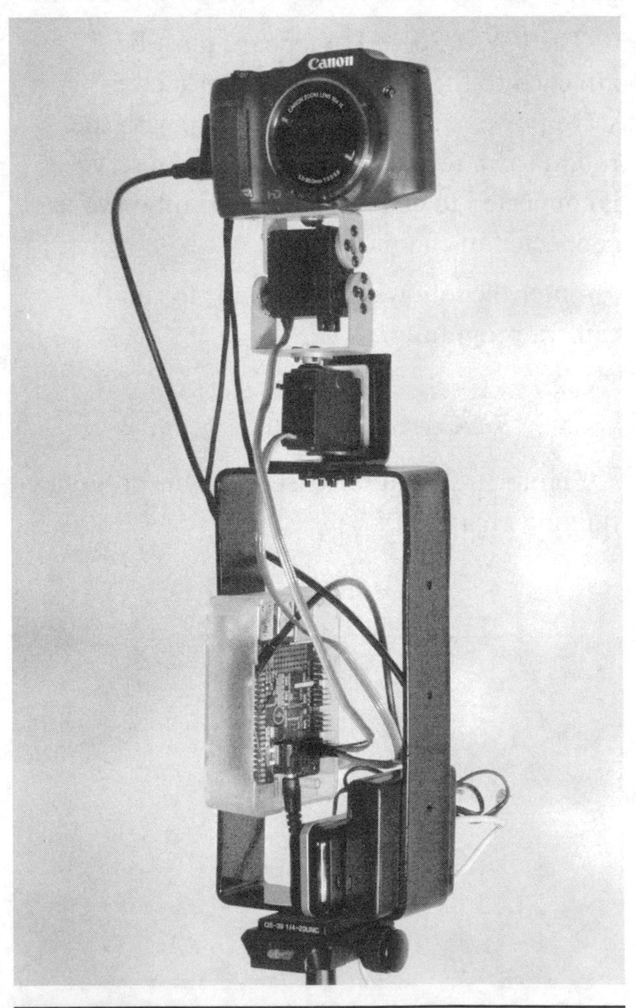

Figure 10-24 Complete point-and-shoot tripod adapter.

resting freely on the tripod adapter makes swapping them a simple task.

The only task left is to create a program that not only modifies the one I already showed you but also automates the panning and picture-taking functions. It will also be started by a real push button because you do not want to have to use a monitor, keyboard, and mouse to access the RasPi.

Automating the Point-and-Shoot Gigapixel Camera System

The changes required to automate the existing point-and-shoot program are extensive. I first removed all the GUI components because they are no longer required for this self-contained program. I next added a forever loop, which, when activated by a user pressing a push button, will cause the camera to pan from left to right in 30° increments, pausing 5 seconds after each motion. The shutter then would be activated after a 2-second delay to allow any vibrations caused by the camera motion to subside. Figure 10-25 shows the schematic for the push-button connection. I recommend that it be wired directly to the HAT board. Any normally open push button can be used. Just don't forget to add the 10-kΩ resistor; otherwise, you will be shorting the 3.3-V supply to ground, which would not be a good thing!

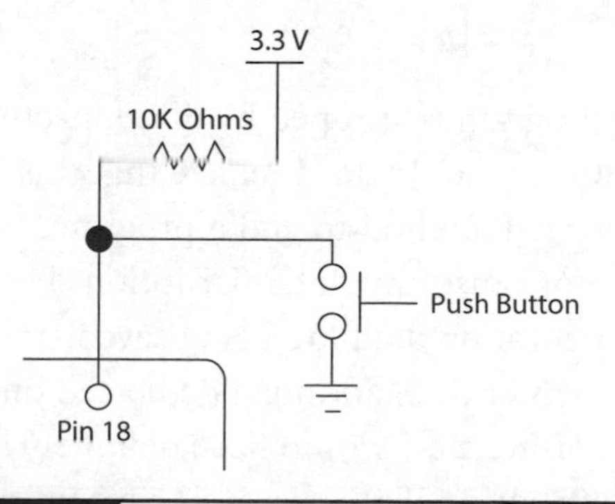

Figure 10-25 Push-button connection.

I mounted the push button on an L-shaped piece of Lexan for easy access on the tripod adapter. Figure 10-26 is a dimensioned sketch of this Lexan piece. Note that I used a hot-air gun to soften the Lexan, which allowed me to bend it into an L shape.

Figure 10-27 shows the push button mounted on the Lexan and connected to the RasPi.

Tilt control was not enabled because you want the camera to be a steady horizontal plane as it is panned. It is easy to gently adjust the tilt angle

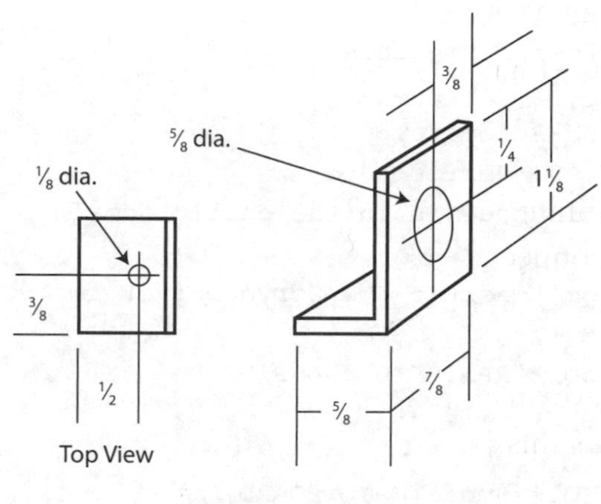

Figure 10-26 Push-button mounting bracket sketch.

Figure 10-27 Mounted push button connected to the RasPi.

because no power is being applied to the tilt servo. Simply leave the tilt angle alone once you have adjusted the camera to the desired plane.

The program also was configured to autostart after the power was applied to the RasPi. I explain how to configure the RasPi for an autostart configuration after the program listing.

This program finally ensures that the camera always starts the picture-taking process facing to the left. I renamed the program `PnS_Automate.py`, and it is listed next. As always, it is also available on this book's companion website.

```
#!/usr/bin/python

#PnS_Automate.py
#D. J. Norris 2015
#This code is in the public domain.

from Adafruit_PWM_Servo_Driver import
PWM
import RPi.GPIO as GPIO
import time

GPIO.setmode(GPIO.BOARD)
#7 is the actual pin number used by the
 RasPi GPIO pin 4
GPIO.setup(7, GPIO.OUT)
GPIO.output(7, False)

#Push button input, physical pin 12 is
 RasPi GPIO pin 18
GPIO.setup(12, GPIO.IN)

#Need a variable to logically latch the
 pb press action
picFlag = 0

pwm = PWM(0x40)
pwm.setPWMFreq(100)

def takePicture():
     GPIO.output(7, True)
     time.sleep(1.0)
     GPIO.output(7, False)

def update0(angle):
     num = float(angle)
     pulseWidth =
int((750*(num+60))/120 + 350)
     pwm.setPWM(0, 0, pulseWidth)

#main code
#set to left position initially
update0(40)
#wait 2 seconds for any vibrations to
 settle
time.sleep(2.0)

while(True):

      #wait for a button press
      #pin 13 is tied high. Pushing the
       pb grounds the input
      pb = GPIO.input(12)
      if(pb == 0):
        picFlag = 1

      if(picFlag == 1):
        takePicture()
        time.sleep(2.0)
        update0(20)
        time.sleep(2.0)
        takePicture()
        time.sleep(2.0)
        update0(0)
        time.sleep(2.0)
        takePicture()
        time.sleep(2.0)
        update0(-20)
        takePicture()
        time.sleep(2.0)
        update0(-40)
        #Need to reset the pb flag to
         await next button press
        picFlag = 0
```

The program is stopped by disconnecting the power to the RasPi. I realize that this is not the preferred method to end a program, but it should not cause any data disruptions because no configuration data are being saved, and it is much easier than adding a dedicated on/off button. However, I would have definitely added an on/off button if this was to be a commercial product.

Configure Program for an Autostart

Be sure that you have installed the push button and resistor before doing this configuration. Otherwise, you will not know if the program is actually running. You will next need to make the following changes to the inittab file:

- `sudo nano /etc/inittab`
- Comment out the following line:

```
1:2345:respawn:/sbin/getty 115200
tty1
```

- Note that the line may appear like this in more recent versions of the Wheezy distro:

```
1:2345:respawn:/sbin/getty --noclear
38400 tty1
```

- Add this line after the line just commented out:

```
1:2345:respawn:/bin/login -f pi tty1
</dev/tty1>/dev/tty1 2>&1
```

- Save the file, and exit the nano editor.

Next, edit the profile file:

- `sudo nano /etc/profile`
- Add the following line to the file (do not add any line returns):

```
sudo python /home/pi/Adafruit-
Raspberry-Pi-Python-Code/Adafruit_
PWM_Servo_Driver/PnS_Automate.py
```

- Save the file, and exit the nano editor.

You should be ready to test the autostart configuration. Just turn on the camera, and connect the servo power supply to the HAT board and the 5-V power supply to the RasPi. You then will need to wait about 45 seconds for the initial boot sequence to finish. Then press the push button, and the camera should move to the left, if not already there, and take a picture. The camera will move again after several seconds and take another picture and continue this sequence three more times. Once at the far right, it will take the final picture and then reposition itself to the far left, ready for another picture-taking sequence. Remember that only one image-capture sequence is needed because there is no way to change the focus on this camera, and no focus stacking can be accomplished.

Creating the actual pano image was discussed in a previous section. All you need to do is to put the camera's SD card into the computer hosting the Panoweaver program and drag and drop the images into the application. It is very easy, and the results are well worth the effort. The only disadvantage to this whole process is that you will need a printer capable of handling large formats if you wish to print the final pano, and that is an expensive purchase.

This last test completes the point-and-shoot portion of the gigapixel camera project. At this point, I will go on to discuss the DSLR pano project.

DSLR Camera

I used a Canon 70D DSLR as the camera for this portion of this chapter's project. It is an excellent camera that, when used with an appropriate lens, is fully capable of taking pictures of excellent quality. I also used a Tamron 24 to 70-mm f/2.8 zoom lens with the camera because it is has a good focal-length range for the pano, and its optical quality is outstanding. The other point of using this lens is that it is not too heavy of a lens, which minimizes any overloading of the pan-and-tilt servo-mechanism. Figure 10-28 shows the 70D with the Tamron lens attached.

DSLR Pan-and-Tilt Servomechanism

The DSLR camera and lens assembly is far too heavy to be used with the same pan-and-tilt servo assembly employed in the point-and-shoot camera project. Instead, I used a heavy-duty unit

Figure 10-28 Canon 70D camera with a Tamron 24- to 70-mm lens.

Figure 10-29 Heavy-duty DSLR pan-and-tilt servo mechanism.

purchased from ServoCity.com, which is shown in Figure 10-29.

The servos used in this mechanism are Hitec Model HS-785HBs, which are 3.5-full-turn winch-style servos. These servos rotate 3.5 turns for a total 1260° range of motion compared with the typical servo range of 120 or 180°. Winch servos are so named because they are used in R/C sailboats to remotely raise and lower model sails. For this pan-and-tilt configuration, the servo turns a small gear, which is, in turn, connected to a much larger gear, as you can readily see in the figure. This arrangement allows the servo to reduce its extended range of motion to match the range appropriate for the pan-and-tilt operation. In addition, the gear ratio provides a significant increase in available torque to easily handle the heavy DSLR-lens combination.

The DSLR is easily mounted to the integral platform on the pan-and-tilt mechanism using a standard ¼-20 tripod mounting screw, which is widely available at camera shops, or you can use a shallow head ¼-20 machine screw, if available. I also added a tripod shoe plate to the bottom of the mechanism, which allowed me to use my Vanguard tripod, just as I did with the point-and-shoot arrangement. Figure 10-30 shows the DSLR mounted on the pan-and-tilt mechanism.

The next part of this chapter introduces the gphoto2 software package that I used to control image capture for the DSLR.

gphoto2

Unlike the point-and-shoot camera, the DSLR used for this part of the pano project is controlled using digital command sequences sent through the USB cable from the RasPi. A

Figure 10-30 DSLR on the pan-and-tilt mechanism mounting plate.

software package named *gphoto2* accomplishes this in a very efficient manner and is able to control a substantial number of the DSLR functions using programmed command sequences. Of course, I will still be using a Python program to control the pan-and-tilt servos because that part of the project is not accommodated by the gphoto2 software. In fact, as you will shortly see, the actual gphoto2 command sequences, which are needed to create the pano, are embedded into the Python program. But first I need to show you how to install and configure the gphoto2 software package.

The RasPi must be connected to the Internet for this procedure to work. Follow these steps to install and configure the gphoto2 software:

1. `sudo apt-get update`
2. `sudo wget raw.github.com/gonzalo/gphoto2-updater/master/gphoto2-updater.sh`
3. `sudo chmod 755 gphoto2-updater.sh`
4. `sudo ./gphoto2-updater.sh` (Be patient; this step takes about 35 minutes on a RasPi 2 and 1 hour and 7 minutes on a RasPi B+.)

You will next need to perform the following steps to ensure that the camera mounts properly:

1. `sudo rm /usr/share/dbus-1/services/org.gtk.Private.GPhoto2VolumeMonitorservice`
2. `sudo rm /usr/share/gvfs/mounts/gphoto2.mount`
3. `sudo rm /usr/share/gvfs/remote-volume-monitors/gphoto2.monitor`
4. `sudo rm /usr/lib/gvfs/gvfs-gphoto2-volume-monitor`

Reboot after completing the preceding steps:

```
sudo reboot
```

Initial Test of gphoto2 with DSLR

You will need to connect the DSLR to the RasPi using an appropriate USB cable. The Canon 70D uses a regular mini-A socket, as was shown in Figure 10-18. Enter the following command once the camera and RasPi are connected and both are turned on:

```
gphoto2 --capture-image-and-download
```

The camera should proceed to take a picture and subsequently download it into the RasPi. The downloaded picture will be in the same directory from which the preceding command was issued. The downloaded picture also would be named *Capt0000.jpg*, assuming that the camera's default setting was to take jpeg images.

Having successfully taken a picture and downloaded it into the RasPi verifies that gphoto2 is working properly and ready for the next steps in this project. These next steps are to create and run a program that will manually control the pan-and-tilt servos and then take and download a picture.

Manual Control Program

This manual control program is a modified version of the PnS_Camera1.py program I used with the point-and-shoot camera. However, that program required the CHDK firmware be installed on the camera, which is not the case for the DSLR camera. The gphoto2 software takes care of all the low-level interactions between the camera and the RasPi control program, which makes the camera interface quite easy to program. I named the program `DSLR_Camera.py` and list it next. It is also available on this book's companion website. You also should note that I added RasPi pin 18 as an input in this code, but it is not used until the next version, which automates the process.

```
#!/usr/bin/python

#DSLR_Camera.py
#D. J. Norris 2015
#This code is in the public domain.
from Tkinter import *
from Adafruit_PWM_Servo_Driver import PWM
import RPi.GPIO as GPIO
import subprocess as call
import time

GPIO.setmode(GPIO.BOARD)
#Push button input, physical pin 12 is RasPi GPIO pin 18
GPIO.setup(12, GPIO.IN)

pwm = PWM(0x40)
pwm.setPWMFreq(100)

def takePicture():
   os.system("gphoto2 --capture-image-and-download -filename %y%m%d%H%M%S")
   time.sleep(2.0)

class Servo:
   def __init__(self, top1):
     win = Frame(top1)
     win.pack()
     label1 = Label(win, text='Pan')
     sc1 = Scale(win, from_=-90, to=30, orient=HORIZONTAL, command=self.update0)
     #Centers the slider tab
     sc1.set(-35)
     sc1.grid(row=0)
     label1.grid(row=1)
     label2 = Label(win, text='Tilt')
     sc2 = Scale(win, from_=-90, to=30, orient=HORIZONTAL, command=self.update1)
     #centers the slider tab
     sc2.set(-32)
     label2.grid(row=3)
     sc2.grid(row=2)
     button = Button(win, text='Take Picture', command=takePicture)
     button.grid(row=4)

   def update0(self, angle):
     num = float(angle)
     pulseWidth = int((750*(num+60))/120 + 350)
     pwm.setPWM(0, 0, pulseWidth)

   def update1(self, angle):
     num = float(angle)
     pulseWidth = int((750*(num+60))/120 + 350)
```

```
        pwm.setPWM(1, 0, pulseWidth)

    top = Tk()
    top.wm_title("DLSR Pan and Tilt Servo Control")
    app = Servo(top)
    top.update_idletasks()
    top.geometry("400x200+0+0")
    top.mainloop()
```

As with all the other similar programs, you must be in X Windows to run this program. Also, be sure that you are in the same directory that contains the Adafruit PWM/Servo library.

I used the same PWM/Servo HAT board that was used with the point-and-shoot system and connected the pan servo to channel 0 and the tilt servo to channel 1. You must also ensure that the 5-V supply connected to the HAT board is capable of providing a minimum of 2 A (peak) because these servos use more current than the point-and-shoot servos.

Be sure that you are in the proper directory, and then enter the following to run this program:

```
sudo python DSLR_Camera.py
```

Figure 10-31 shows the resulting GUI display after starting this program.

I clicked the button and repositioned the servo slider controls to confirm that the camera and servos responded as they should.

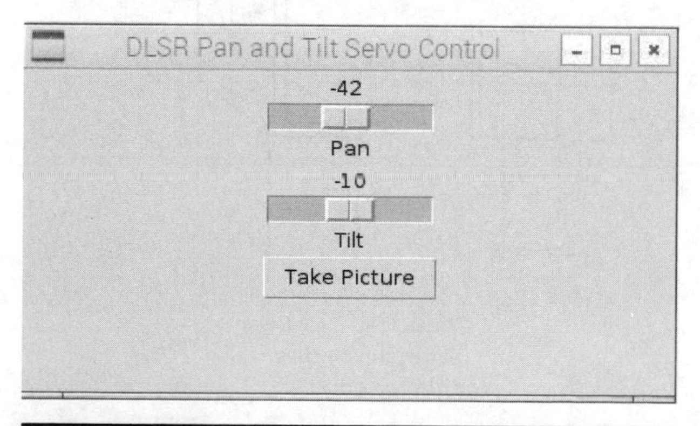

Figure 10-31 DSLR_Camera GUI.

Image-Processing Sequence

All image captures are stored on the RasPi SD card instead of the camera, which I believe will simplify the postprocessing workflow. You can easily change this feature by making the following edit to the program: change

```
os.system("gphoto2 --capture-image-and-
download -filename %y%m%d%H%M%S.jpg")
```

to

```
os.system("gphoto2 --capture-image
-filename %y%m%d%H%M%S.jbg")
```

All the images will now be stored on the camera once you change the capture command. Also notice that the images are all automatically named using the date and time to the second when the picture was taken. This ensures that all the image files are unique, and you should be readily able to determine the sequence of when they were taken.

The DLSR focus cannot be programmed using gphoto2 for image capture. It is possible to use gphoto2 to change the focus while in the live view mode, which is the situation where the camera mirror is locked-up and the camera's viewfinder cannot be consequently used. Unfortunately, the mirror must be unlocked and the camera in a non–live view mode in order to take a picture. This means that you will need to take a series of pictures using different focus points to have the input images to do both focus stacking and stitching.

I would suggest that you start the whole process by manually focusing at a nearby point and taking an automated panning sequence. Next, readjust the focus point to a middle area between close-by and distant objects and repeat the automated panning sequence. Finally, focus at infinity, and do one more automated panning sequence. You will need to review all the images and select the ones that first will be focus stacked and then stitch all the processed stacked images together. All this means is that a total of 12 images will be processed, three each for stacking and then four for stitching, assuming that there are four pan positions. This is not too much work considering how easy it is to use the stack and stitch applications.

I will next discuss how I mounted the RasPi along with the activation push button to the heavy-duty pan-and-tilt mechanism.

Mounting the RasPi to the Pan-and-Tilt Mechanism

I mounted the RasPi and the activation push button on a piece of clear Lexan, which is, in turn, mounted on the pan-and-tilt mechanism. Figure 10-32 is a sketch of this mounting plate. You can use material other than Lexan if that is what you have available. Just be sure that it is sufficiently sturdy to support the RasPi and any battery packs hanging from it.

The push button is also mounted on an L-shaped piece of Lexan, as was the case for the point-and-shoot system. Figure 10-26 is a sketch of the push-button L-shaped mounting bracket. Fortunately, the rotating base leg is approximately 2.5 inches above the mounting base, which allows plenty of room for the RasPi and push button to be positioned without interfering with leg motion.

The RasPi is attached to the mounting plate with a simple L-shaped bracket, as shown in Figure 10-33. The clear plastic case I used has a

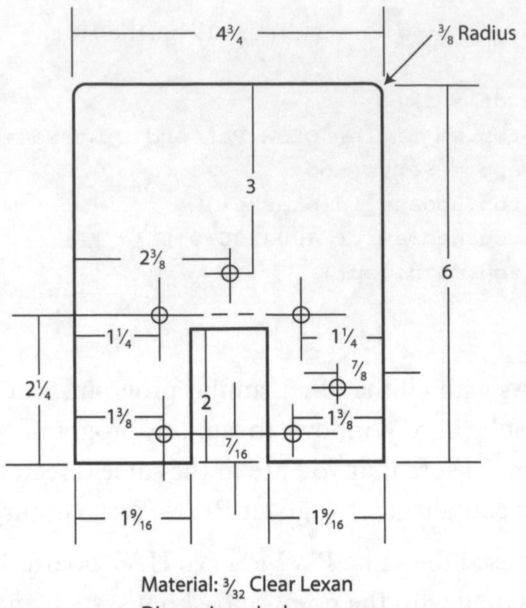

Figure 10-32 RasPi mounting plate.

slot in the long wall opposite the wall with the HDMI cutout, which is a convenient feature to hold a small 4-40 machine screw.

Figure 10-34 shows the complete assembly with camera, RasPi, and supporting battery packs mounted on my Vanguard tripod.

This whole assembly is heavy, and I would caution anyone against trying to carry all of it

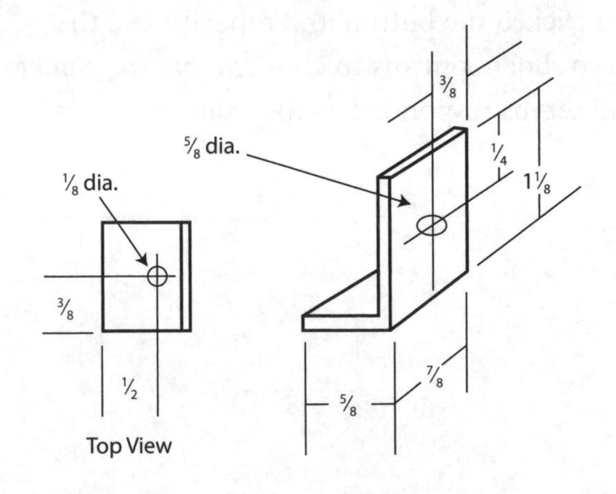

Figure 10-33 RasPi mounting bracket.

Figure 10-34 Complete DSLR assembly mounted on tripod.

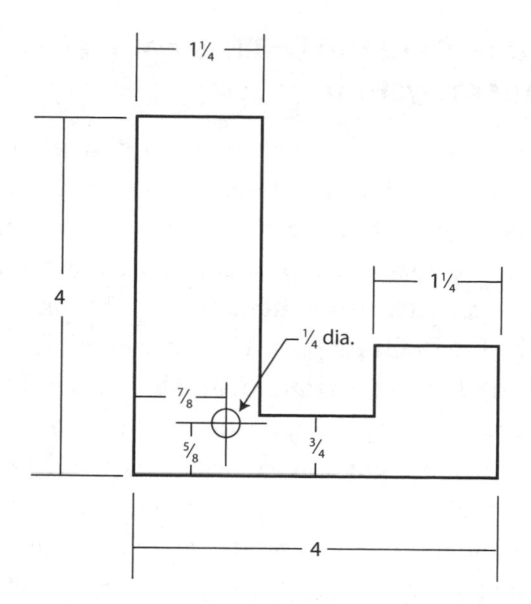

Figure 10-35 Lexan stop plate.

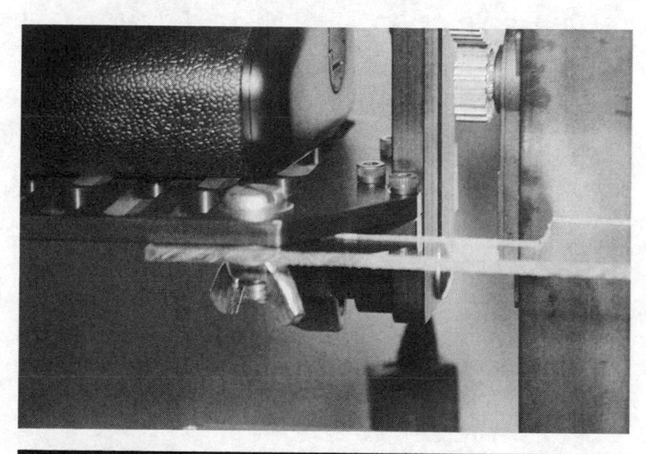

Figure 10-36 Stop plate attached to the camera mounting plate.

as a single package. The assembly also seems to have a stable center of gravity (CG), meaning that it should not tip over once set up on a firm tripod, which is, in turn, set on level ground.

The camera, when secured to the pan-and-tilt mounting plate, will swing down unless the tilt servo is powered on. However, I did not want to continually power this servo because the camera and lens combination puts a considerable bending load on the servo. I found that it started to get warm if I left it powered on with the camera in place. Because there is no need for an automated tilt operation, I configured a simple stop plate, which I made out of a piece of Lexan. Figure 10-35 is a sketch of this stop plate.

The plate was held in place by a ¼-20 machine screw secured with a wing nut. The machine screw went through one of the camera mounting plate slots, as you can see in Figure 10-36.

This arrangement may not be elegant, but it is effective, and it literally saves the tilt servo from overheating and perhaps failing.

Automating the DSLR Gigapixel Camera System

At this point, I will show you the automation program, which functions almost identically to the point-and-shoot version. The only difference is that the gphoto2 application is used to capture an image instead of a digital pulse from one of the RasPi GPIO pins. There are no GUI features in this program, but there is a check for the user pressing the activation button, which is identical in function to the one used for the point-and-shoot system. The program is named `DSLR_Automate.py` and is available on the companion website.

```
#!/usr/bin/python

#DSLR_Automate.py
#D. J. Norris 2015
#This code is in the public domain.

from Adafruit_PWM_Servo_Driver import
PWM
import RPi.GPIO as GPIO
import time
import os

GPIO.setmode(GPIO.BOARD)

#Push button input, physical pin 12 is
 RasPi GPIO pin 18
GPIO.setup(12, GPIO.IN)

#Need a variable to logically latch the
 pb press action
picFlag = 0

pwm = PWM(0x40)
pwm.setPWMFreq(100)

def takePicture():
    os.system("gphoto2 --capture-image
-and-download -filename %y%m%d%H%M%S")
    time.sleep(5.0)

def update0(angle):
    num = float(angle)
    pulseWidth = int((750*(num+60))/120
+ 350)
    pwm.setPWM(0, 0, pulseWidth)

#main code
#set to left position initially
update0(30)
#wait 2 seconds for any vibrations to
 settle
time.sleep(2.0)

while(True):

    #wait for a button press
    #pin 12 is tied high. Pushing the pb
     grounds the input
    pb = GPIO.input(12)
    if(pb == 0):
      picFlag = 1

    if(picFlag == 1):
      takePicture()
      time.sleep(2.0)
      update0(24)
      time.sleep(2.0)
      takePicture()
      time.sleep(2.0)
      update0(18)
      time.sleep(2.0)
      takePicture()
      time.sleep(2.0)
      update0(12)
      takePicture()
      time.sleep(2.0)
      #Reset back to left position
      update0(30)
      #Need to reset the pb flag to await
       next button press
      picFlag = 0
```

I did find that when I initially ran the program, the RasPi didn't connect or discover the DSLR. This is simply remedied by turning the camera off and then on again. My research on this problem seems to indicate that it is a known issue with the gphoto2 USB implementation. Other than that problem, everything went smoothly with the first test of this program. I also kept the RasPi connected in a workstation configuration because I knew that I had to adjust the servo angle parameters to match the pan requirements for that servo. The numbers in the program listing reflect the correct angles for the initial and subsequent pan positions. Figure 10-37 is a screenshot of the RasPi terminal display for a series of pictures.

If you examine the date/time stamp on the image files, you should notice that it takes about 16 seconds to pan, take a picture, download it to the RasPi, and finally delete the image from the camera itself. You should plan on taking about a minute to progress through a single panning operation.

It is now time to configure the autostart after confirming that the pan operation works properly and the camera takes the pictures as commanded. I will follow the exact same procedure I laid out for the point-and-shoot system.

Configure Program for an Autostart

Just follow the exact procedure I described in the point-and-shoot section, except that the following line should be put in the profile file:

```
sudo python /home/pi/Adafruit-Raspberry-
Pi-Python-Code/Adafruit_PWM_Servo_
Driver/DSLR_Automate.py
```

You should be all ready for the field once you have completed the changes and rebooted the RasPi.

I have finally included Figure 10-38, which is a pano image showing the impressive capabilities of this system after all the individual images

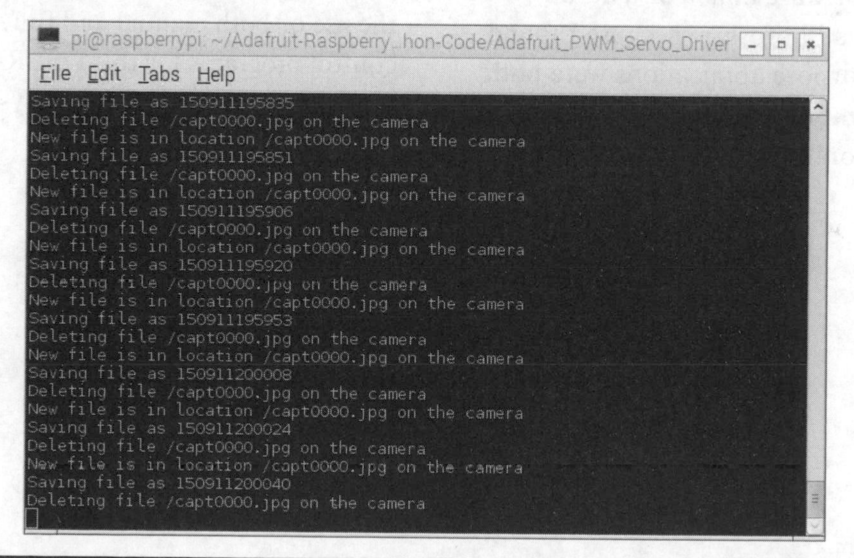

Figure 10-37 Terminal display after several pan operations completed.

Figure 10-38 Sample stacked and stitched pano image.

have been stacked and stitched. The book's monochrome presentation really doesn't do it justice, which is why I have asked that a full-color image be put in this book's companion website. Please take a look at it to get an appreciation of this gigapixel camera system.

Summary

This chapter's project demonstrated how to build two variations of a camera system that is capable of creating a panoramic (pano) image composed of multiple images. One variation was designed for a point-and-shoot type of camera and the other for a digital single-lens reflex (DSLR) camera. The beginning portions of this chapter concerned the stack and stitch postprocessing software that is used to create the ultimate pano. I used specialty software in lieu of a do-all application such as Photoshop because I found that the special-purpose applications were both easier to use and more efficient than the general-purpose application.

The next sections concerned how to assemble and program the point-and-shoot system. I noted that my design required the use of a Canon camera because CHDK firmware needed to be installed in the camera to enable remote shutter activation. The RasPi was used to control the pan and image-capture operations to generate the required number of images for the pano generation. I demonstrated both manual control and fully automated programs for this system.

The concluding chapter sections basically repeating the same point-and-shoot discussions except that they were changed to handle the DSLR camera and its requirements. The two biggest changes were use of a very heavy-duty pan-and-tilt mechanism capable of moving the DSLR and use of the gphoto2 application to control the camera image capture. I demonstrated the use of both manual and fully automated control programs for this DSLR system, as I had done for the point-and-shoot system.

CHAPTER 11

Nighttime Garden Monitor

IN THIS CHAPTER, I WILL SHOW YOU how to build a video surveillance system capable of operating in the dark. I will show you a solution to a common aggravating situation that many home gardeners have encountered, namely, nighttime marauders. It is unsettling to look at the remains of your well-kept garden after some critter has ravaged it during the evening hours. Use of this system will enable you to literally see in the dark and allow you to identify the miscreants and take appropriate actions to prevent any further nighttime raids. Let's begin with the device that allows this wondrous ability to see in the dark.

Pi Noir Camera

The Pi Noir camera is sensitive to nonvisible light, which in this case is light in the infrared (IR) spectrum. It turns out that the Pi Noir camera is just a regular RasPi camera with its IR-blocking filter removed. This means that the Noir camera can operate in regular lighting conditions, but it could have artifacts introduced into a daytime image because of the normally filtered IR light waves being displayed. This situation is not a concern in this project because the Pi Noir camera will be operated only at night with an IR illuminator module as a light source.

Chapter 11 Parts List

Item	Model	Quantity	Source
Raspberry Pi 2 Model B	2358	1	adafruit.com
Pi Noir camera	1567	1	adafruit.com
Adjustable Pi camera mount	1434	1	adafruit.com
Univivi IR illuminator	U03R	1	amazon.com
5-V relay	Various	1	digikey.com mouser.com
Laser Trip Sensor Kit	LTS1	1	ramseyelectronics.com
5-V cell phone external battery	Generic	1	amazon.com
12-V sealed lead-acid battery	Commodity	1	amazon.com
0.125 inch of Lexan sheet material	Commodity	See build drawings	Local home-improvement store
Wood screws, machine screws, and nuts	Commodity	Various	Local home-improvement store
Wood stock	Commodity	2 ft 2 × 4 9 × 14 × 1	Local home-improvement store

Figure 11-1 Pi Noir camera mounted on a stand.

Figure 11-1 shows the Pi Noir camera mounted in a holder specifically designed for it. The stand itself is mounted on a magnet, which I used to facilitate quick repositioning of the camera.

Table 11-1 lists some key specifications for the Pi Noir camera.

TABLE 11-1 Key Technical Pi Noir Camera Specifications

Feature	Specification
Board size	25 × 24 × 9 mm
Native resolution	5 megapixels
Pixel size	2592 × 1944
Lens	Fixed focus
Video modes	1080p30, 720p60, 640×480p60/90

Installing the Camera

The following installation and configuration steps are based on the steps provided by the Raspberry Foundation.

NOTE **The camera can be easily damaged by static electricity. Before removing the camera from its antistatic bag, ensure that you have discharged any static electricity buildup by touching a grounded object such as a water faucet.**

Figure 11-2 is a close-up of the camera serial interface (CSI) flex socket, which is located directly behind the RJ45 Ethernet connector.

You first need to gently pull up on the black plastic clamp, which is designed to hold the flex cable in place. Next, insert the camera's flex cable with the silver finger connectors pointing away from the Ethernet connector. Finally, gently push down on the black plastic clamp to secure the cable. Ensure that the cable is aligned properly and not skewed or the camera will not function. Figure 11-3 shows a properly inserted and secured cable for your reference.

The camera may come with a small piece of translucent blue plastic film covering the lens. This is only present to protect the lens and needs to be removed by gently peeling it off.

This camera must be set up with appropriate drivers before it can be used, which is the topic of the next section.

Installing and Configuring the Camera Driver Software

It is relatively easy to set up the basic software that will enable the camera to be used with the RasPi. You will need to have the RasPi connected to the Internet to be successful with this procedure. Please follow these four steps in the order presented:

1. `sudo apt-get update`
2. `sudo apt-get upgrade` (Be patient; this could take a long time.)
3. `sudo raspi-config` (Navigate to Camera and select Enable.)
4. Select Finish, and reboot.

The next section describes some basic camera operation commands that will allow you to test the camera and confirm that it is operating properly.

Figure 11-2 CSI connector.

Figure 11-3 Properly seated flex cable.

Initial Camera Tests

The Pi camera software supports two primary operational modes: still photographs or videos. The two applications implementing these modes are

- **raspistill.** Captures still images.
- **raspivid.** Captures videos in one of several available video formats.

Both these applications have options that you can use to configure the output to suit your requirements. To view the possible options for raspivid or raspistill, enter the following:

```
raspivid | less
raspistill | less
```

I summarize the key options next with some examples:

- `-o` or `--output` specifies the output filename.
- `-t` or `--timeout` specifies the amount of time that the preview will be displayed in milliseconds. The default is 5000, or 5 seconds.
- `-d` or `--demo` runs a demo mode, which will cycle through all the available image effects.

Example Commands

- Capture an image in jpeg format:

```
raspistill -o myPic.jpg
```

- Capture a 5-second video in h264 format:

```
raspivid -o myVideo.h264
```

- Capture a 30-second video in h264 format:

```
raspivid -o myVideo.h264 -t 30000
```

- Capture a 20-second video in h264 format while in the demo mode (type `q` to exit the demo mode):

```
raspivid -o myVideo.h264 -t 20000 -d
```

You will next need to download an application to convert the h264 format to mp4, which, in turn, can be viewed using the mplayer application. Follow these next steps to download and install the converter and viewer applications:

- `sudo apt-get update`
- `sudo apt-get install gpac`
- `sudo apt-get install mplayer`

Enter the following commands to test the Pi Noir camera using X Windows:

- `startx`
- Open a lxTerminal window and type in

```
raspistill -o TestImage.jpg
```

- Use the File Manager application to find and open the test image with the image application.
- Examine the image to confirm that the Pi Noir still capture operates properly. Recall that a daytime image may contain IR artifacts.
- Next, enter this command in the terminal window to check the video capability:

```
raspvid -o myVideo.h264 -t 30000
```

- Use the MP4Box app part of the gpac package to convert the video to mp4 format.

```
MP4Box -add myVideo.h264 myVideo.mp4
```

- Use the mplayer application to view the video.

It is time to work on the physical monitoring system now that the basic video software has been installed and confirmed to function properly.

Physical Monitoring System

The first step in any reasonable system design is to articulate the requirements that the system must meet or fulfill to be deemed acceptable.

After some consideration, I set the following system requirements for this nighttime monitoring system:

- Video surveillance using an infrared illuminator
- Video recording automatically initiated
- Survey a 10 × 10 ft area
- Weather resistant
- Battery operated (temporary installation)
- Raspberry Pi used as the primary controller

Figure 11-4 shows a conceptual diagram for this monitoring system with all its main components.

The following sections explain each of the main system components.

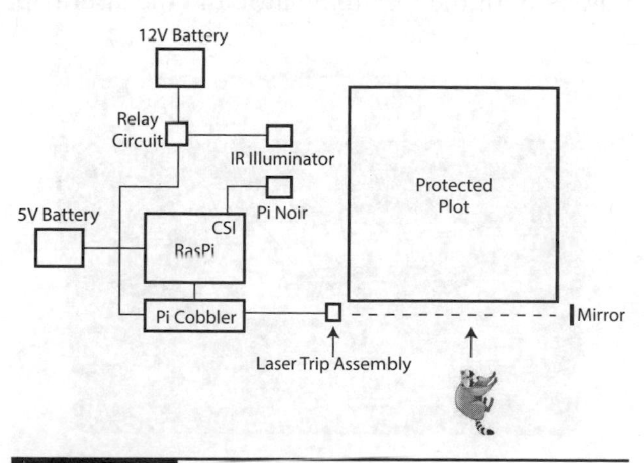

Figure 11-4 Conceptual system diagram.

Laser Trip Assembly

I also elected to use the Ramsey Electronics' Laser Trip Sensor Kit, Model LTS1, to act as the intruder detector. This kit is relatively inexpensive and includes both a laser pen (pointer) and a light detector. The detector requires assembly, and Figure 11-5 shows both items mounted on a small piece of 2.5 × 6 inch Lexan.

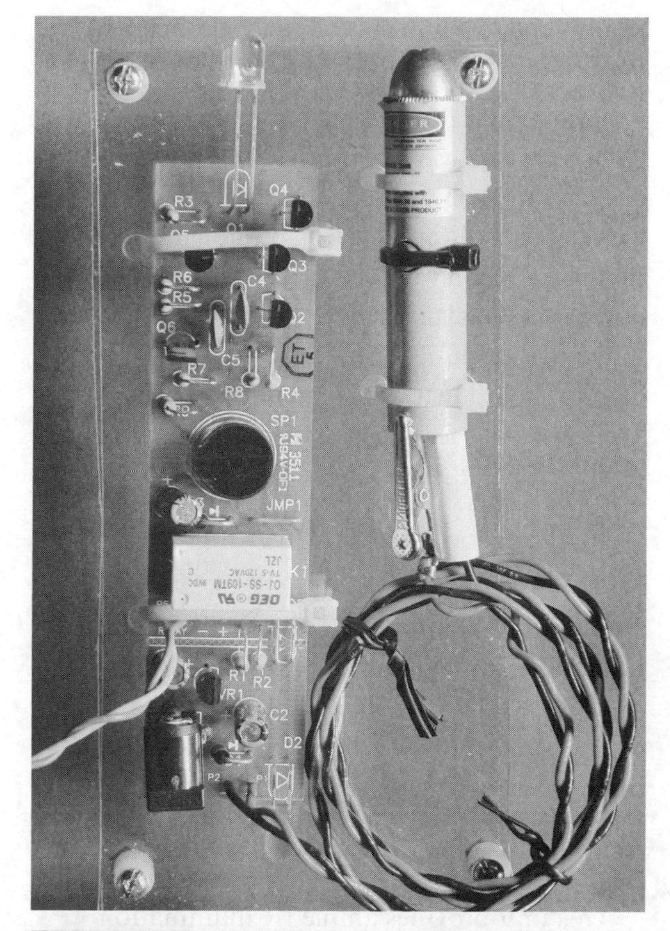

Figure 11-5 Laser pen and light-detector module.

There are two assemblies, not counting the 12-V power supply, that make up this sensor module. The first is the red laser pointer that acts as the light source, and the second is the light sensor that consists of a phototransistor with some associated analog signal-processing circuitry. The laser pointer is a very common low-power device that is typically battery powered. In this case, there are no batteries because it is powered by 4.3 V supplied by the detector board. The laser pointer's power is delivered via a pair of small alligator clips, visible in the figure.

The light detector's sensor is a phototransistor, which is normally encased in an ambient light-blocking tube at the front of the assembly, which I removed in Figure 11-5. The phototransistor, which looks like a normal LED, is really a transistor with only the collector and emitter leads externally connected. It will transition into a conducting state when the laser light strikes the base-emitter region. The black tube light shield, which I made out of black duct tape, helps to prevent any false triggers when in place.

The module also has a relay with contacts that are normally closed when light is not being detected and will open when the laser light strikes the phototransistor. This will mean that the Python program will need to detect a low-to-high transition. I explain this in more detail in the software section. There is also a red LED on the detector board that will illuminate when the relay is energized. This feature is very handy when you are trying to align the laser beam with the detector.

IR Illuminator

Another key component is the IR illuminator, which is required in order for the Pi Noir camera to capture any video in the darkness. The illuminator I used was a very inexpensive unit that I purchased online. It is powered by 12 V and has four IR LEDs installed in it, as may be seen in Figure 11-6.

This unit provides ample IR illumination for the target 10 × 10 plot that I set forth in the systems requirement. The illuminator does need an external relay circuit to provide an effective interface between the RasPi's 3.3-V GPIO output and the 12 V needed for the illuminator

Figure 11-6 IR illuminator.

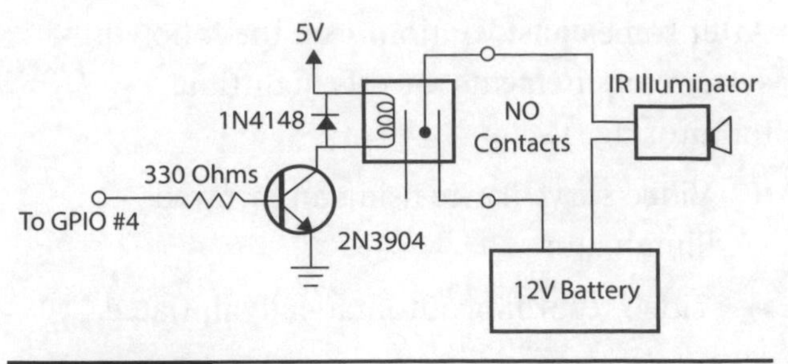

Figure 11-7 Relay interface schematic.

module. Figure 11-7 is a schematic of this simple circuit, which easily enables this interface.

Note that I happened to have a 5-V reed relay in my collection of surplus parts, which made the circuit quite simple to design and build. You should be able to find a similar relay online, but if not, you certainly could use an opto-isolator chip to accomplish the same function as the relay.

Pi Noir Camera

This was introduced at the start of this chapter, and its function will not be further discussed.

12- and 5-V Power Supplies

I used the same external cell phone battery supply that I used in Chapter 10's project to power the RasPi. It will provide sufficient power for at least one evening's monitoring, if not more.

A sealed lead-acid battery, as shown in Figure 11-8, provided the 12 V for the illuminator. This powers both the IR illuminator and the laser trip

Figure 11-8 12-V sealed lead-acid battery.

assembly, which includes both the laser pointer and the photodetector module.

This battery is rated at over 9000 mAh, which is more than adequate to last for several monitoring sessions before being depleted. You certainly could use two alternating-current (AC) main-powered supplies if you do not want to constantly be recharging the system batteries. This would necessitate running an AC extension cord to wherever the system is located, but you are free of the battery-charging requirement, especially if you wanted to make it a permanent installation instead of a temporary setup.

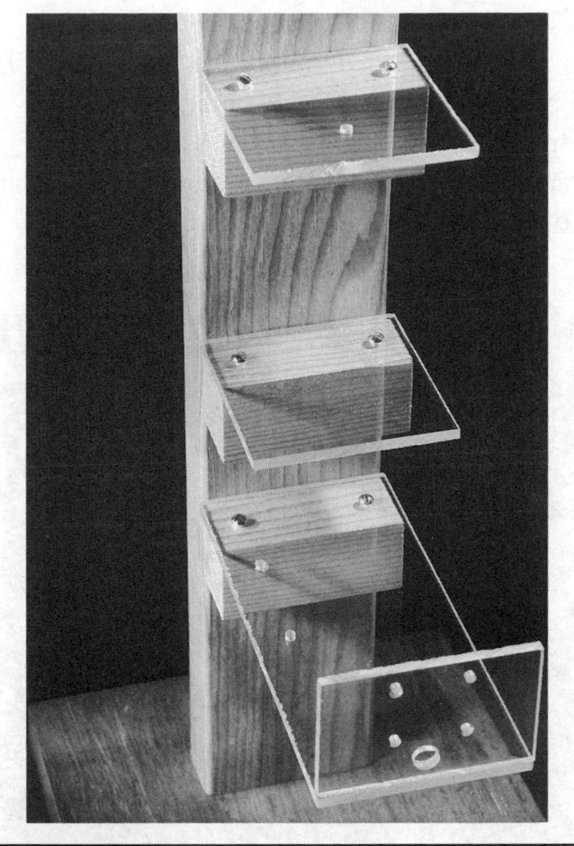

Figure11-10 Complete mounting assembly.

Mounting All the System Components

I designed a simple mounting assembly that has three acrylic mounting plates attached to a short piece of 2 × 4, which, in turn, is attached to a wooden mounting plate. Figure 11-9 is a dimensioned drawing of this mounting assembly.

A complete mounting assembly without any attached components is shown in Figure 11-10.

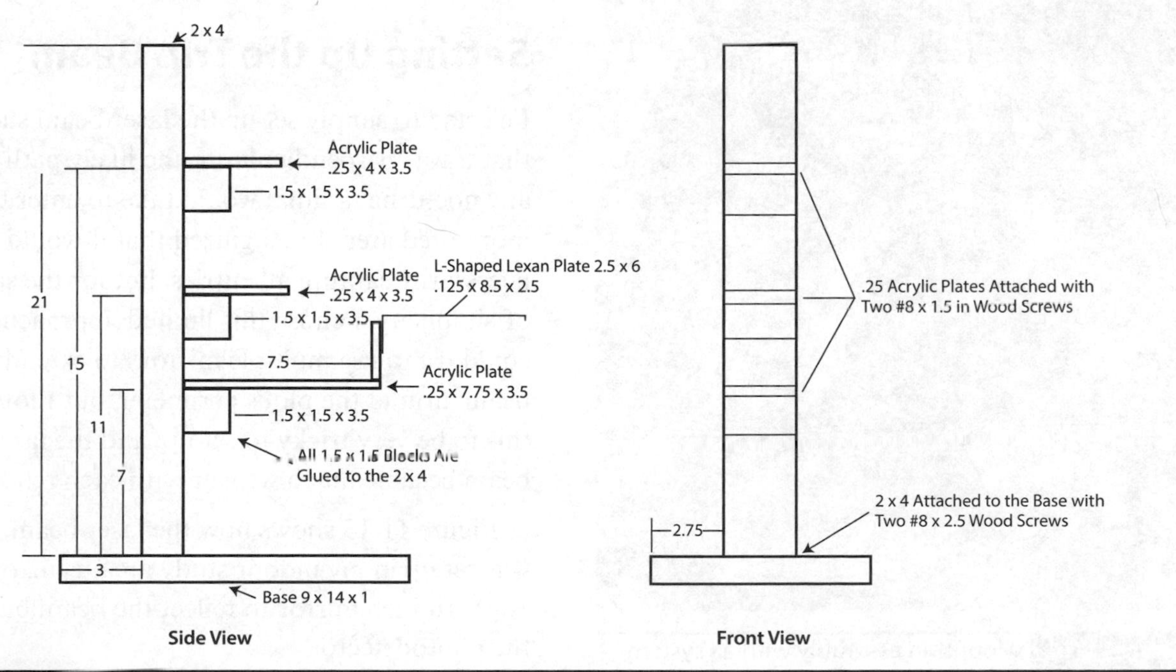

Figure 11-9 Mounting assembly drawing.

The three mounting plates from top to bottom are designed to hold the IR illuminator, Pi Noir camera, and RasPi–laser trip assembly, respectively. Figure 11-11 shows all the components mounted in the mounting assembly. Notice that the 12-V battery is simply placed

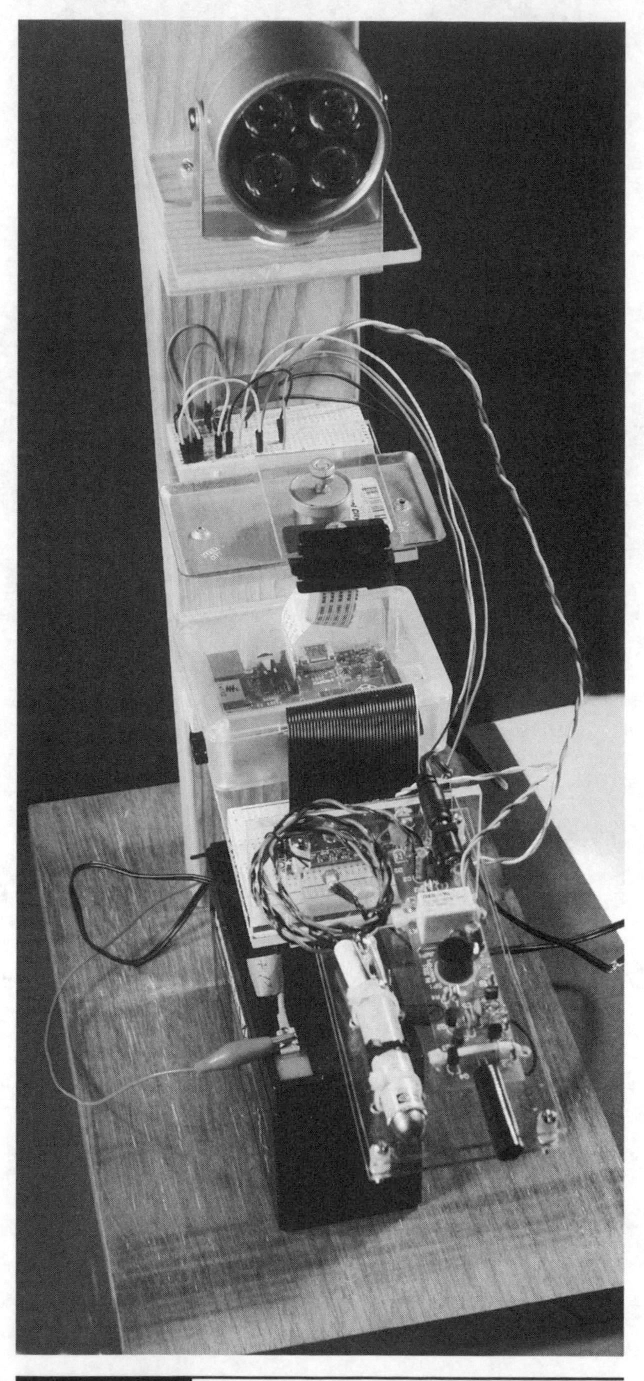

Figure11-11 Mounting assembly with all system components attached.

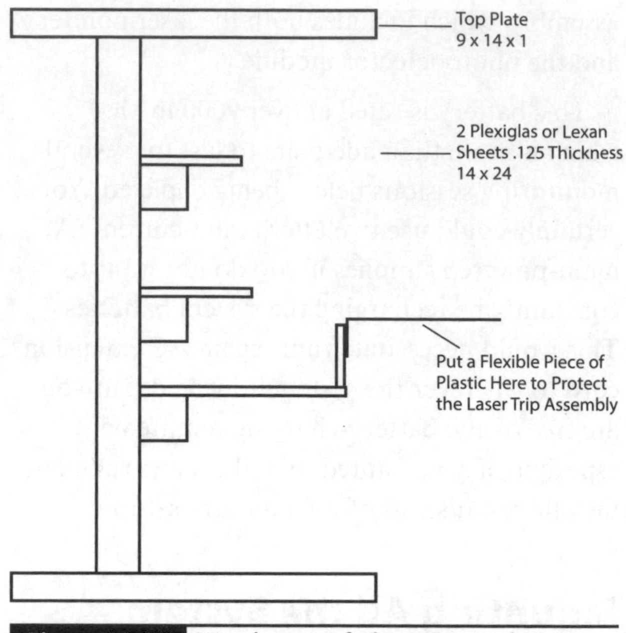

Figure11-12 Weatherproofed system enclosure.

on the bottom wooden support plate. It is fairly heavy and helps stabilize the overall assembly.

I also sized the main 2 × 4 support to allow for a top wooden cover such that sheet Plexiglas or Lexan can be used to enclose the entire assembly to help weatherproof the system. Figure 11-12 shows this weatherproofed system.

Setting Up the Trip Beam

I elected to simply set up the laser beam such that it was perpendicular to the likely path that any nocturnal animal would take to enter the monitored area. I recognized that it would not necessarily capture all entries, but for the sake of simplicity, I chose this limited approach. You could try to use multiple mirrors to extend the beam around the plot's perimeter, but I found this to be very tricky to set up, and the laser beam became much weaker with each reflection.

Figure 11-13 shows how the laser beam was set up in my indoor study area using one 16- × 16-inch mirror to reflect the beam back to the photodetector.

Figure 11-13 Test setup.

I purchased an inexpensive 16- × 16-inch mirror that has a ½-inch-thick plastic frame that neatly fits into ½-inch channels that are cut into vertical wood supports. Figure 11-14 is a build diagram for this mirror holder, and Figure 11-15 shows the complete mirror assembly.

Note that I applied polyurethane to all the wood pieces in the assembly to help protect it from the weather because it may be placed outdoors.

This last assembly completes the physical build, which means that it's time for the software installation and configuration.

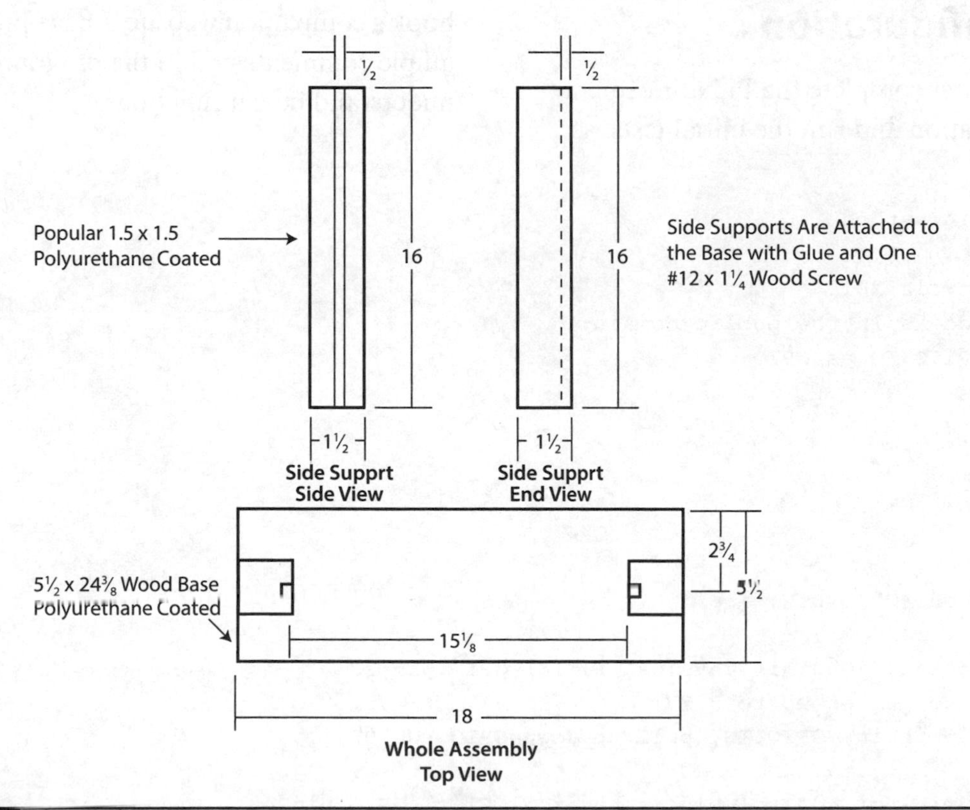

Figure 11-14 Mirror holder.

Figure 11-15 Complete mirror assembly.

Software Installation and Configuration

You should first complete the Pi Noir camera driver installation and run the initial tests, as I discussed in an earlier section. It is time to enter the control program, once you are certain that the camera functions as expected. I created the control program using Python because this language makes any modifications quite easy to accomplish. The program requirements are fairly simple and are listed next:

- Autostart when the RasPi is powered on.
- Continuously check for a relay activation from the photodetector module.
- Turn on the IR illuminator once the relay is activated.
- Start recording video for 60 seconds after illuminator is turned on.
- Ensure that videos are uniquely named.
- After 60 seconds has elapsed, turn the IR illuminator off.
- Reset and wait for the next activation.

The complete program is named nv1.py and is listed next. It is also available from this book's companion website. I have provided ample comments within the program to help you understand how it functions.

```
#!/usr/bin/python
#nv1.py
#D. J. Norris 2015
#This code is in the public domain.
import RPi.GPIO as GPIO
import os
import time

inPin = 23
outPin = 4

GPIO.setmode(GPIO.BCM)

#Relay contact input is physical P1 16, GPIO pin 23
#It is also pulled up to 3.3V
GPIO.setup(inPin, GPIO.IN, pull_up_down=GPIO.PUD_UP)

#The illuminator control pin is physical P1 x, GPIO pin 4
```

```
GPIO.setup(outPin. GPIO.OUT)

#Used to uniquely number the capture videos
index = 0

#Forever loop, stop the program by pressing ctl-C
while(1):

    #Check to see if the relay is activated
    if(GPIO.input == False):
    #Increment the file index
    index = index + 1
    #Turn on the IR illuminator
    GPIO.output(outPin, GPIO.HIGH)
    #Create the video commandline string
    vidStr = "raspivid -o vid" + str(index) + ".h264 -t 60000"
    #execute an os commandline
    os.system(vidStr)
    #Turn off the IR illuminator
    GPIO.output(outPin, GPIO.LOW)
    #short delay before starting another relay check
    time.sleep(2)
```

The autostart feature is done in exactly the same manner as described in Chapter 10. The next section details how to set up the autostart.

Configure the Program for an Autostart

You will first need to make the following changes to the inittab file:

- `sudo nano /etc/inittab`
- Comment out the following line:

```
1:2345:respawn:/sbin/getty 115200
tty1
```

- Note that the line may appear as follows in more recent versions of the Wheezy distro:

```
1:2345:respawn:/sbin/getty --noclear
38400 tty1
```

- Add this line after the line just commented out:

```
1:2345:respawn:/bin/login -f pi tty1
</dev/tty1>/dev/tty1 2>&1
```

- Save the file, and exit the nano editor.

  Next, edit the profile file.

- `sudo nano /etc/profile`
- Add the following line to the file:

```
sudo python /home/pi/nv1.py
```

- Save the file, and exit the nano editor.

You should be ready to test the autostart configuration simply by powering on the RasPi and checking whether a video is recorded when the laser beam is interrupted. The easiest way to see if the video is present is to SSH into the RasPi and check whether a video named vid1.h264 is present in the home directory /home/pi. If so, everything worked as planned, and you are ready to deploy this system. Just note that each time the system is rebooted, any existing videos will be automatically written over, starting with vid1.h264.

Sample from a Capture Video

Figure 11-16 is a screen capture from a capture video in which I used a giant fake spider as the intruder in the monitored space. Note that there is absolutely no visible illumination source used in this video, just the IR illuminator.

I created the screenshot by first transferring all the captured videos from the RasPi home directory to a Windows computer. I next used a free application named VLC from www.videolan.org that played one of the h264 video files from which I generated the screenshot.

This last section completes this chapter's project. On a final note, it should be readily apparent that this system could easily be incorporated into a regular security surveillance system with some minor modifications to the control program. The major change would be the need to have the RasPi send an alert signal on intruder detection. This could be done in different ways, including changing the state of a GPIO pin or sending a text message via WiFi.

Figure 11-16 Sample screenshot from a capture video.

Summary

This chapter was focused on building a nighttime surveillance system. Its purpose was to detect and make a video recording of any nocturnal animal visitor to a monitored garden plot. Of course, the system may be easily extended to any nighttime surveillance application, including home and business security.

The Pi Noir camera was the primary means to capture the video in the subject area. The Pi Noir needs IR light waves to illuminate the area to be videoed. A compact IR illuminator provided the IR, which was fully capable of illuminating the target area. A laser trip assembly was used to signal the RasPi to commence video capture when an intruder interrupted the laser beam covering an approach path to the monitored site. A uniquely identified video recording was then started and stored on the RasPi's SD card, which could be viewed at a later date. I also discussed how to install a viewer application that would display the h264 encoded video.

Index

A

ACARS, 175–176
  downloading and installing a multichannel ACARS decoder, 176
  sample results, 179–183
  using the Kalibrate application, 176–179
acarsdec program, 178
acknowledgment packets (ACKs), 60
ad hoc networks, 61
Adafruit Industries, 7
  capacitive touchscreens, 28
Aircraft Communications Addressing and Reporting System. *See* ACARS
amplitude modulation (AM), 163
analog-to-digital converters (ADCs), 25–26
  MCP3008 ADC, 141–142
aperture, 238, 239
Arduino Uno
  development board, 65–67
  integrated development environment (IDE), 67–69
  and Lidar, 73–74
arrays, 85

B

baud rate, 56
Bitcoin network, 110
Blink program, 67–69
blocking, 135
BrickPi controller
  motor ports, 193
  overview, 191, *192*
  sensor ports, 192–193
  specialized I/O board, 193–195
broadcast messages, 113, 134–135
bus network. *See* I2C serial protocol

C

callback methods, 152
camera. *See* gigapixel camera system
cancel, 15
Canon Hack Development Kit (CHDK), 244
  firmware, 246–249
canvas, 98
capacitive touchscreens, 26–28
  See also touchscreens
CasterBot, 191
  obstacle-avoidance algorithm, 203–207
  obstacle-avoidance demonstration, 207–208
  overview, 195–196
  power supply, 196–199
  software installation and configuration, 199–202
  ultrasonic sensor, 202–203
  WiFi dongles, 199
centralized structure, 108
chip enable, 57
clean flags, 139
client ID, 139
cloning SD cards, 119
cluster operations
  monitoring cluster network traffic, 135
  unique functions for, 130–132
  See also RasPi cluster
collective communication operations, 113, 134–135
communication implementation techniques, 55–56
  additional communications link hardware, 59
  GPIO interrupt, 58
  GPIO poll/respond, 57–58
  I2C serial protocol, 56–57
  specialized with external hardware, 58–65
  SPI serial protocol, 57
  UART serial protocol, 56

compartmentalization, 108
continuous wave (CW), 163
Conway's game of life, 94–96
coprocessors, defined, 55
cPickle, 115
CuteCom program, 80–82
cyclic redundancy checks (CRCs), 60

D

damping effect, 220
dead reckoning, 204
deadband, 220
degrees of freedom (DOFs), 211–212
  free-body diagram (FBD), 212
  translation, 212
  workspace, 212–213
demodulation, 164–165
depth-map method, 241
deserialization, 115
development stations, 7
  headless setup, 9–10
  stand-alone setup, 8–9
deviation, 164
DHCP, 17
differential mode, 141
diffraction, 238
Digi International, 63
digi-peating, 60
discrete Fourier transform (DFT), 174
distributed architecture, 108
DMap, 241
DSLR camera, 253–262
Dynamic Host Configuration Protocol (DHCP), 119

E

Eclipse.org, 137–139
EDIMAX EW-7811Un, 9
error pulse, 219

F

fast Fourier transforms (FFTs), 109, 173–174
focal length, 238
focus stacking, 238, 240–242
  See also stacking
folding@home project, 109
Fourier, Jean-Baptiste, 173
four-wire serial bus, 57
four-wire touchscreens, 26
  See also touchscreens
framebuffer, 41
framebuffer image (FBI) viewer, 41–42
frames, 56
free-body diagram (FBD), 212
Freescale transceiver, 63–64
frequency modulation (FM), 163–164
Fry, Eric, 167
f-stops, 238
full-duplex communication, 29, 142
fundamental harmonic, 174

G

general-purpose input-output. *See* GPIO
gigaFLOPS, 109
gigapixel camera system
  automating, 251–253
  automating the DSLR camera system, 260–262
  building the tripod adapter, 249–251
  Canon Hack Development Kit (CHDK), 244, 246–249
  configuring for an autostart, 253, 261–262
  DSLR camera, 253–262
  gphoto2, 254–255
  manual control program for DSLR camera, 255–258
  mounting the RasPi to the pan-and-tilt mechanism, 258–259
  overview, 237
  pan-and-tilt servomechanism, 244–246, 253–254
  point-and-shoot camera, 244–253
  stacking, 238–242
  stitching, 242–244
  USB control cable, 246–249
GIMP, 100–101
gphoto2, 254–255
GPIO, 4–7
GPIO interrupt, 58
GPIO poll/respond, 57–58

H

HAT specification, 90
HDMI-to-VGA converter modules, 8
heterodyning, 164–165
hostnames, assigning, 121

I

I2C serial protocol, 56–57
image stacking. *See* stacking
integrated development environment (IDE), 67–69

interchip circular buffer arrangement, 142
interface board, 89–92
interintegrated circuit interface. *See* I2C serial protocol
interleaving, 87
intermediate frequency (IF), 164
interrupts, 58
IR illuminator, 268
ITO layers, 24–27

K

Kalibrate, 176–179
keygen, 121–126
Kolokowsky, Steve, 24

L

laser trip assembly, 267–268
Lidar
  Arduino software, 74–80
  and Arduino Uno, 73–74
  building the Lidar project, 72–80
  demonstration project, 69, 82–84
  laser emitter, 70
  main components, 70–72
  photo detector and signal-processing electronics, 70–71
  project parts and components, 72
  RasPi software, 80–82
  scanning and optics, 70
  technology overview, 69–70
LIDAR-Lite, 69, *70*, 71, 73
  portable Arduino-LIDAR-Lite system, 83–84
Light, Roger, 150
light-detection and ranging system. *See* Lidar
linear motion, 214
Linux, users, privileges and permissions, 21–22
LiPo battery, 196–197
Logitech K400 keyboard/mouse devices, 9
logout, 15
low null bits, 143
low-noise amplifiers (LNAs), 165
LXDE Desktop, 14–15

M

machinefiles, 121–122
machine-to-machine (M2M) communications, 137
macrophotography, 239
manufactured jumper wires, 7
matrix, 85
  See also RGB LED matrix
MCM Electronics, 7
MCP3008 ADC, 141–142
  connecting and testing with the RasPi, 143–147
merge function, 242
Message Passing Interface (MPI), 112–115
  basic MPI operations, 132–135
  MPI for Python (MPI4PY), 115
Message Queuing Telemetry Transport (MQTT), 137, 138
  adding MQTT features to an application, 147–150
  MQTT Brokers, 138, 150–151
  subscriber client, 152–155
  two-phase thermostats, 155–162
METARS, 183
middleware applications, 138
Mindstorm sensors, 192–193
Model A, 4
Model A+, 4
Model B, 3–4
Mode-S transponders, 174–175
modulation, 163–164
modules, 85
Morse code, 163
Mosquito Project, 150
MQTT. *See* Message Queuing Telemetry Transport (MQTT)
MQTT Brokers, 138, 150–151
multidrop. *See* I2C serial protocol

N

Netpbm, 100
nighttime garden monitor
  configuring for an autostart, 273
  installing and configuring the camera driver software, 264
  installing the camera, 264
  IR illuminator, 268
  laser trip assembly, 267–268
  mounting system components, 269–270
  power supplies, 268–269
  Pi Noir camera overview, 263–264
  sample from a capture video, 274
  setting up the trip beam, 270–272
  software installation and configuration, 272–273
  system requirements, 267
  testing, 266
nmap, 120
nonblocking, 134
NOOBS file, 11–12

O

obstacle avoidance
  algorithm, 203–207
  dead reckoning, 204
  demonstration, 207–208
OS, 12–15
  setting up the RasPi OS using an image file, 12–15
  updating and upgrading the Raspian image, 16–17
output, 89

P

packets, 59
Paho Project, 137–139
Palosaari, Antii, 167
pan-and-tilt servomechanism, 244–246
panos, 242
Panoweaver v9.1 Standard Edition, 242
parallel computing, 107
  problem types suitable for, 108–109
parts lists
  Chapter 1, 1–2
  Chapter 2, 24
  Chapter 3, 55
  Chapter 4, 86
  Chapter 5, 107
  Chapter 6, 137
  Chapter 7, 163
  Chapter 8, 191
  Chapter 9, 211
  Chapter 10, 237
  Chapter 11, 263
payload, 153
permissions, 21–22
personal area network (PAN), 58–59
  See also ZigBee protocol
petaFLOPS, 109
phase modulation (PM), 164
pi, 21
pi calculations, 126–127
  using the RasPi cluster, 128–130
Pi Cobbler, 6–7, 29–31, 143
Pi Noir camera
  installing, 264
  installing and configuring the camera driver software, 264
  overview, 263–264
  testing, 266
Pi4J library, 156–159
pick-and-place, 226
pickling, 115
PiTFT touchscreens, 23–24
  interface, 28–29
  logical names, 42
  See also touchscreens
Pluggable powered hubs, 9
PMax, 241
point-and-shoot camera, 244–253
point-to-point communication, 113, 132–134
polling, 58
portable bitmap format (PBM), 100
portable pixel maps. *See* ppm files
Poskanzer, Jef, 100
powered hubs, 9
ppm files, 100
  converting an existing image file to ppm, 102–103
  creating, 102–104
  creating an original ppm file, 103–104
  file format, 101–102
  ppm to jpeg file size test, 100–101
pressure-sensitive adhesive (PSA), 27
printed circuit boards (PCBs), 7
processes, defined, 107–108
processors, defined, 107–108
projected capacitive sensing, 28
proportional gain, 221
pulsing color, 94, *95*
PuTTY, 17–18
pyramid method, 241
Python
  and Message Passing Interface (MPI), 115
  using with the RGB LED matrix display, 104–106
Python libraries
  Pi4J library, 156–159
  Python Imaging Library (PIL), 104
  pytho-rpi.gpio, 195

Q

quality of service (QoS), 138–139

R

radial-theta angle robotic arm. *See* R-Theta
radio
  amplitude modulation (AM), 163
  basic concepts, 163–164
  continuous wave (CW), 163
  demodulation, 164–165
  deviation, 164
  frequency modulation (FM), 163–164
  heterodyning, 164–165
  intermediate frequency (IF), 164

low-noise amplifiers (LNAs), 165
modulation, 163–164
Morse code, 163
phase modulation (PM), 164
Tayloe detector, 165
See also software-defined radio
Raspberry Pi
features, 3
interface board, 89–92
Model A, 4
Model A+, 4
Model B, 3–4
overview, 1
Raspberry Pi Foundation (RPF), 4
HAT specification, 90
RasPi. *See* Raspberry Pi
RasPi cluster, 110–112
calculating Pi using, 128–130
monitoring cluster network traffic, 135
software, 112–115
software setup, 115–126
unique functions for cluster operations, 130–132
Raspian OS, 12–15
raspi-config menu, 12–15
real-time clock (RTC), 90
reboot, 15
reconnecting, 139
resistance temperature detectors (RTDs), 57–58
resistive touchscreens, 24–26
See also touchscreens
RGB LED matrix
32 x 64, 85–87
how it works, 87–88
HUB75 connection, 88–89
minimal example demonstration program, 96–97
software, 92–104
text-example demonstration program, 98–100
using Python with, 104–106
robotic arms
arm positioning, 214–215
base, 229–230
degrees of freedom (DOFs), 211, 212–213
initial test program, 222–223
linear-to-rotary translation, 214
overview, 211–212
R-Theta, 213
SainSmart 3-DOF, 215–216
SainSmart 6-DOF, 216–217
SCARA, 213
servo control interface board, 217–218
servo control pulses, 221
software, 221–235
training, 214
X-Y-Z coordinate (Cartesian Coordinate), 213–214
rotating square, 94, *95*
routing, 60
R-Theta, 213

S

SainSmart 3-DOF robotic arm, 215–216
servo control program, 223–225
testing, 225–226
SainSmart 6-DOF robotic arm, 216–217
mounting the RasPi/HAT, 228–229
robotic arm base, 229–230
servo control program, 226–228
testing, 230
X-Y-Z coordinate (Cartesian Coordinate) application, 230–235
Savage, Robert, 156
SCARA, 213
SD cards, 10, 17
cloning, 119
SDR. *See* software-defined radio
SDR dongles, 167–169
selectively compliant articulated robot arm. *See* SCARA
self-healing mesh, 61
sensor ports, 192–193
serial data transmission. *See* UART serial protocol
serial peripheral interface (SPI), 28–29, 57, 142–143
servo control pulses, 221
servo motors, 218–221
SETI@home project, 109
shoe plate, 249
shutdown, 15
sine waves, 173–174
single-ended mode, 141
sketches, 65
software, Arduino Uno, 67–69
software setup, 11–15
headless configuration, 17–19
headless operation with graphics, 19–21
setting up the RasPi OS using an image file, 15–16
updating and upgrading the Raspian image, 16–17
software-defined radio, 163
basic radio concepts, 163–164
demodulation, 164–165
GNU radio software package installation, 172–173
heterodyning, 164–165

software-defined radio (*continued*)
- monitoring ACARS, 175–183
- monitoring Mode-S transponders, 174–175
- receiving aviation data signals, 174–183
- rtl-sdr software package installation, 169–171
- SDR dongles, 167–169
- signal reconstruction using I and Q waveforms, 165–166
- spectrum analyzer, 183–188

spectrum analyzer, 183–184
- making it portable, 186–188
- software installation, 184–186

SPI serial protocol, 57
square test, 94, *95*
SSH Protocol, 17–19
stacking, 238–240
- focus stacking, 240–242

start bits, 143
stitching, 242–244
subscriber client, 152–155
subtopics, 148
super users, 21
supercomputers
- attributes, 109–110
- overview and history, 107–108

synchronous serial interface (SSI), 57

T

Tayloe detector, 165
Tianhe-2, 109
TMP36 temperature sensor, 139–141
toggling, 44
topics, 148
touchscreens
- automagic calibration script, 37–38
- button installation, 31–32
- calibration, 35–41
- capacitive, 26–28
- configuration, 34–35
- demonstration project, 43–53
- evtest, 36–37
- four-wire, 26
- framebuffer, 41–42
- hardware installation, 29–32
- helper file software installation, 33–34
- logical names, 42
- manual calibration, 38–41
- PiTFT touchscreens, 23–24
- resistive, 24–26
- software installation, 32–33
- ts_test, 39–40
- udev rule, 35–36
- video player, 42–43

transcendental numbers, 126
translation, 212
- linear-to-rotary translation, 214

U

UART serial protocol, 56
ultrasonic sensor, 202–203
universal asynchronous receive transmit. *See* UART serial protocol
Upton, Eben, 1

V

video player, 42–43
virtual networking connections (VNC), 19–20
volume bars, 96

W

Walker-Morgan, D.J., 137
WiFi dongles, 199
Williams, Mark, 44
wills, 139
Win32 Disk Imager, 15
WiringPi, 157
workspace, 212–213

X

XBee hardware, 62–65
XBee implementation, 61–62
xrdp, 19–20
X-Y-Z coordinate (Cartesian Coordinate) robotic arm, 213–214
- application, 230–235

Z

Zeller, Henner, 92
ZigBee protocol, 58–61
zombie fingers, 28